AF411963

Advances in Neurobiology

Volume 19

Series Editor
Arne Schousboe

More information about this series at http://www.springer.com/series/8787

Liliana Letra • Raquel Seiça
Editors

Obesity and Brain Function

 Springer

Editors
Liliana Letra
Institute of Physiology, Institute
  for Biomedical Imaging and Life
  Sciences—IBILI
Faculty of Medicine, University of Coimbra
Coimbra, Portugal

Neurology Department
Centro Hospitalar do Baixo Vouga
Aveiro, Portugal

Raquel Seiça
Institute of Physiology, Institute
  for Biomedical Imaging and Life
  Sciences—IBILI
Faculty of Medicine, University of Coimbra
Coimbra, Portugal

ISSN 2190-5215   ISSN 2190-5223   (electronic)
Advances in Neurobiology
ISBN 978-3-319-63259-9   ISBN 978-3-319-63260-5   (eBook)
DOI 10.1007/978-3-319-63260-5

Library of Congress Control Number: 2017952609

© Springer International Publishing AG 2017
This work is subject to copyright. All rights are reserved by the Publisher, whether the whole or part of the material is concerned, specifically the rights of translation, reprinting, reuse of illustrations, recitation, broadcasting, reproduction on microfilms or in any other physical way, and transmission or information storage and retrieval, electronic adaptation, computer software, or by similar or dissimilar methodology now known or hereafter developed.
The use of general descriptive names, registered names, trademarks, service marks, etc. in this publication does not imply, even in the absence of a specific statement, that such names are exempt from the relevant protective laws and regulations and therefore free for general use.
The publisher, the authors and the editors are safe to assume that the advice and information in this book are believed to be true and accurate at the date of publication. Neither the publisher nor the authors or the editors give a warranty, express or implied, with respect to the material contained herein or for any errors or omissions that may have been made. The publisher remains neutral with regard to jurisdictional claims in published maps and institutional affiliations.

Printed on acid-free paper

This Springer imprint is published by Springer Nature
The registered company is Springer International Publishing AG
The registered company address is: Gewerbestrasse 11, 6330 Cham, Switzerland

Preface

The concept of this book arose from an increasing interest in one of the greatest epidemics of modern societies—obesity. Just a few years ago, adipose tissue function and dysfunction were not even included in Physiology books while currently they became the focus of extensive and exciting investigation.

Obesity and Brain Function emerges, then, by the recognition of the influence of adipose tissue and particularly of its derived products (adipokines) on brain structure and function as well as their role in the development and progression of neurological diseases. It is written by talented basic researchers and skilled clinical neurologists who were gathered by their interest on this particular topic and whom the editors thank for their dedication and diligent work.

This book is expected to provide a comprehensive, though concise and practical, review of adipose tissue biology in health and in neurological disease, also comprising hot topics such as bariatric surgery and functional neuroimaging, and ultimately serve as a useful resource to researchers and/or physicians interested in obesity. Enjoy.

Coimbra, Portugal

Liliana Letra
Raquel Seiça

Contents

Part III   Brain Function After Bariatric Surgery

Part IV   Neuroimaging in Obesity

Part I
Adipose Tissue Biology and Crosstalks

Chapter 1
Function and Dysfunction of Adipose Tissue

Paulo Matafome and Raquel Seiça

Abstract Adipose tissue is an endocrine organ which is responsible for postprandial uptake of glucose and fatty acids, consequently producing a broad range of adipokines controlling several physiological functions like appetite, insulin sensitivity and secretion, immunity, coagulation, and vascular tone, among others. Many aspects of adipose tissue pathophysiology in metabolic diseases have been described in the last years. Recent data suggest two main factors for adipose tissue dysfunction: accumulation of nonesterified fatty acids and their secondary products and hypoxia. Both of these factors are thought to be on the basis of low-grade inflammatory activation, further increasing metabolic dysregulation in adipose tissue. In turn, inflammation is involved in the inhibition of substrate uptake, alteration of the secretory profile, stimulation of angiogenesis, and recruitment of further inflammatory cells, which creates an inflammatory feedback in the tissue and is responsible for long-term establishment of insulin resistance.

Keywords Nutrient storage • Adipokines • Lipid intermediates • Hypoxia • Inflammation • Angiogenesis

1.1 Adipose Tissue Structure

Adipose tissue is a complex and heterogeneous tissue composed by cells with lipid storage functions, called adipocytes, and a stromovascular function, composed by endothelial and mesenchymal stem cells, preadipocytes, fibroblasts, and resident cells from the immune system (Rajala and Scherer 2003; Juge-Aubry et al. 2005a;

P. Matafome (✉)
Institute of Physiology, Institute for Biomedical Imaging and Life Sciences—IBILI,
Faculty of Medicine, University of Coimbra, Coimbra, Portugal

Department of Complementary Sciences, Coimbra Health School (ESTeSC), Instituto
Politécnico de Coimbra, Coimbra, Portugal
e-mail: paulomatafome@gmail.com

R. Seiça
Institute of Physiology, Institute for Biomedical Imaging and Life Sciences—IBILI,
Faculty of Medicine, University of Coimbra, Coimbra, Portugal

© Springer International Publishing AG 2017
L. Letra, R. Seiça (eds.), *Obesity and Brain Function*, Advances in
Neurobiology 19, DOI 10.1007/978-3-319-63260-5_1

Guilherme et al. 2008b; Christiaens and Lijnen 2010). During the embryonic development, the vascular network develops before adipocytes and the extracellular matrix which supports blood vessels is the first to be deposited, showing the crucial role of the vascular system in adipose tissue development (Neels et al. 2004; Christiaens and Lijnen 2010). In fact, during the embryonic development, a close communication between the stromovascular fraction and the adipocytes results in a mutual control between angiogenesis and adipogenesis. Recent data show that adipocytes may develop from capillary networks as the progenitor cells respond to pro-angiogenic stimuli in association with the expanding capillaries (Min et al. 2016).

In adult life, a well-developed vascular network is observable at the microscope and each adipocyte is surrounded by at least one capillary (Neels et al. 2004; Rutkowski et al. 2009; Christiaens and Lijnen 2010). The capillaries of the adipose tissue are fenestrated and are rich in trans-endothelial channels, which allow a close communication with the adipocytes (Christiaens and Lijnen 2010). Moreover, even in the adult life, this vascular network is very dynamic and is continuously adapting to changing nutritional fluxes. However, the mechanisms governing such remodeling are still far from being understood. Interestingly, the dynamics of vascular remodeling apparently influence adipocyte behavior during expansion. Adipocyte hypertrophy is usually associated with the formation of aberrant capillaries, while adipocyte hyperplasia is usually associated with increased angiogenesis and development of new capillaries. Hyperplasia is considered a harmless form of adipose tissue expansion, given the formation of smaller well-irrigated adipocytes with lower inflammatory activity than hypertrophic ones (Christiaens and Lijnen 2010). When the tissue is forced to expand, the formation of local phenomena of hypoxia leads to the expression of angiogenic factors which stimulate angiogenesis, including several cytokines and adipokines. The balance between these factors determines vessel density and permeability, and thus the "good" physiological or the "bad" pathophysiological expansion of the adipose tissue. The mechanisms will be detailed in the following sections.

1.2 Metabolic Functions of the Adipose Tissue

Although its important endocrine functions, the primary function of the adipose tissue is to store energy in the form of lipids, mainly in intracellular triglycerides droplets, also regulating lipid catabolism in different tissues due to the actions of adipokines (Rajala and Scherer 2003; Juge-Aubry et al. 2005a; Guilherme et al. 2008b). Lipid droplets are coated by a group of proteins, which the main one is Perilipin A (Per A), which prevent the contact between the stored triglycerides and the cytoplasm. In times of energetic need, such triglycerides are quickly hydrolyzed into free fatty acids and glycerol, a process called lipolysis, and released to the blood in order to feed other organs demands (Arner 2005; Guilherme et al. 2008b; Galic et al. 2010). Thus, the adipose tissue is able to recognize the metabolic state of the organism, not only by local energetic sensors, but also through different inputs coming mainly from the gut after and between meals. Moreover, adipocytes also regulate cholesterol metabolism, as they are able to produce high-density lipoproteins (HDL) in order to send excessive cholesterol to the liver.

When adipocytes accumulate excessive amounts of nonesterified fatty acids, for reasons that are currently under investigation, their metabolism and endocrine function are shifted in order to produce a broad range of proinflammatory factors, while adiponectin secretion is decreased. Such factors include cytokines and chemokines, growth factors, tumour necrosis factor (TNF)-α, interleukin (IL)-6, monocyte chemoattractant factor (MCP)-1, vascular endothelial graowth factor (VEGF), leptin, and resistin (Wellen and Hotamisligil 2005; Tilg and Moschen 2006; Guilherme et al. 2008b). In fact, a strong relationship between metabolism and innate immunity is present at the adipose tissue. Many of the intracellular pathways involved in metabolic signaling are recruited as well during an immune response and most of the adipokines and adipose tissue-derived factors, besides regulating metabolism, also have paracrine and endocrine functions in regulating the immune response. Many authors support the existence of a metabolism—immunity axis, as any metabolic change immediately induces alterations in the immune response and the regulation of metabolic fluxes usually involves the activation of intracellular inflammatory and stress pathways (Rajala and Scherer 2003; Juge-Aubry et al. 2005a; Goossens 2008; Guilherme et al. 2008b; Rutkowski et al. 2009).

1.2.1 Mechanisms of Nutrient Uptake and Storage in Adipocytes

Triglycerides synthesis, a process called esterification, occurs from one molecule of glucose-derived glycerol and three fatty acyl chains. Adipocytes have a limited ability to store glycogen and thus all the glucose that is not consumed in the adipocyte metabolism is transformed into glycerol and stored in the triglycerides pool (Tilg and Moschen 2006; Goossens 2008; Guilherme et al. 2008b). Esterification is mainly stimulated by insulin, which induces tyrosine kinase activity in its receptor (Fig. 1.1). This in turn leads to the tyrosine phosphorylation and activation of the insulin receptor substrate-1 (IRS-1) and initiates a signaling pathway which involves PI$_3$K and Akt/PKB activation. Among other actions, the activation of this signaling cascade leads to the translocation to the membrane of GLUT4-containing vesicles, allowing glucose uptake (Wellen and Hotamisligil 2005; Bugianesi et al. 2005). Besides inducing glucose uptake, insulin is also responsible for lipolysis inhibition, through the inhibition of adenylate cyclase, the main enzyme involved in AMPc synthesis. AMPc activates PKA, which in turn phosphorylates and activates hormone-sensitive lipase (HSL), the main enzyme involved in triglycerides hydrolyzation (Fig. 1.1). On the other hand, in times of nutrient demand, contra-regulatory hormones, like glucagon, cortisol, growth hormone, and adrenaline, increase AMPc levels, leading to the activation of HSL and thus the release of fatty acids into the circulation (Fig. 1.1) (Tilg and Moschen 2006; Goossens 2008; Guilherme et al. 2008b). Increased lipolysis is also observed in obese individuals due to the development of insulin resistance and the increased secretion of proinflammatory cytokines that promote lipolysis (Langin 2006; Guilherme et al. 2008b). The subsequent flux of free fatty acids from the adipose tissue to the circulation may in turn cause their

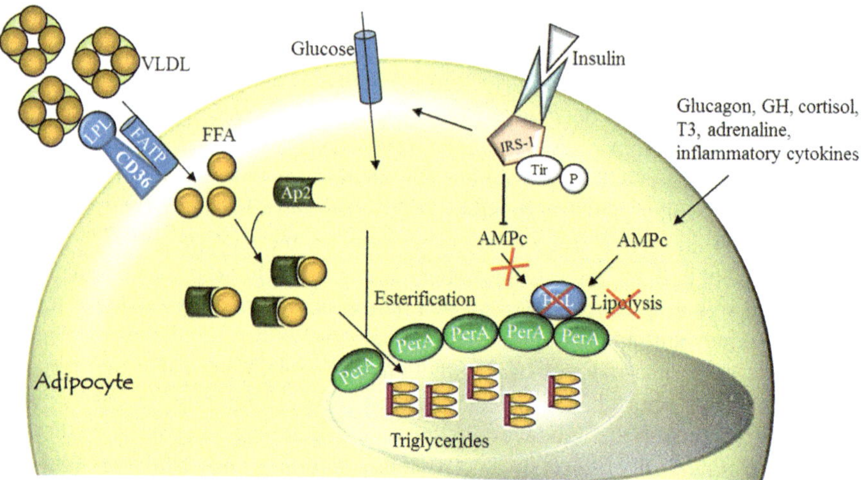

Fig. 1.1 Mechanisms of lipid storage in adipocytes and their mobilization from lipid droplets. Lipolysis is inhibited by insulin signaling and promoted by other hormones like glucagon, GH, cortisol, T3, or adrenaline, due to the stimulation of AMPc and HSL. *cAMP* cyclic adenosine monophosphate, *FATP* fatty acid binding protein, *FFA* free fatty acids, *HSL* hormone-sensitive lipase, *IRS-1* insulin receptor substract-1, *LPL* lipoprotein lipase, *PerA* perilipin A, *VLDL* very-low density lipoprotein

ectopic accumulation in other tissues, such as the liver and the skeletal muscle, mechanism which will be described in the chapter dedicated to the pathophysiology of adipose tissue.

Circulating lipids are derived from hepatic incorporation into VLDL or from intestinal absorption and included in chylomicrons. Such lipoproteins bind to the CD36 receptor present at the adipocyte membrane. Lipoproteins are then hydrolyzed by the lipoprotein lipase (LPL) and the fatty acids are transported to the cytoplasm through the fatty acid transporter protein (FATP/CD36). Once at the cytoplasm the nonesterified fatty acids are captured by the protein aP2, which prevent their free circulation in the cell and the consecutive activation of inflammatory and stress pathways (Ram 2003; Wellen and Hotamisligil 2005). Fatty acids are finally esterified into triglycerides and stored in lipid droplets (Fig. 1.1).

Alternatively to esterification, fatty acids may be metabolized in several mediators of intracellular signaling pathways, namely eicosanoids, which are important activators of the peroxissome proliferation activated receptor-gamma (PPARγ). This nuclear receptor controls the events involved in lipid esterification, being also activated by the pharmacological class of tiazolidinediones (TZD) (Fig. 1.2). The genes controlled by the activation of PPARγ include proteins involved in fatty acid uptake (FATP, CD36 and LPL), metabolism (PEPCK and SCD-1), storage (perilipin A), and oxidation (Adiponectin and UCP-1). PPARγ activation also inhibits cellular inflammatory pathways, including NF-κB, which will be discussed in the chapter dedicated to adipose tissue pathophysiology (Wellen and Hotamisligil 2003; Ram 2003). Thus, PPARγ promotes insulin sensitivity due to the reduction of cytoplasmatic free fatty acids and due to the inhibition of inflammatory pathways (Ram 2003; Guilherme et al. 2008b).

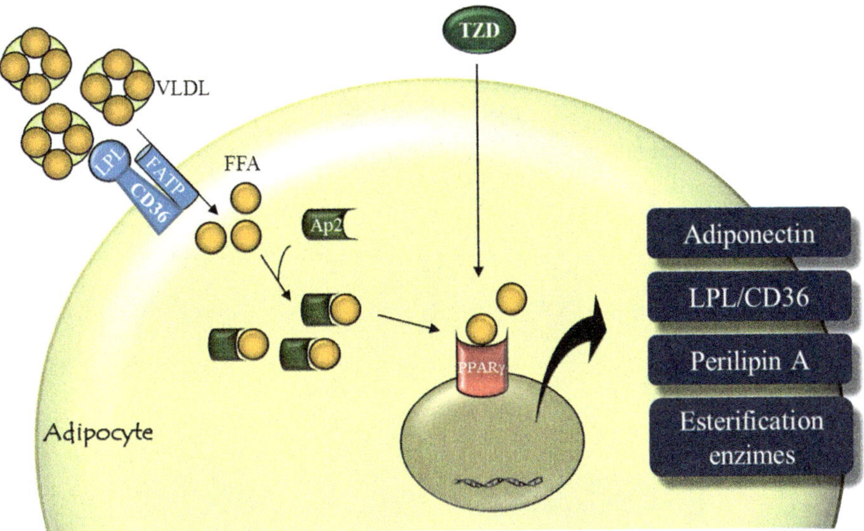

Fig. 1.2 Mechanisms of PPARγ activation in response to fatty acids and their metabolites, stimulating the expression of adiponectin and proteins involved in fatty acid esterification and storage. The activity of PPARγ can be improved by the pharmacological class of TZDs. *FATP* fatty acid binding protein, *FFA* free fatty acids, *LPL* lipoprotein lipase, *PPARγ* peroxisome proliferating-activated receptor γ, *VLDL* very-low density lipoprotein

PPARγ is expressed in adipocytes and macrophages, controlling lipid uptake, but also preadipocyte differentiation into mature adipocytes, a process called adipogenesis (Tamori et al. 2002; Lee et al. 2011). PPARγ inhibition in obesity conducts to the inhibition of adipogenesis and to the hypertrophy of the existing adipocytes. Hypertrophic adipocytes have shown to be hypoxic and metabolically dysregulated (Trayhurn et al. 2008a; Trayhurn 2014). PPARα controls the expression of genes involved in lipid oxidation and is expressed in tissues with catabolic activity like the liver and the skeletal muscle. It is regulated by adiponectin and by drugs like fibrates and metformin (indirectly), increasing the oxidation of fatty acids on mitochondria.

1.2.2   Cholesterol Fluxes

The adipocyte is also an important regulator of cholesterol storage and mobilization. Excessive intracellular cholesterol is incorporated into HDL/Apo-A1 particles, being transported to the liver where it used in the synthesis of biliary acids or excreted in bile (Yin et al. 2010). Two of the most important proteins in this transport are ABCA1 and ABCG1 ("ATP-Binding Membrane Cassete Transporter A1 e G1"), being involved in the transport of cholesterol, phospholipids, and other lipophilic substances to the HDL particles (Yin et al. 2010). The expression of ABC proteins is regulated by the PPAR and by the AMPc/PKA pathway, which is important in the mobilization of stored lipids (cholesterol). However, decreased PPAR activity in dysfunctional adipocytes

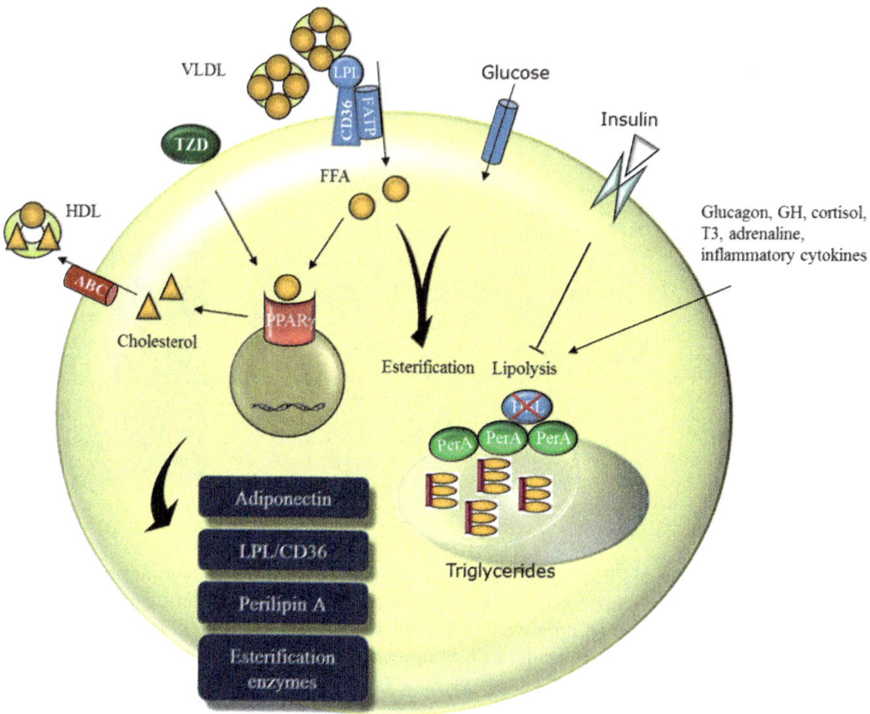

Fig. 1.3 Integration of the mechanisms involved in the uptake of lipids and glucose into the adipocyte, namely the role of insulin in inducing glucose uptake and in inhibiting lipolysis, as well as the role of PPARγ in promoting fatty acid esterification and cholesterol efflux to HDL particles. *FATP* fatty acid binding protein, *FFA* free fatty acids, *HDL* high-density lipoprotein, *HSL* hormone-sensitive lipase, *LPL* lipoprotein lipase, *PerA* perilipin A, *PPARγ* peroxisome proliferating-activated receptor γ, *VLDL* very-low density lipoprotein

may compromise cholesterol mobilization into the liver (Fig. 1.3) (Yin et al. 2010). Knockout models for ABCA1 have more infiltration of inflammatory cells in several tissues, due to decreased cholesterol transport to the liver and excessive deposition in the tissues. On the other hand, ABCA1 overexpression was shown to prevent cholesterol accumulation and atherosclerosis progression. Thus, ABCA1 prevents the accumulation of cholesterol in tissues, which is known to induce the formation of foam cells from recruited macrophages. The effect of ABCA1 and ABCG1 knockout is cumulative, suggesting that they have distinct roles in regulating cholesterol efflux. It is believed that ABCA1 and ABCG1 have distinct affinities for different HDL proteins (Yin et al. 2010).

1.3 Endocrine Function of the Adipose Tissue

The adipose tissue secretes a broad range of factors, including adipokines, cytokines, chemokines, angiogenic factors, coagulation factors, and vasoactive factors, among others. More than 600 different factors have been identified as produced and secreted by the adipose tissue, affecting glucose and lipid metabolism, appetite, vascular function, inflammation, coagulation, or the cardiovascular function. However, the complete list of adipokines and their functions is not yet completely known (Wellen and Hotamisligil 2003; Trayhurn and Wood 2004; Guilherme et al. 2008b; Galic et al. 2010). Moreover, there are differences in the secretory profile between the different fat depots, with the visceral ones being more susceptible to nutritional signals due to their proximity to the intestine and liver.

1.3.1 Leptin

Leptin was the first factor originally identified as a product of adipose tissue and its discovery changed the view about this tissue, from a mere fat reserve to an endocrine organ able to control energy homeostasis. Leptin is almost exclusively produced by the adipocyte (95%) and its levels are proportional to the fat mass (Rajala and Scherer 2003; Meier and Gressner 2004; Golay and Ybarra 2005; Lorincz and Sukumar 2006; Vona-Davis and Rose 2007). Differences in leptin secretion from different fat depots were also shown, being the subcutaneous one more active (Nielsen et al. 2009). The main function of leptin is to inform the hypothalamus about nutrient availability. Leptin activates afferent nervous fibers and acts directly on the hypothalamus in order to suppress appetite and modulate energy expenditure (Fig. 1.4) (Kiess et al. 2008). Moreover, it accutely supresses insulin secretion, but increases long-term β cell survival and function. Moreover, leptin increases fatty acid uptake and oxidation in the skeletal muscle and liver, also acting as a growth factor for endothelial cells (Rajala and Scherer 2003; Meier and Gressner 2004; Bugianesi et al. 2005; Lorincz and Sukumar 2006).

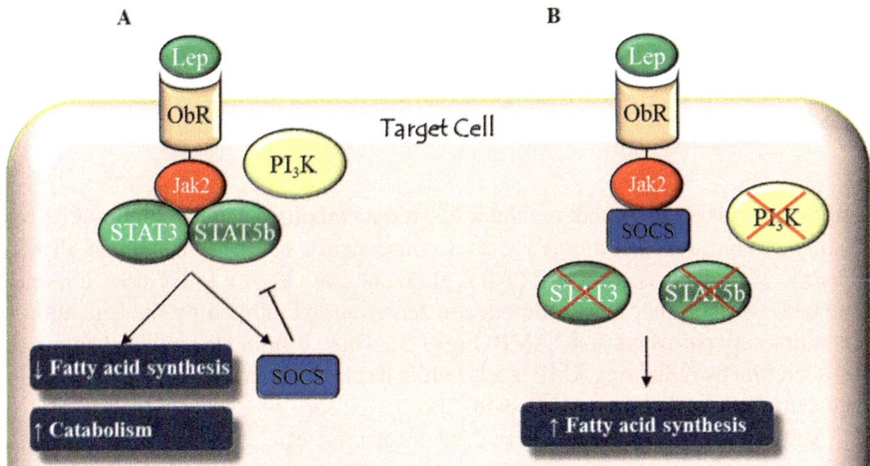

Fig. 1.4 Mechanisms of leptin signaling in target cells (**a**), as well the alterations occurring in leptin resistance (**b**). *Lep* Leptin, *ObR* leptin receptor, *SOCS* supressor of cytokine signalling, *STAT* signal transducer and activator of transcription

Although leptin levels are proportional to fat mass, its secretion may be stimulated by insulin and inhibited by increased levels of AMPc, i.e., it increases after meals and decreases in times of energetic demand (Meier and Gressner 2004). When leptin levels are lower, appetite is stimulated and thyroid hormones, thermogenesis, and immune system function are inhibited (Münzberg et al. 2005). Mutations of the leptin or leptin receptor genes are known to induce obesity. On the other hand, leptin administration leads to an increase of fatty acid oxidation and a reduction of their circulating levels. Such effects are not observed in obese patients, which are known to have a resistance to the action of the hormone (Rajala and Scherer 2003; Meier and Gressner 2004). Other functions of leptin include the stimulation of angiogenesis, the modulation of the immune response, and the expression of the key enzyme involved in estrogen synthesis, aromatase (Catalano et al. 2003, Juge-Aubry et al. 2005a, b Lorincz and Sukumar 2006).

Leptin receptor is linked to a Jak/STAT pathway, which first protein is Jak2 and leads to the activation of STAT3 and STAT5. STAT proteins are transcription factors that stimulate catabolic processes and insulin secretion in β cells (Ueki et al. 2004a, b; Laubner et al. 2005; Kaneto et al. 2010; Blüher and Mantzoros 2015) (Fig. 1.4). STAT proteins also activate SOCS proteins, involved in the negative feedback. Thus, hyperleptinemia leads not only to increased signaling but also to increased negative feedback, contributing to the leptin resistance observed in obese patients mainly in the hypothalamus (Laubner et al. 2005). SOCS proteins also inhibit insulin signaling, contributing to insulin resistance. In fact, leptin and insulin signaling have a common pathway, namely the activation of IRS-1, PI3K, and Akt (Ahima and Flier 2000; Ueki et al. 2004a; Münzberg et al. 2005; Ahima 2005; Imrie et al. 2010). Leptin also activates AMPK inducing lipid oxidation and promoting insulin sensitivity (Rajala and Scherer 2003; Meier and Gressner 2004; Ahima 2005; Juge-Aubry et al. 2005a). Moreover, leptin inhibits fatty acid synthesis, due to the suppression of the transcription factor SREBP-1c and the key proteins of the biosynthetic pathway ACC and FAS. On the other hand, such mechanisms were shown to be activated by SOCS proteins (Fig. 1.4) (Ueki et al. 2004a, b).

1.3.2 Adiponectin

Adiponectin is an important regulator of lipid metabolism and stimulator of its oxidation produced in the adipocyte as a consequence of the activation of PPARγ. Pharmacological activators of PPARγ, glitazones, are known to increase adiponectinemia. On the other hand, adiponectin secretion is inhibited by the activation of inflammatory pathways and cAMP (Fig. 1.5). Thus, insulin also induced adiponectin secretion by reducing cAMP levels, while its elevation in times of energy demand prevents fatty acid consume (Xu et al. 2003).

Adiponectin was shown to have cardioprotective effects by promoting cell viability and inhibiting apoptosis during ischemia (Ding et al. 2012; Park and Sweeney 2013; Smekal and Vaclavik 2017). As well, similar effects were demonstrated in β

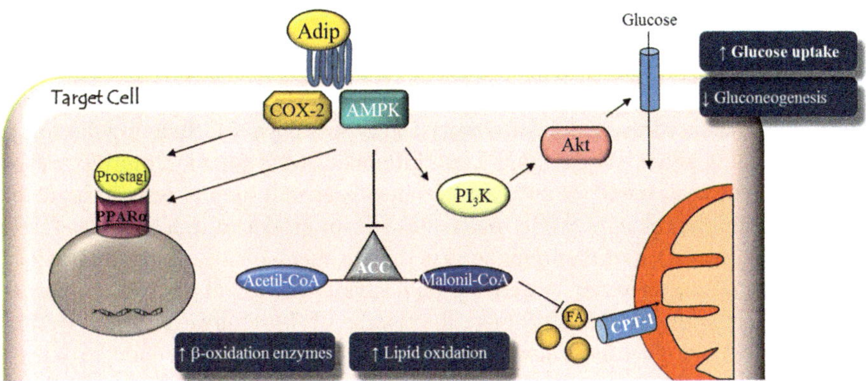

Fig. 1.5 Mechanisms of adiponectin signaling in target cell, including AMPK-mediated glucose uptake and inhibition of ACC. Inhibition of ACC prevents fatty acids synthesis and prevents inhibition of fatty acid uptake by the ACC product malonyl-CoA. Activation of PPARα increases the expression of fatty acids oxidation enzymes. *ACC* acetyl-CoA carboxylase, *Adip* adiponectin, *AMK* AMP-activated protein kinase, *COX-2* ciclooxygenas-2, *PPARγ* peroxisome proliferating-activated receptor γ

cells, improving insulin secretion. Adiponectin was shown to exert anti-inflammatory effects on the vessel wall preventing atherosclerosis through the inhibition of adhesion molecules expression and smooth muscle cell proliferation (Fig. 1.5) (Juge-Aubry et al. 2005a). Protective effects of adiponectin have also been reported in different pathologies like cancer and neurodegenerative diseases (Vona-Davis and Rose 2007; Pais et al. 2009; Matafome 2013; Letra et al. 2014).

Adiponectin circulates in three molecular forms (trimeric, hexametric, and high-molecular weight isoform (HMW) composed of several hexamers). Reduced levels of the HMW isoform have been specifically associated with metabolic disorders (Rajala and Scherer 2003; Vona-Davis and Rose 2007, 2009; Pais et al. 2009). The globular region of the adiponectin chain binds to two membrane receptors, AdipoR1 and AdipoR2. While AdipoR1 is found mainly in the skeletal muscle, AdipoR2 is more abundant in the liver (Meier and Gressner 2004). A third receptor, T-cadherin, was recently identified as an adiponectin receptor, but without intracellular signaling and mainly involved in the anchorage of the HMW isoform to the membrane (Takeuchi et al. 2007). Adiponectin binding to AdipoR1 and AdipoR2 leads to AMPK and PPARα activation, which in turn leads to increased uptake and oxidation of glucose and lipids (Fig. 1.5) (Kamon et al. 2003; Yamauchi et al. 2003; Lorincz and Sukumar 2006; Pais et al. 2009). Moreover, AMPK is involved in promoting glucose uptake through GLUT4 translocation to the membrane and in inhibiting gluconeogenesis (Juge-Aubry et al. 2005a; Nawrocki et al. 2006). Similar to leptin, adiponectin directly inhibits the enzymes involved in the fatty acids synthesis pathway, through the inhibition of acctyl-coA carboxylase (ACC), the key enzyme of such pathway (Fig. 1.5) (Ouchi et al. 2001; Xu et al. 2003; Nawrocki et al. 2006). On the other hand, the activation of PPARα increases the expression of the enzymes involved in lipid oxidation and uptake to the mitochondria (Fig. 1.5) (Kamon et al. 2003; Yamauchi et al. 2003; Gealekman et al. 2008, 2012).

1.3.3   Resistin

Resistin is an adipokine associated with the establishment of insulin resistance as its levels are increased in models of obesity and type 2 diabetes. Studying the role of human resistin is not easy as significant differences were found to murine resistin. Resistin gene was found in different chromosomes and human resistin is mainly produced in macrophages while murine resistin in produced in adipocytes (Lazar 2007). The activation of the immune system after metabolic dysregulation induces resistin secretion, in order to activate pathways which block nutrient uptake by hypertrophic adipocytes and promote the release of those already stored. Resistin also induces angiogenesis aiming to increase blood flow and thus adipocyte oxygenation. Importantly, resistin increases lipid uptake by macrophages, also inhibiting cholesterol efflux from these cells, in order to store adipocyte-derived lipids. However, the chronic activation of such mechanisms leads to the formation of foam cells, common in atherosclerotic lesions (Fig. 1.6) (Lazar 2007; Robertson et al. 2009).

Resistin is directly involved in blocking insulin signaling in adipocytes, but also in the liver and skeletal muscle. Resistin knockout mice were shown to be protected of obesity-related insulin resistance, suggesting that it may be an important link between the activation of the immune system and glucose metabolism (Lazar 2007; Qatanani and Szwergold 2009; Robertson et al. 2009). Moreover, resistin neutralization increases insulin sensitivity and decreases hepatic glucose production, while

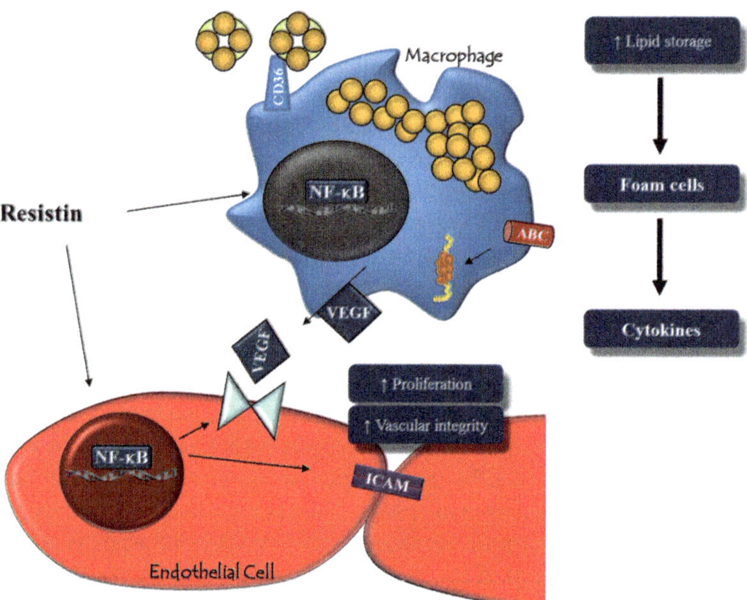

Fig. 1.6 Mechanisms of resistin action in macrophages and endothelial cells, promoting the formation of foam cells and endothelial proliferation. *ICAM* intercellular adhesion molecule, *NF-kB* nuclear factor-kappa B, *VEGF* vascular endothelial growth factor

resistin administration induces severe insulin resistance (Bugianesi et al. 2005; Juge-Aubry et al. 2005a; Lazar 2007). The expression of human resistin specifically in the macrophages of resistin knockout mice was able to increase adipose tissue lipolysis and fatty acid ectopic accumulation in the skeletal muscle, which led to insulin resistance (Qatanani and Szwergold 2009).

Resistin is a cysteine-rich protein which forms a dimer responsible for its biological activity (Bugianesi et al. 2005; Juge-Aubry et al. 2005a; Lazar 2007; Robertson et al. 2009; Smith and Yellon 2011). Human resistin was shown to induce the proinflammatory activity of macrophages and endothelial cells, by activating the Nuclear Factor-kB (NF-kB) and inhibiting the insulin signaling (Fig. 1.6). Resistin promotes the proliferation and adhesion of endothelial cells through NF-κB activation and increased expression of the VEGF receptors (VEGFR1 and VEGFR2), matrix metalloproteinases (MMP-1 and MMP-2), and adhesion proteins (VCAM and ICAM) (Fig. 1.6) (Juge-Aubry et al. 2005a).

In macrophages, resistin promotes adipocyte-like mechanisms of lipid uptake resulting in the formation of foam cells. Such mechanisms include the upregulation of scavenger receptors (SR-A and CD36), also promoting the degradation of lipid efflux proteins in the proteasome (ABCA1, ABCG1). Foam cells are known for their proinflammatory and proangiogenic potential which is thus promoted by resistin actions (Fig. 1.6) (Lazar 2007; Lee et al. 2009; Qatanani and Szwergold 2009).

1.3.4 Other Adipose Tissue Products

Recently, a broad range of other adipokines has been identified, including apelin, visfatin, vaspin, omentin, and others involved in canonical mechanisms like coagulation, vascular tone, and immunity.

Visfatin was initially discovered as a factor controlling B lymphocytes development. On the other hand, an enzyme called Nampt (nicotinamide 5-fosforibosyl-1-pyrophosfate transferase) catalyzing the formation of nicotinamide adenine dinucleotide (NAD) was also described as product from the same gene (Kiess et al. 2008; Gallí et al. 2010; Saddi-Rosa et al. 2010). Thus, visfatin appears to be a factor with a simultaneous enzymatic and endocrine activity. Its circulating concentrations are correlated with the amount of visceral adipose tissue and increases during weight gain. Moreover, its synthesis was identified mostly in macrophages residing and infiltrating the tissue (Kiess et al. 2008; Gallí et al. 2010; Saddi-Rosa et al. 2010). Increased circulating visfatin levels were found in a number of pathological situations like portal inflammatory infiltration, steatohepatitis, coronary accumulation of oxidized LDL and atherosclerotic plaque instability, higher incidence of ischemic events and decreased kidney function (Saddi-Rosa et al. 2010). On the other hand, visfatin was observed to produce a rapid insulin-independent hypoglycemic effect after its administration, while animal models lacking visfatin have shown hypoinsulinemia and lower insulin-stimulated glucose uptake (Kiess et al. 2008; Gallí et al. 2010; Saddi-Rosa et al. 2010). Visfatin was also observed to increase adipocyte and myocyte glucose uptake and to suppress hepatic glucose production (Saddi-Rosa et al. 2010). In vitro and in vivo studies also demonstrated protective effects of visfatin in ischemic events and its ability to increase endothelial cell activation, promoting angiogenesis (Mocan Hognogi and Simiti 2016). Thus, contradictory results about visfatin effects have been observed. Apparently, it is a compensatory mechanism for hyperglycemia with protective effects at multiple levels (Saddi-Rosa et al. 2010). Such effects were recently attributed to its ability to synthesize NAD, which improves the activity of several enzymes involved in glucose metabolism and mitochondrial function (Gallí et al. 2010; Mocan Hognogi and Simiti 2016).

Apelin was only recently identified, after the identification of its receptor APJ. Its functions are not very clear but apparently it regulates distinct biological processes like angiogenesis, vascular tone, heart function, and insulin secretion. Apelin was shown to antagonize Angiopoietin (Ang)-2 effects of endothelial cells, increasing NO-dependent vasodilation, but it was also shown to increase smooth muscle contraction. Although these are contradictory results, apelin also has the ability to bind to antidiuretic hormone (ADH) producing neurons in the hypothalamus, decreasing its production. Thus, together with its ionotropic actions in the heart, apelin is thought to be involved in the control of blood pressure (Glassford et al. 2007; Pitkin et al. 2010; Fasshauer and Blüher 2015). Moreover, insulin was observed to increase Apelin secretion, but apelin apparently has inhibitory functions on insulin secretion, suggesting that it might be involved in a negative feedback to insulin secretion (Glassford et al. 2007; Pitkin et al. 2010; Fasshauer and Blüher 2015). Nevertheless, little is yet known about apelin and more research is needed in this field.

Vaspin is another adipose-derived hormone, initially identified as inhibitor of serine proteases, but with benefic effects on insulin sensitivity. Although the mechanisms of vaspin stimulation and secretion are unknown, it was shown to be secreted by visceral fat mass and nutritionally regulated, due to its circadian release after meals (Jeong et al. 2010; Fasshauer and Blüher 2015). Vaspin levels are decreased in type 2 diabetes and its administration was shown to improve glucose tolerance and insulin sensitivity (Fig. 1.7) (Klöting et al. 2006; Jeong et al. 2010; Fasshauer and Blüher 2015). Moreover, diet or surgery-induced weight loss leads to a decrease of vaspin levels, which may result from decreased fat mass (Klöting et al. 2006; Handisurya et al. 2010).

Omentin is mainly produced in the omental adipose tissue, but it is expressed in vascular cells and not in adipocytes (Schäffler et al. 2005; Yang et al. 2006; Yamawaki et al. 2010). Omentin was recently suggested to increase insulin-stimulated glucose uptake, being correlated with insulin sensitivity. Its concentrations were observed to be decreased in obese patients and to increase after weight loss (Fig. 1.7) (Yang et al. 2006; De Souza Batista et al. 2007; Moreno-Navarrete et al. 2010). Importantly, omentin was shown to induce eNOS-dependent vessel relaxation, although the mechanisms are still unknown (Yamawaki et al. 2010).

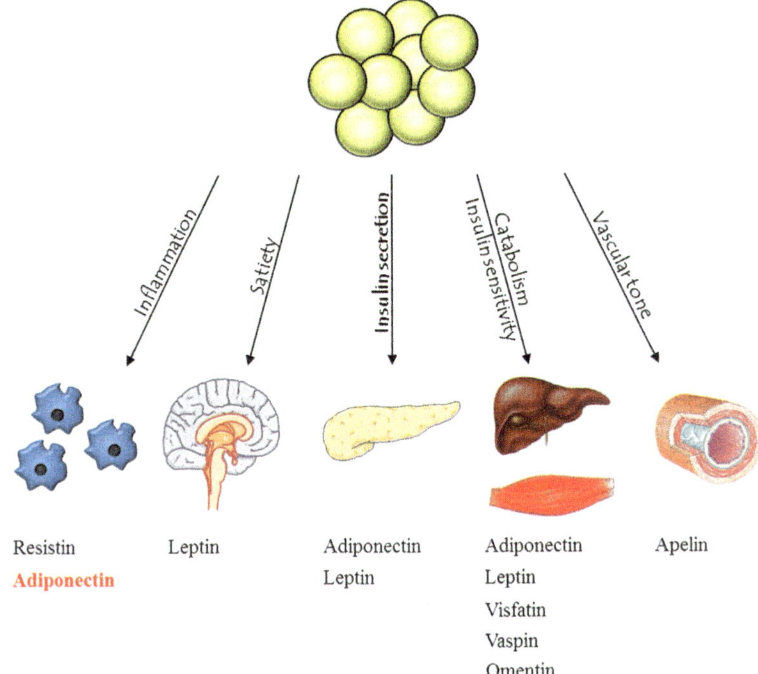

Fig. 1.7 Effects of adipose tissue-derived endocrine factors on inflammation, appetite, sensitivity and secretion of insulin and vascular tone

1.4 Dysregulation of Adipose Tissue Function in Obesity

Adipocyte hypertrophy is a process occurring after excessive nutrient uptake, especially in visceral adipose tissue, which is more susceptible to nutrient fluxes. Hypertrophy has been recently associated with the development of hypoxic regions, with a strong influence on the metabolic and secretory functions of the adipose tissue.

Several studies demonstrated inflammation as the link between metabolic dysregulation and hypoxia, changing the secretory profile to create a proangiogenic environment (Fig. 1.8). Inflammation blocks nutrient uptake (insulin resistance and PPARγ inhibition) in order to prevent uncontrolled cell and tissue expansion and promotes lipolysis. Dysfunctional adipose tissue secretes a broad range of cytokines, but also chemokines, which recruit circulating monocytes to the tissue. Infiltrating macrophages are then important in maintaining the inflammatory environment and in uptake of excessive lipids, leading to the formation of adipose tissue foam cells. Many authors claim the existence of a metabolism—immunity axis, given that all metabolic changes lead to an immune response and many of the inflammatory factors simultaneously regulate metabolism and the immune system (Fig. 1.8) (Rajala and Scherer 2003; Golay and Ybarra 2005; Juge-Aubry et al.

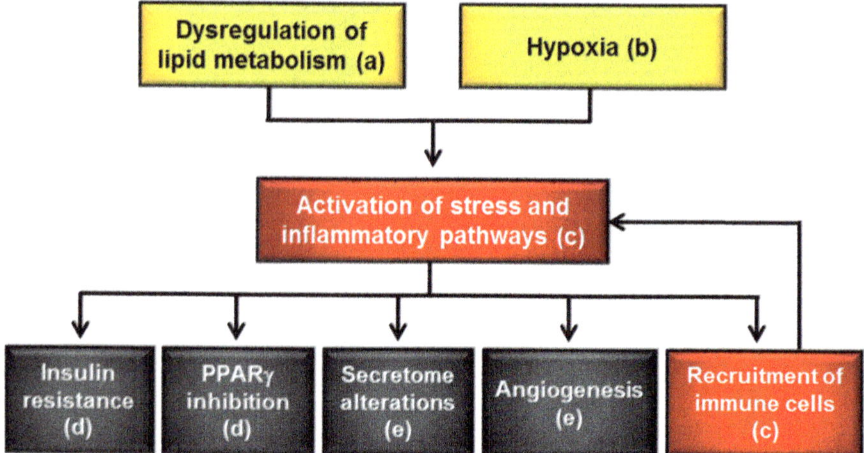

Fig. 1.8 The role of hypoxia (*b*) and dysregulation of the lipid metabolism (*a*) in activating inflammatory mechanisms (*c*), aiming to restore homeostasis. Inflammation inhibits nutrient uptake and promotes lipolysis (*d*), also increasing angiogenesis (*e*) and the secretion of inflammatory adipokines (*e*). The mechanisms involved will be discussed in the following sections according to the letters in each figure's box

2005a; Ye et al. 2007; Goossens 2008; Guilherme et al. 2008b; Rutkowski et al. 2009; Maury and Brichard 2010).

1.4.1 Dysregulation of Lipid Metabolism

During the process of fat accumulation, adipocytes continuously accumulate triglycerides, increasing the expression of enzymes involved in fatty acid esterification. When adipose tissue is not able anymore to store all the lipids from the diet, adipocytes release fatty acids to the circulation, which then accumulate in tissues like skeletal muscle and liver. Type 2 diabetic patients usually have hepatic and muscle steatosis, leading to insulin-resistance and morphological alterations in such tissues (Rajala and Scherer 2003; Guilherme et al. 2008b; Kawano and Cohen 2013; Lonardo et al. 2015). The increasing availability of fatty acids also increases their esterification into triglycerides and oxidation in the mitochondria (Bugianesi et al. 2005; Golay and Ybarra 2005). However, lipid oxidation is a limited process and in such conditions intermediates of fatty acid metabolism like ceramides and diacylglycerols accumulate. Such intermediates inhibit insulin signaling and glucose uptake (Fig. 1.9). In physiological condition, such mechanisms prevent excessive nutrient uptake, which would cause cell death by oxidative stress. However when such mechanisms are chronically activated they conduce to insulin resistance and contribute to the onset of type 2 diabetes (Golay and Ybarra 2005; Guilherme et al. 2008a).

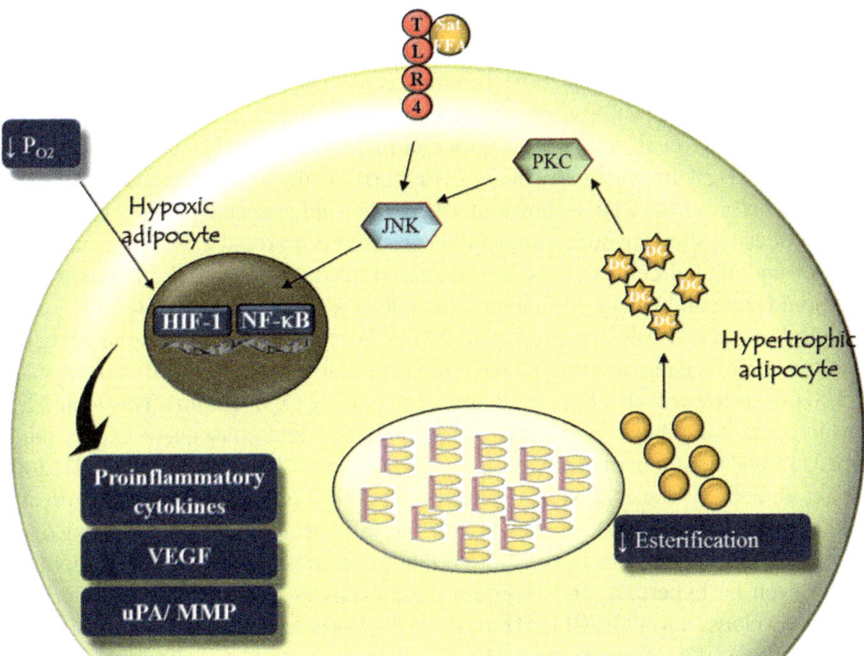

Fig. 1.9 Main mechanisms conducting to the activation of inflammatory pathways in adipocytes, namely the cytoplasmatic accumulation of lipid mediators and activation of membrane receptors by saturated fatty acids. *FFA-Sat* saturated free fatty acids, *HIF-1* hypoxia inducible factor-1, *JNK c-jun n-terminal kinase, MMP* matrix metalloproteinase, P_{O2} oxygen pressure, *TLR4* toll-like receptor 4, *TNF-α* tumor necrosis factor alpha, *uPA* plasminogen activator urokinase-type, *VEGF* vascular endothelial growth factor, *VEGFR* VEGF receptor

The accumulation of diacylglycerols directly activates several PKC isoforms which besides inhibiting insulin receptor activation through serine phosphorylation, also promotes the activation of the IKKβ (Fig. 1.9) (Moeschel et al. 2004; Boden et al. 2005; Guilherme et al. 2008b). IKKβ is also a serine kinase which directly inhibits the insulin receptor and leads to the activation of the NF-κB. Moreover, JNK is also activated by the PKC and by fatty acid-activated toll-like receptor 4 (TLR4), contributing to the same final objective of NF-κB activation (Qatanani and Lazar 2007; Zhang et al. 2010). In turn, NF-κB activated the expression of proinflammatory and proangiogenic molecules (MCP-1, TNF-α, VEGF, adhesion molecules, etc.), which further contribute to the inhibition of substrate uptake (Fig. 1.9).

1.4.2 Adipose Tissue Hypoxia

It has been demonstrated that angiogenesis is a vital process in adipose tissue expansion and proper nutrient storage. Tissue expansion is believed to lead to the formation of physiological hypoxic regions, which induce the secretion of angiogenic factors and the reestablishment of tissue homeostasis. However, chronic hypoxia has been suggested to dramatically change adipokine secretion, with a decrease of adiponectin and an increase of proinflammatory adipokines (Hosogai et al. 2007; Wang et al. 2007; Goossens 2008; Trayhurn et al. 2008b; Wood et al. 2009). Despite the reasons to the development of chronic hypoxia are not known, it was described that obese individuals lack the postprandial increase of blood supply to the adipose tissue observed in lean individuals (Trayhurn et al. 2008b; Wood et al. 2009). Hypoxic regions have been shown in several diet-induced and genetic animal models of obesity, showing deficient angiogenesis, oxygen pressure, with decreased levels of endothelial cell markers and increased accumulation of the hypoxia probe pimonidazole and lactate formation (Hosogai et al. 2007; Ye et al. 2007; Goossens 2008; Pang et al. 2008; Trayhurn et al. 2008b; Wood et al. 2009). Despite being currently controversial, a current view supports that the oxygen diffusion distance (100 μm) is lower than adipocyte diameter in obesity (150–20 μm). This would lead to hypoxia-inducible factor-1 (HIF-1) stabilization and consequent expression of GLUT1 and secretion of angiogenic factors like the VEGF (Wood et al. 2007, 2009; Ye et al. 2007; Wang et al. 2007; Rausch et al. 2008; Goossens 2008; Trayhurn et al. 2008a; He et al. 2011). However, this view is recently being abandoned and Goossens and colleagues have reported that the adipose tissue of obese individuals may even be hyperoxic under certain circumstances due to decreased metabolic activity (Goossens et al. 2011). Thus, the real causes for impaired vascular function and hypoxia (if confirmed) are currently unknown.

1.4.3 Activation of Stress and Inflammatory Pathways by Lipid Mediators and Hypoxia: Consequences for Inflammatory Cell Recruitment

There is no consensus about the role of adipocytes and preadipocytes in tissue response to hypoxia as current models does not discriminate the secretome change occurring in each cell type. On the other hand, cell lines do not mimic all the events observed in vivo. Nevertheless, several authors have suggested NF-κB and HIF-1 activation leading to the expression of proinflammatory cytokines like IL-6, MCP-1, plasminogen activator inhibitor-1 (PAI-1), transforming growth factor (TGF)-β, and matrix metalloproteinases (MMP), but also leptin (Ye et al. 2007; Trayhurn et al. 2008b; Halberg et al. 2009; Wood et al. 2009). NF-κB activation in hypoxia is well documented in different models of hypoxia like tumors, interacting with HIF-1 to increase the expression of inflammatory cells (Fig. 1.9) (van Uden et al. 2008). It has been described that cultured hypoxic adipocytes secrete higher levels of MCP-1, TNF-α, and VEGF and increase GLUT1 expression. Preadipocytes apparently are more susceptible to hypoxic stimuli, acquiring the ability to secrete leptin. However, hypoxia inhibits preadipocytes differentiation to adipocytes and promotes their transformation to macrophages. The release of chemoattractant factors like MCP-1 further increases the number of macrophages in the tissue (Trayhurn et al. 2008b; Wood et al. 2009).

1.4.3.1 Adipose Tissue Immune System: Recruitment of Inflammatory Cells and Secretion of Inflammatory Cytokines

Changes in the metabolic status are detected by the immune system, which acts in order to establish homeostasis. The activation of the immune system is an important mechanism in times of excessive nutritional fluxes, avoiding excessive nutrient uptake by the cells and maintaining cell viability and nutrient distribution by the different cells of the body. However, the low-grade inflammation observed in obesity resulting from chronic nutrient excess leads to insulin resistance in adipocytes (Galic et al. 2010). Such mechanisms include MCP-1 expression, which is involved in recruiting cells from the immune system to the tissue. Such cells, mainly monocytes-derived macrophages, further support the inflammatory feedback in the tissue, with the release of inflammatory mediators like TNF-α. In turn, inflammatory mediators further inhibit nutrient accumulation and tissue growth, showing the existence of a metabolism—immunity continuous (Schäffler et al. 2007; Galic et al. 2010).

A large number of circulating monocytes can be recruited to the adipose tissue through the action of proinflammatory cytokines and chemokines and develop into macrophages (Schäffler et al. 2007; Guilherme et al. 2008a; Surmi and Hasty 2010; Olefsky and Glass 2010). Macrophages may be derived from recruited monocytes and from differentiation from preadipocytes or mesenchymal cells and can become 50% of the total number of cells in the adipose tissue in obesity (Schäffler et al. 2007; Guilherme et al. 2008a; Surmi and Hasty 2010; Olefsky and Glass 2010). MCP-1, a major chemokine, was shown to be increased in obese patients and animal models. Its inhibition in diabetic and obese animal models improves insulin resistance and glucose tolerance (Guilherme et al. 2008a; Olefsky and Glass 2010).

The reason why preadipocytes can differentiate into macrophage is because they share the same cell lineage, sharing several intracellular mechanisms of lipid uptake, storage, and metabolism (Schäffler et al. 2007). Macrophages are known to surround death adipocytes forming crown-like structures, which are typically found in the adipose tissue of obese patients and animal models (Trayhurn et al. 2008a). Such macrophages uptake and store large amounts of fat released from their lipid droplets, leading to the formation of foam cells. In turn, foam cells have an increased inflammatory activity with increased activation of the NF-κB and secretion of pro-inflammatory factors (Ye et al. 2007; Rutkowski et al. 2009; Maury and Brichard 2010; Galic et al. 2010). Adipose tissue hypoxic regions were shown to be strongly populated by macrophages, being strongly positive for their membrane marker F4/80, and by CD4+ and CD8+ T cells, although their role in metabolic syndrome is currently unknown (Ye et al. 2007; Rausch et al. 2008; Goossens 2008; Trayhurn et al. 2008a; Wood et al. 2009).

Two populations of macrophages may be found at the adipose tissue: M1 macrophages recruited to the tissue by the gradient of chemokines like the MCP-1 typically have a more pronounced proinflammatory secretome, originating most of the foam cells. Tissue resident M2 macrophages secreting a broad range of angiogenic and tissue remodeling factors like metalloproteinases are involved in tissue adaptation to hypoxia and have a more modest inflammatory activity (Maury and Brichard 2010; Olefsky and Glass 2010; Galic et al. 2010).

Activation of endothelial cells by inflammatory cytokines leads to the overexpression of adhesion molecules, which are important for macrophage binding and migration to the tissue matrix (Rutkowski et al. 2009; Maury and Brichard 2010). Moreover, the increased HIF-1-dependent expression of the migration inhibition factor (MIF-1) in hypoxia inhibits macrophage exit from the tissue, contributing to their accumulation (Ye et al. 2007). Most of the cytokines released by the adipose tissue in obesity, namely TNF-α, IL-6, and resistin, are derived from the infiltrating macrophages. TNF-α acts on adipocytes in order to inhibit insulin receptor and the PPARγ and to activate NF-κB, completing the inflammatory feedback in the tissue aiming to block further nutrient uptake and to increase angiogenesis and tissue remodeling (Wellen and Hotamisligil 2003; Moeschel et al. 2004; Sandu et al. 2005; Qatanani and Lazar 2007; Guilherme et al. 2008a; Olefsky and Glass 2010; Min et al. 2016). On the other hand, TNF-α and MCP-1 inhibition in models of diet-induced insulin resistance was shown to improve insulin sensitivity (Wellen and Hotamisligil 2003; Bugianesi et al. 2005; Guilherme et al. 2008b).

1.4.4 Inhibition of Nutrient Uptake by Inflammatory Stimuli: Insulin Resistance and PPARγ Inhibition

Insulin resistance is characterized by a deficient cell response to physiological insulin concentrations, leading to increased β-cell insulin secretion and compensatory hyperinsulinemia (Yki-Järvinen 2005, 2015; Yki-Järvinen and Westerbacka 2005). Insulin resistance is not associated with the body mass index (BMI), but depends on impaired adipose tissue lipid storage and activation of inflammatory pathways (Yki-Järvinen 2005, 2015; Yki-Järvinen and Westerbacka 2005). Anti-inflammatory molecules like PPARγ agonists (thiazolidinediones and polyunsaturated fatty acids) and salicylates are known to improve insulin sensitivity. As well, physical exercise is known to decrease inflammation and to improve insulin sensitivity (Bugianesi et al. 2005; Guilherme et al. 2008b; Oliveira et al. 2011).

Fatty acids are known to directly cause insulin resistance. Even nondiabetic individuals were shown to decrease insulin sensitivity after an acute exposure to fatty acids, which is a mechanism important to prevent excessive accumulation of nonesterified fatty acids in the cytoplasm (Golay and Ybarra 2005; Einstein et al. 2010). Circulating free fatty acids are also strong inducers of insulin resistance acting through the membrane receptor TLR4 (Qatanani and Lazar 2007; Guilherme et al. 2008a; Zhang et al. 2010; Olefsky and Glass 2010).

Hypoxia is a strong inhibitor of nutrient uptake. In cultured adipocytes hypoxia-dependent HIF-1 activation leads to the upregulation of stress pathways that blunts insulin signaling (Regazzetti et al. 2009; Ye 2009; Wood et al. 2009; Zhang et al. 2010). As well, the same mechanisms also inhibit PPARγ activity, leading to decreased secretion of adiponectin (Gentil et al. 2006; Chen et al. 2006; Goossens

2008; Ye 2009). However, it was shown that the overexpression of an inactive form of HIF-1 during a high-fat diet challenge aggravates insulin resistance, suggesting distinct acute and chronic effects of hypoxia in adipose tissue. Such evidences support the idea that acute activation of hypoxia response mechanisms are important regulators of nutrient uptake be adipocytes, but long-term activation of such mechanisms is important to the regulation of mitochondrial biogenesis and angiogenesis (Zhang et al. 2010; He et al. 2011). In fact, adipose tissue vascular density was shown to correlate with insulin ability to suppress lipolysis, i.e., insulin sensitivity (Pasarica et al. 2010).

A main contributor to the long-term activation of such mechanisms is macrophage-derived inflammatory signals. Activation of inflammatory signals results not only in the inhibition of nutrient uptake but also the release of those already accumulated, namely through the stimulation of lipolysis. Cytokines like the TNF-α and IL-6 are known to be mostly produced in adipose tissue infiltrating macrophages. These cytokines were shown to activate NF-κB through serine kinases like JNK and IKK, leading to the transcription of inflammatory factors and proteins involved in stress and inflammatory pathways (Fig. 1.10) (Tilg and Moschen 2006; Kiess et al. 2008; Olefsky and Glass 2010). Moreover, serine kinases phosphorylate

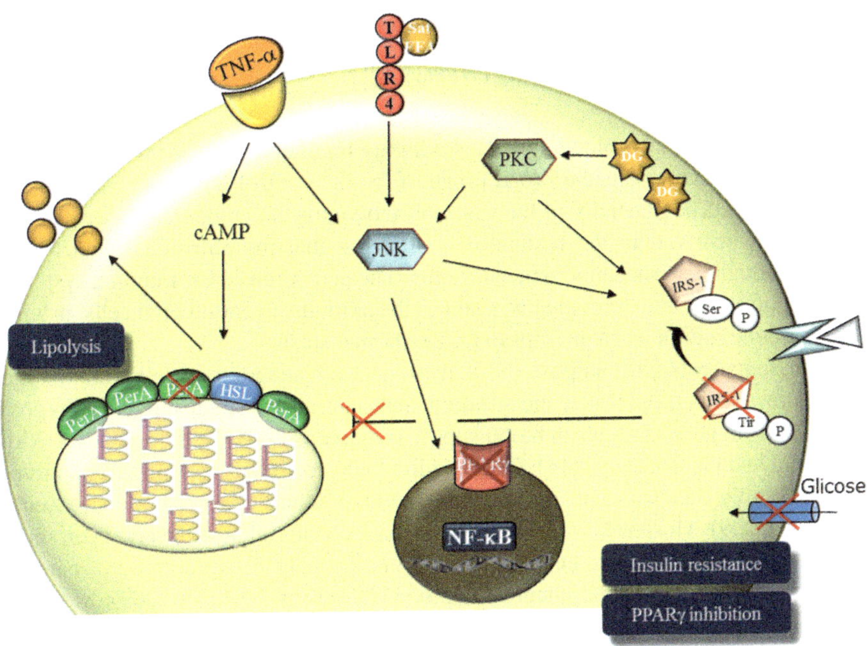

Fig. 1.10 Consequences of adipocyte inflammatory mechanisms for inhibition of insulin signaling and PPARγ activity, inhibiting nutrient uptake and promoting lipolysis. *FFA-Sat* saturated free fatty acids, *IRS-1* insulin receptor substract-1, *JNK* c-jun n-terminal kinase, *NF-kB* nuclear factor kappa B, *PerA* perilipin A, *PKC* protein kinase C, *PPARγ* peroxisome proliferating-activated receptor γ, *Ser* serine, *TLR4* toll-like receptor 4, *TNF-α* tumor necrosis factor alpha

the insulin receptor and its substrate (IRS-1) in serine residues instead of the stimulatory tyrosine phosphorylation, leading to the inactivation of the pathway (Fig. 1.10) (Moeschel et al. 2004; Boden et al. 2005; Wellen and Hotamisligil 2005; Qatanani and Lazar 2007; Guilherme et al. 2008a; Wang et al. 2009; Kaneto et al. 2010; Maury and Brichard 2010; Olefsky and Glass 2010; Galic et al. 2010). JNK was also shown to increase the expression of SOCS proteins, which are known to bind to and signalize IRS-1 to proteasome degradation (Ueki et al. 2004a; Howard and Flier 2006). Such mechanisms perpetuate the inhibition of insulin signaling in response to inflammatory cytokines.

Regarding lipid storage, TNF-α was shown to inhibit PPARγ activity, causing the downregulation of key enzymes in fatty acid uptake (LPL and FATP) and esterification, as well as adiponectin (Boden et al. 2005; Juge-Aubry et al. 2005a; Qatanani and Lazar 2007; Guilherme et al. 2008b). TNF-α was also shown to increase intracellular levels of cAMP, which is known to activate hormone-sensitive lipase (HSL) and thus to induce lipolysis (Fig. 1.10) (Wellen and Hotamisligil 2003; Juge-Aubry et al. 2005a; Guilherme et al. 2008a; Maury and Brichard 2010; Galic et al. 2010). Interestingly, hypoxia is also associated with decreased levels of ATP and increased cAMP levels, also contributing to lipolysis (Ye 2009).

1.4.5 Activation of Angiogenesis

The formation of new blood vessels is a highly regulated process involving several cell types, which produce a broad range of factors controlling vessel permeability and stability. The plasticity of the vascular network is also modulated by a series of hormones involved in the metabolism, showing that the nutritional status has a major influence in the angiogenic stimulus. The activation of the immune system in metabolic syndrome dramatically changes the stimulus to endothelial cells, promoting vascularization as an attempt to restore homeostasis.

Obese patients with adipose tissue hypoxia were shown to have normal oxygen pressure and hemoglobin concentration and saturation, suggesting that hypoxia derives from local changes in blood supply (Goossens 2008; Rutkowski et al. 2009; Corvera and Gealekman 2014). Obese individuals may have a 30–40% reduction of blood supply than lean ones and a similar reduction was observed in obese Zucker rats (Ye 2009). Moreover, obese and type 2 diabetic animal models have been shown to have a decrease of the endothelial cells marker CD31, which may possibly be associated with the development of hypoxia (Goossens 2008; Ye 2009). Moreover, recent studies also demonstrated a reduction of endothelial progenitor cells, which may contribute to impaired angiogenic function (Neels et al. 2004; Gealekman et al. 2008; Tam et al. 2009; Corvera and Gealekman 2014).

Inhibition of angiogenesis was initially suggested as an effective strategy to prevent adipose tissue growth (Lijnen 2008). However, this hypothesis has been abandoned given that the vascular network was shown to be essential for proper tissue oxygenation and storage and metabolic functions and blood supply was shown to correlate with insulin sensitivity (Mannerås-Holm and Krook 2012). Moreover, it was shown that adipose tissue expansion is necessary for nutrient storage and to prevent ectopic lipid deposition in liver and skeletal muscle.

The formation of hypoxic regions activates the secretion of angiogenic factors which ultimately restore homeostasis (Cao 2007; Goossens 2008; Rutkowski et al. 2009). However, if angiogenesis is not enough, hypoxia is maintained and HIF-1 levels increase in order to stimulate the expression of VEGF, angiopoietins, leptin, and matrix-remodeling factors (Minet et al. 2001; Yamakawa et al. 2003; Cao 2007; Hosogai et al. 2007; Goossens 2008). However, HIF-1 chronic activation also leads to the secretion of fibrotic factors (Rutkowski et al. 2009; Wood et al. 2009; Suga et al. 2010).

1.4.5.1   Pro and Antiangiogenic Factors

Adipose tissue angiogenesis is regulated by the balance between pro and antiangiogenic factors derived from all the cells of the tissue (Neels et al. 2004; Cao 2007; Rutkowski et al. 2009; Tinahones et al. 2012; Corvera and Gealekman 2014).

VEGF is the main endothelial cell growth factor, being associated with the development of obesity (Cao 2007; Rutkowski et al. 2009; Tinahones et al. 2012; Corvera and Gealekman 2014). Animal models of obesity and obese individuals have increased VEGF levels, which were shown to decrease after weight loss (Gómez-Ambrosi et al. 2010). VEGF is mostly produced by stroma cells and by mature adipocytes, acting in adjacent capillaries (Fig. 1.11) (Neels et al. 2004; Christiaens and Lijnen 2010). During adipose tissue expansion, a significant part of VEGF is produced by infiltrating inflammatory cells. VEGFR2 inhibition prevents diet-induced tissue expansion and preadipocyte differentiation, showing that angiogenesis is crucial to the development of new adipocytes (Tam et al. 2009; Tran et al. 2012; Min et al. 2016). The major regulators of VEGF expression are hypoxia and insulin, while inflammatory cytokines can also induce its expression. VEGF inhibition as a strategy to prevent obesity is still controversial. VEGF was shown to increase in genetic models of obesity, but not in models of metabolic syndrome, suggesting that hypoxia-induced VEGF overexpression may be lost in hyperglycemia (Trayhurn et al. 2008b; Wood et al. 2009). Such observations derive from impaired HIF-1 stabilization in hyperglycemic models, resulting in impaired VEGF expression and dysregulation of the angiogenic process (Bento et al. 2010b). Moreover, insulin resistance may also contribute to impaired VEGF expression in type 2 diabetes. Thus, the initial idea that inhibiting VEGF could be an effective strategy to prevent obesity is currently being changed and is now believed that VEGF in necessary for proper adipose tissue angiogenesis and function (Treins et al. 2002; Hausman and Richardson 2004; Ye 2009; Vona-Davis and Rose 2009).

VEGF effects on the vascular network depend on its interaction with angiopoietins (Ang). Ang-1 and Ang-2 have opposite effects on the vasculature, although both were shown to be necessary for embryonic development. While Ang-1 is involved in vessel stabilization, increasing cell-cell adhesion, Ang-2 induces destabilization of cell-cell and cell-matrix interactions in order to allow cell proliferation and migration (Cao 2007; Christiaens and Lijnen 2010; Corvera and Gealekman 2014).

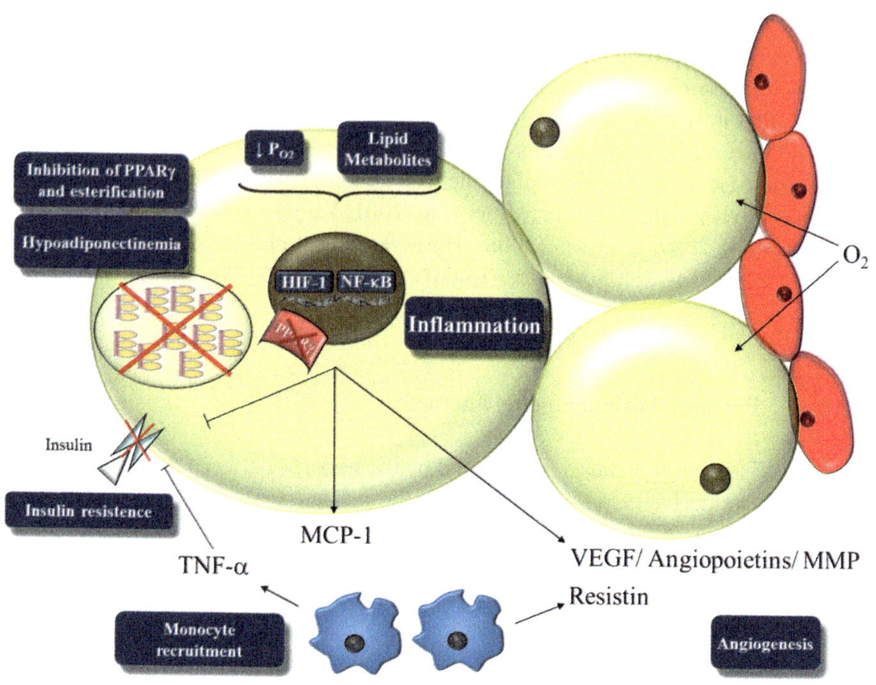

Fig. 1.11 Causes and consequences of inflammatory processes in adipose tissue. Inflammation directly inhibits excessive nutrient uptake by adipocytes and their continuous hypertrophy. Moreover, such mechanisms also stimulate angiogenic function in order to reestablish homeostasis. *HIF-1* hypoxia inducible factor-1, *MCP-1* monocyte chemoattractant protein-1, *MMP* matrix metalloproteinases, *NF-kB* nuclear factor-kappa B, P_{O2} oxygen pressure, *PPARγ* peroxisome proliferating-activated receptor γ, *TNF-α* tumor necrosis factor alpha, *VEGF* vascular endothelial growth factor

Ang-1 promotes vessels stiffness through modulation of TGFβ activity and deposition of extracellular matrix proteins (Cao 2007; Rutkowski et al. 2009; Christiaens and Lijnen 2010). On the other hand, Ang-2 expression is induced by hypoxia like VEGF and its receptors (Tie-2) are often co-localized with VEGFR2 (Christiaens and Lijnen 2010; Bento et al. 2010a). Ang-2 action in order to induce destabilization of cell interactions is necessary for VEGF-induced proliferation (Christiaens and Lijnen 2010; Bento et al. 2010a). The absence of VEGF after Ang-2 induced destabilization of endothelial cells leads to the formation of aberrant blood vessels. Such mechanisms have been suggested to be involved in the development of diabetic vascular complications and to be present in adipose tissue during the process of metabolic dysregulation (Bento et al. 2010a; Matafome et al. 2012).

Leptin was also shown to be proangiogenic, sharing many mechanisms with insulin, namely MAPK activation (Cao 2007; Vona-Davis and Rose 2007; Lijnen 2008; Christiaens and Lijnen 2010; Bento et al. 2010a). Leptin and insulin promote endothelial cells migration, proliferation, and survival, causing the increased VEGF secretion and VEGFR2 expression. Moreover, leptin was shown to increase MMP-dependent matrix remodeling (Cao 2007; Vona-Davis and Rose 2007; Lijnen 2008; Christiaens and Lijnen 2010; Bento et al. 2010a). On the other hand, adiponectin was shown to counteract all these effects, by inhibiting VEGF actions in endothelial cells (Vona-Davis and Rose 2009). Adiponectin is mostly produced by small and non-inflamed adipocytes. Thus, when adipose tissue blood supply is enough, adiponectin prevents excessive blood vessel growth. On the other hand, leptin is produced proportionally to fat mass and thus its angiogenic functions are necessary to sustain tissue angiogenesis.

Other factors that have been suggested to be involved in the regulation of adipose tissue angiogenesis are fibroblast growth factor (FGF) (Lijnen 2008; Christiaens and Lijnen 2010), hepatic growth factor (HGF), and platelet-derived growth factor (PDGF) (Lijnen 2008; Pang et al. 2008; Christiaens and Lijnen 2010). As well, fibrotic and inflammatory factors like TGF-β and TNF-α have been shown to exert direct effects on adipose tissue endothelial cells. While TNF-α is involved in endothelial cell proliferation, TGF-β is important for the deposition of the extracellular matrix and vessel stabilization (Lijnen 2008; Niu et al. 2008; Christiaens and Lijnen 2010).

Angiogenesis is not only regulated by endothelial cell proliferation and migration, but also by matrix remodeling. Extracellular matrix degradation is a highly regulated process by the plasminogen system and by MMPs. The plasminogen system is known for its role on blood clotting but it is also shown to be involved in the regulation of extracellular matrix degradation (Lijnen 2008; Carter and Church 2009; Christiaens and Lijnen 2010). Animal models of obesity and metabolic syndrome have been shown to have increased levels of the inhibitor of plasminogen activator (PAI-1). Transgenic models of adipocyte PAI-1 expression have shown lack of visceral adipose tissue, showing that PAI-1 inhibits matrix degradation and vessel growth. However, although PAI-1 is apparently involved in preventing matrix degradation, its expression is increased by hypoxia and its levels are correlated with the body mass index (BMI) and insulin resistance, suggesting that its function in adipose tissue may be far more complex. Regarding MMPs, they are involved in matrix degradation, allowing endothelial cell migration. Their levels are increased in obesity and their overexpression results in decreased capillary density in adipose tissue. However, the mechanisms of their regulation and their real impact in adipose tissue pathophysiology are currently unknown (Lijnen 2008; Christiaens and Lijnen 2010; Corvera and Gealekman 2014).

1.4.6 *Inflammatory Feedback in Adipose Tissue: Unifying Mechanism*

The dysregulation of adipocyte metabolism occurs with the accumulation of fatty acids and their intermediates and with the development of hypoxia. All these conditions lead to the activation of stress pathways resulting in inhibition of nutrient uptake and increased expression of chemokines involved in the recruitment of inflammatory cells to the tissue (Galic et al. 2010). PPARγ inhibition also leads to decreased adiponectin secretion, which contributes to the inflammatory feedback in the tissue. Monocyte recruitment and their differentiation into macrophages further increase the release of inflammatory chemokines, cytokines like TNF-α, and angiogenic factors. Such events further chronically inhibit insulin signaling and fatty acid uptake, also increasing lipolysis (Fig. 1.11) (Qatanani and Lazar 2007; Olefsky and Glass 2010).

References

Ahima RS (2005) Central actions of adipocyte hormones. Trends Endocrinol Metab 16:307–313

Ahima RS, Flier JS (2000) Leptin. Annu Rev Physiol 62:413–437

Arner P (2005) Human fat cell lipolysis: biochemistry, regulation and clinical role. Best Pract Res Clin Endocrinol Metab 19:471–482

Bento CF, Fernandes R, Matafome P, Sena C, Seiça R, Pereira P (2010a) Methylglyoxal-induced imbalance in the ratio of vascular endothelial growth factor to angiopoietin 2 secreted by retinal pigment epithelial cells leads to endothelial dysfunction. Exp Physiol 95:955–970

Bento CF, Fernandes R, Ramalho J, Marques C, Shang F, Taylor A, Pereira P (2010b) The chaperone-dependent ubiquitin ligase CHIP targets HIF-1α for degradation in the presence of methylglyoxal. PLoS One 5:1–13

Blüher M, Mantzoros CS (2015) From leptin to other adipokines in health and disease: facts and expectations at the beginning of the 21st century. Metabolism 64(1):131–145

Boden G, She P, Mozzoli M, Cheung P, Gumireddy K, Reddy P, Xiang X, Luo Z, Ruderman N (2005) Free fatty acids produce insulin resistance and activate the proinflammatory nuclear factor-κb pathway in rat liver. Diabetes 54:3458–3465

Bugianesi E, McCullough AJ, Marchesini G (2005) Insulin resistance: a metabolic pathway to chronic liver disease. Hepatology 42:987–1000

Cao Y (2007) Angiogenesis modulates adipogenesis and obesity. J Clin Invest 117:2362–2368

Carter JC, Church FC (2009) Obesity and breast cancer: the roles of peroxisome proliferator-activated receptor-gamma and plasminogen activator inhibitor-1. PPAR Res 2009:345320

Catalano S, Marsico S, Giordano C, Mauro L, Rizza P, Panno ML, Andò S (2003) Leptin enhances, via AP-1, expression of aromatase in the MCF-7 cell line. J Biol Chem 278:28668–28676

Chen B, Lam KSL, Wang Y, Wu D, Lam MC, Shen J et al (2006) Hypoxia dysregulates the production of adiponectin and plasminogen activator inhibitor-1 independent of reactive oxygen species in adipocytes. Biochem Biophys Res Commun 341:549–556

Christiaens V, Lijnen HR (2010) Angiogenesis and development of adipose tissue. Mol Cell Endocrinol 318:2–9

Corvera S, Gealekman O (2014) Adipose tissue angiogenesis: impact on obesity and type-2 diabetes. Biochim Biophys Acta 1842:463–472

De Souza Batista CM, Yang RZ, Lee MJ, Glynn NM, DZ Y, Pray J et al (2007) Omentin plasma levels and gene expression are decreased in obesity. Diabetes 56:1655–1661

Ding M, Rzucidlo EM, Davey JC, Xie Y, Liu R, Jin Y, Stavola L, Martin KA (2012) Adiponectin in the heart and vascular system. Vitam Horm 90:289–319

Einstein FH, Huffman DM, Fishman S, Jerschow E, Heo HJ, Atzmon G, Schechter C, Barzilai N, Muzumdar RH (2010) Aging per se increases the susceptibility to free fatty acid-induced insulin resistance. J Gerontol A Biol Sci Med Sci 65(8):800–808

Fasshauer M, Blüher M (2015) Adipokines in health and disease. Trends Pharmacol Sci 36(7):461–470

Galic S, Oakhill JS, Steinberg GR (2010) Adipose tissue as an endocrine organ. Mol Cell Endocrinol 316(2):129–139

Gallí M, Van Gool F, Rongvaux A, Andris F, Leo O (2010) The nicotinamide phosphoribosyltransferase: a molecular link between metabolism, inflammation, and cancer. Cancer Res 70(1):8–11

Gealekman O, Burkart A, Chouinard M, Nicoloro SM, Straubhaar J, Corvera S (2008) Enhanced angiogenesis in obesity and in response to PPARgamma activators through adipocyte VEGF and ANGPTL4 production. Am J Physiol Endocrinol Metab 295(5):E1056–E1064

Gealekman O, Guseva N, Gurav K, Gusev A, Hartigan C, Thompson M, Malkani S, Corvera S (2012) Effect of rosiglitazone on capillary density and angiogenesis in adipose tissue of normoglycaemic humans in a randomised controlled trial. Diabetologia 55:2794–2799

Gentil C, Le Jan S, Philippe J, Leibowitch J, Sonigo P, Germain S, Piétri-Rouxel F (2006) Is oxygen a key factor in the lipodystrophy phenotype? Lipids Health Dis 5:1–11

Glassford AJ, Yue P, Sheikh AY, Chun HJ, Zarafshar S, Chan DA, Reaven GM, Quertermous T, Tsao PS (2007) HIF-1 regulates hypoxia- and insulin-induced expression of apelin in adipocytes. Am J Physiol Endocrinol Metab 293(6):E1590–E1596

Golay A, Ybarra J (2005) Link between obesity and type 2 diabetes. Best Pract Res Clin Endocrinol Metab 19(4):649–663

Gómez-Ambrosi J, Catalán V, Rodríguez A, Ramírez B, Silva C, Gil MJ, Salvador J, Frühbeck G (2010) Involvement of serum vascular endothelial growth factor family members in the development of obesity in mice and humans. J Nutr Biochem 21(8):774–780

Goossens GH (2008) The role of adipose tissue dysfunction in the pathogenesis of obesity-related insulin resistance. Physiol Behav 94(2):206–218

Goossens GH, Bizzarri A, Venteclef N, Essers Y, Cleutjens JP, Konings E, Jocken JW, Cajlakovic M, Ribitsch V, Clément K, Blaak EE (2011) Increased adipose tissue oxygen tension in obese compared with lean men is accompanied by insulin resistance, impaired adipose tissue capillarization, and inflammation. Circulation 124(1):67–76

Guilherme A, Virbasius J, Puri V, Czech MP (2008a) Adipocyte dysfunctions linking obesity to insulin resistance and type 2 diabetes. Nat Rev Mol Cell Biol 9(5):367–377

Guilherme A, Virbasius JV, Puri V, Czech MP (2008b) Adipocyte dysfunctions linking obesity to insulin resistance and type 2 diabetes. Nat Rev Mol Cell Biol 114:715–728

Halberg N, Khan T, Trujillo ME, Wernstedt-Asterholm I, Attie AD, Sherwani S et al (2009) Hypoxia-inducible factor 1alpha induces fibrosis and insulin resistance in white adipose tissue. Mol Cell Biol 29(16):4467–4483

Handisurya A, Riedl M, Vila G, Maier C, Clodi M, Prikoszovich T et al (2010) Serum vaspin concentrations in relation to insulin sensitivity following RYGB-induced weight loss. Obes Surg 20(2):198–203

Hausman GJ, Richardson RL (2004) Adipose tissue angiogenesis. J Anim Sci 82:925–934

He Q, Gao Z, Yin J, Zhang J, Yun Z, Ye J (2011) Regulation of HIF-1{alpha} activity in adipose tissue by obesity-associated factors: adipogenesis, insulin, and hypoxia. Am J Physiol Endocrinol Metab 300(5):E877–E885

Hosogai N, Fukuhara A, Oshima K, Miyata Y, Tanaka S, Segawa K et al (2007) Adipose tissue hypoxia in obesity and its impact on adipocytokine dysregulation. Diabetes 56:901–911

Howard JK, Flier JS (2006) Attenuation of leptin and insulin signaling by SOCS proteins. Trends Endocrinol Metab 17(9):365–371

Imrie H, Abbas A, Kearney M (2010) Insulin resistance, lipotoxicity and endothelial dysfunction. Biochim Biophys Acta 1801(3):320–326

Jeong E, Youn BS, Kim DW, Kim EH, Park JW, Namkoong C et al (2010) Circadian rhythm of serum vaspin in healthy male volunteers: relation to meals. J Clin Endocrinol Metab 95(4):1869–1875

Juge-Aubry CE, Henrichot E, C a M (2005a) Adipose tissue: a regulator of inflammation. Best Pract Res Clin Endocrinol Metab 19(4):547–566

Juge-Aubry CE, Somm E, Pernin A, Alizadeh N, Giusti V, Dayer J-M, Mcier C (2005b) Adipose tissue is a regulated source of interleukin-10. Cytokine 29:270–274

Kamon J, Yamauchi T, Terauchi Y, Kubota N, Kadowaki T (2003) The mechanisms by which PPARgamma and adiponectin regulate glucose and lipid metabolism. Nihon yakurigaku zasshi Folia pharmacologica Japonica 122(4):294–300

Kaneto H, Katakami N, Matsuhisa M, Matsuoka T (2010) Role of reactive oxygen species in the progression of type 2 diabetes and atherosclerosis. Mediat Inflamm 2010:1–11

Kawano Y, Cohen DE (2013) Mechanisms of hepatic triglyceride accumulation in non-alcoholic fatty liver disease. J Gastroenterol 48(4):434–441

Kiess W, Petzold S, Töpfer M, Garten A, Blüher S, Kapellen T, Körner A, Kratzsch J (2008) Adipocytes and adipose tissue. Best Pract Res Clin Endocrinol Metab 22(1):135–153

Klöting N, Berndt J, Kralisch S, Kovacs P, Fasshauer M, Schön MR, Stumvoll M, Blüher M (2006) Vaspin gene expression in human adipose tissue: association with obesity and type 2 diabetes. Biochem Biophys Res Commun 339(1):430–436

Langin D (2006) Adipose tissue lipolysis as a metabolic pathway to define pharmacological strategies against obesity and the metabolic syndrome. Pharmacol Res 53(6):482–491

Laubner K, Kieffer TJ, Lam NT, Niu X, Jakob F, Seufert J (2005) Inhibition of preproinsulin gene expression by leptin induction of suppressor of cytokine signaling 3 in pancreatic beta-cells. Diabetes 54:3410–3417

Lazar MA (2007) Resistin- and obesity-associated metabolic diseases. Horm Metab Res 39(10):710–716

Lee TS, Lin CY, Tsai JY, YL W, KH S, KY L et al (2009) Resistin increases lipid accumulation by affecting class a scavenger receptor, CD36 and ATP-binding cassette transporter-A1 in macrophages. Life Sci 84:97–104

Lee J-Y, Hashizaki H, Goto T, Sakamoto T, Takahashi N, Kawada T (2011) Activation of peroxisome proliferator-activated receptor-α enhances fatty acid oxidation in human adipocytes. Biochem Biophys Res Commun 407(4):818–822

Letra L, Santana I, Seiça R (2014) Obesity as a risk factor for Alzheimer's disease: the role of adipocytokines. Metab Brain Dis 29:563–568

Lijnen HR (2008) Angiogenesis and obesity. Cardiovasc Res 78:286–293

Lonardo A, Ballestri S, Marchesini G, Angulo P, Loria P (2015) Nonalcoholic fatty liver disease: a precursor of the metabolic syndrome. Dig Liver Dis 47(3):181–190

Lorincz AM, Sukumar S (2006) Molecular links between obesity and breast cancer. Endocr Relat Cancer 13(2):279–292

Mannerås-Holm L, Krook A (2012) Targeting adipose tissue angiogenesis to enhance insulin sensitivity. Diabetologia 55:2562–2564

Matafome P (2013) Common mechanisms of dysfunctional adipose tissue and obesity-related cancers. Diabetes Metab Res Rev 29(4):285–295

Matafome P, Santos-Silva D, Crisóstomo J, Rodrigues T, Rodrigues L, Sena CM, Pereira P, Seiça R (2012) Methylglyoxal causes structural and functional alterations in adipose tissue independently of obesity. Arch Physiol Biochem 118(2):58–68

Maury E, Brichard SM (2010) Adipokine dysregulation, adipose tissue inflammation and metabolic syndrome. Mol Cell Endocrinol 314(1):1–16

Meier U, Gressner AM (2004) Endocrine regulation of energy metabolism: review of pathobiochemical and clinical chemical aspects of leptin, ghrelin, adiponectin, and resistin. Clin Chem 50(9):1511–1525

Min SY, Kady J, Nam M, Rojas-Rodriguez R, Berkenwald A, Kim JH et al (2016) Human "brite/beige" adipocytes develop from capillary networks, and their implantation improves metabolic homeostasis in mice. Nat Med 22(3):312–318

Minet E, Michel G, Mottet D (2001) Transduction pathways involved in hypoxia-inducible factor-1 phosphorylation and activation. Free Radic Biol Med 31(7):847–855

Mocan Hognogi LD, Simiti LV (2016) The cardiovascular impact of visfatin—an inflammation predictor biomarker in metabolic syndrome. Clujul Medical 89:322

Moeschel K, Beck A, Weigert C, Lammers R, Kalbacher H, Voelter W, Schleicher ED, Häring H-U, Lehmann R (2004) Protein kinase C-zeta-induced phosphorylation of Ser318 in insulin receptor substrate-1 (IRS-1) attenuates the interaction with the insulin receptor and the tyrosine phosphorylation of IRS-1. J Biol Chem 279(24):25157–25163

Moreno-Navarrete JM, Catalán V, Ortega F, Gómez-Ambrosi J, Ricart W, Frühbeck G, Fernández-Real JM (2010) Circulating omentin concentration increases after weight loss. Nutr Metab (Lond) 7:27

Münzberg H, Björnholm M, Bates SH, Myers MG (2005) Leptin receptor action and mechanisms of leptin resistance. Cell Mol Life Sci 62(6):642–652

Nawrocki AR, Rajala MW, Tomas E, Pajvani UB, Saha AK, Trumbauer ME et al (2006) Mice lacking adiponectin show decreased hepatic insulin sensitivity and reduced responsiveness to peroxisome proliferator-activated receptor gamma agonists. J Biol Chem 281(5):2654–2660

Neels JG, Thinnes T, Loskutoff DJ (2004) Angiogenesis in an in vivo model of adipose tissue development. FASEB J 18:983–1002

Nielsen NB, Højbjerre L, Sonne MP, Alibegovic AC, Vaag A, Dela F, Stallknecht B (2009) Interstitial concentrations of adipokines in subcutaneous abdominal and femoral adipose tissue. Regul Pept 155(1–3):39–45

Niu J, Azfer A, Zhelyabovska O, Fatma S, Kolattukudy PE (2008) Monocyte chemotactic protein (MCP)-1 promotes angiogenesis via a novel transcription factor, MCP-1-induced protein (MCPIP). J Biol Chem 283(21):14542–14551

Olefsky JM, Glass CK (2010) Macrophages, inflammation, and insulin resistance. Annu Rev Physiol 72:219–246

Oliveira AG, Carvalho BM, Tobar N, Ropelle ER, Pauli JR, R a B, Guadagnini D, Carvalheira JBC, Saad MJA (2011) Physical exercise reduces circulating lipopolysaccharide and TLR4 activation and improves insulin signaling in tissues of DIO rats. Diabetes 60:784–796

Ouchi N, Kihara S, Arita Y, Nishida M, Matsuyama A, Okamoto Y et al (2001) Adipocyte-derived plasma protein, adiponectin, suppresses lipid accumulation and class a scavenger receptor expression in human monocyte-derived macrophages. Circulation 103:1057–1063

Pais R, Silaghi H, Silaghi AC, Rusu ML, Dumitrascu DL (2009) Metabolic syndrome and risk of subsequent colorectal cancer. World J Gastroenterol 15(41):5141–5148

Pang C, Gao Z, Yin J, Zhang J, Jia W, Ye J (2008) Macrophage infiltration into adipose tissue may promote angiogenesis for adipose tissue remodeling in obesity. Am J Physiol Endocrinol Metab 295(2):E313–E322

Park M, Sweeney G (2013) Direct effects of adipokines on the heart: focus on adiponectin. Heart Fail Rev 18(5):631–644

Pasarica M, Rood J, Ravussin E, Schwarz J-M, Smith SR, Redman LM (2010) Reduced oxygenation in human obese adipose tissue is associated with impaired insulin suppression of lipolysis. J Clin Endocrinol Metab 95(8):4052–4055

Pitkin SL, Maguire JJ, Bonner TI, Davenport AP (2010) International Union of Basic and Clinical Pharmacology. LXXVIII. Lysophospholipid receptor nomenclature. Pharmacol Rev 62(3):331–342

Qatanani M, Lazar MA (2007) Mechanisms of obesity-associated insulin resistance: many choices on the menu. Genes Dev 21(12):1443–1455

Qatanani M, Szwergold N (2009) Macrophage-derived human resistin exacerbates adipose tissue inflammation and insulin resistance in mice. J Clin Invest 119(3):531–539

Rajala MW, Scherer PE (2003) Minireview: the adipocyte—at the crossroads of energy homeostasis, inflammation, and atherosclerosis. Endocrinology 144:3765–3773

Ram VJ (2003) Therapeutic significance of peroxisome proliferator-activated receptor modulators in diabetes. Drug Today 39:609–632

Rausch ME, Weisberg S, Vardhana P, Tortoriello D V (2008) Obesity in C57BL/6J mice is characterized by adipose tissue hypoxia and cytotoxic T-cell infiltration. Int J Obes 32:451–463

Regazzetti C, Peraldi P, Grémeaux T, Najem-Lendom R, Ben-Sahra I, Cormont M et al (2009) Hypoxia decreases insulin signaling pathways in adipocytes. Diabetes 58:95–103

Robertson SA, Rae CJ, Graham A (2009) Resistin: TWEAKing angiogenesis. Atherosclerosis 203:34–37

Rutkowski JM, Davis KE, Scherer PE (2009) Mechanisms of obesity and related pathologies: the macro- and microcirculation of adipose tissue. FEBS J 276:5738–5746

Saddi-Rosa P, Oliveira CS, Giuffrida FM, Reis AF (2010) Visfatin, glucose metabolism and vascular disease: a review of evidence. Diabetol Metab Syndr 2:21

Sandu O, Song K, Cai W, Zheng F, Uribarri J, Vlassara H (2005) Insulin resistance and type 2 diabetes in high-fat—fed mice are linked to high glycotoxin intake. Diabetes 54(8):2314–2319

Schäffler A, Neumeier M, Herfarth H, Fürst A, Schölmerich J, Büchler C (2005) Genomic structure of human omentin, a new adipocytokine expressed in omental adipose tissue. Biochim Biophys Acta 1732:96–102

Schäffler A, Schölmerich J, Salzberger B (2007) Adipose tissue as an immunological organ: toll-like receptors, C1q/TNFs and CTRPs. Trends Immunol 28:393–399

Smekal A, Vaclavik J (2017) Adipokines and cardiovascular disease: a comprehensive review. Biomed Pap 161:31–40

Smith CCT, Yellon DM (2011) Adipocytokines, cardiovascular pathophysiology and myocardial protection. Pharmacol Ther 129:206–219

Suga H, Eto H, Aoi N, Kato H, Araki J, Doi K, Higashino T, Yoshimura K (2010) Adipose tissue remodeling under ischemia: death of adipocytes and activation of stem/progenitor cells. Plast Reconstr Surg 126:1911–1923

Surmi BK, Hasty AH (2010) The role of chemokines in recruitment of immune cells to the artery wall and adipose tissue. Vasc Pharmacol 52:27–36

Takeuchi T, Adachi Y, Ohtsuki Y, Furihata M (2007) Adiponectin receptors, with special focus on the role of the third receptor, T-cadherin, in vascular disease. Med Mol Morphol 40:115–120

Tam J, Duda DG, Perentes JY, Quadri RS, Fukumura D, Jain RK (2009) Blockade of VEGFR2 and not VEGFR1 can limit diet-induced fat tissue expansion: role of local versus bone marrow-derived endothelial cells. PLoS One 4:e4974

Tamori Y, Masugi J, Nishino N, Kasuga M (2002) Role of peroxisome proliferator-activated receptor-gamma in maintenance of the characteristics of mature 3T3-L1 adipocytes. Diabetes 51:2045–2055

Tilg H, Moschen AR (2006) Adipocytokines: mediators linking adipose tissue, inflammation and immunity. Nat Rev Immunol 6:772–783

Tinahones FJ, Coín-Aragüez L, Mayas MD, Garcia-Fuentes E, Hurtado-Del-Pozo C, Vendrell J et al (2012) Obesity-associated insulin resistance is correlated to adipose tissue vascular endothelial growth factors and metalloproteinase levels. BMC Physiol 12:4

Tran K, Gealekman O, Frontini A, Zingaretti M, Morroni M, Giordano A et al (2012) The vascular endothelium of the adipose tissue gives rise to both white and brown fat cells. Cell Metab 15:222–229

Trayhurn P (2014) Hypoxia and adipocyte physiology: implications for adipose tissue dysfunction in obesity. Annu Rev Nutr 34:207–236

Trayhurn P, Wood IS (2004) Adipokines: inflammation and the pleiotropic role of white adipose tissue. Br J Nutr 92:347–355

Trayhurn P, Wang B, Wood IS (2008a) Hypoxia in adipose tissue: a basis for the dysregulation of tissue function in obesity? Br J Nutr 100:227–235

Trayhurn P, Wang B, Wood IS (2008b) Hypoxia and the endocrine and signalling role of white adipose tissue. Arch Physiol Biochem 114:267–276

Treins C, Giorgetti-Peraldi S, Murdaca J, Semenza GL, Van Obberghen E (2002) Insulin stimulates hypoxia-inducible factor 1 through a phosphatidylinositol 3-kinase/target of rapamycin-dependent signaling pathway. J Biol Chem 277:27975–27981

van Uden P, Kenneth NS, Rocha S, Van UP (2008) Regulation of hypoxia-inducible factor-1alpha by NF-kappaB. Biochem J 412:477–484

Ueki K, Kondo T, Kahn C (2004a) Suppressor of cytokine signaling 1 (SOCS-1) and SOCS-3 cause insulin resistance through inhibition of tyrosine phosphorylation of insulin receptor substrate proteins. Mol Cell Biol 24:5434–5446

Ueki K, Kondo T, Tseng Y-H, Kahn CR (2004b) Central role of suppressors of cytokine signaling proteins in hepatic steatosis, insulin resistance, and the metabolic syndrome in the mouse. Proc Natl Acad Sci U S A 101:10422–10427

Vona-Davis L, Rose DP (2007) Adipokines as endocrine, paracrine, and autocrine factors in breast cancer risk and progression. Endocr Relat Cancer 14:189–206

Vona-Davis L, Rose DP (2009) Angiogenesis, adipokines and breast cancer. Cytokine Growth Factor Rev 20:193–201

Wang B, Wood IS, Trayhurn P (2007) Dysregulation of the expression and secretion of inflammation-related adipokines by hypoxia in human adipocytes. Pflugers Arch: Eur J Physiol 455:479–492

Wang Y, Nishina P, Naggert J (2009) Degradation of IRS1 leads to impaired glucose uptake in adipose tissue of the type 2 diabetes mouse model TALLYHO/Jng. J Endocrinol 203:65–74

Wellen K, Hotamisligil G (2003) Obesity-induced inflammatory changes in adipose tissue. J Clin Investig 112:1785–1788

Wellen K, Hotamisligil G (2005) Inflammation, stress, and diabetes. J Clin Investig 115:1111–1119

Wood IS, Wang B, Lorente-cebrián S, Trayhurn P (2007) Hypoxia increases expression of selective facilitative glucose transporters (GLUT) and 2-deoxy-D-glucose uptake in human adipocytes. Biochem Biophys Res Commun 361:468–473

Wood IS, Heredia FP, Wang B, Trayhurn P (2009) Cellular hypoxia and adipose tissue dysfunction in obesity. Proc Nutr Soc 68:370–377

Xu A, Wang Y, Keshaw H, Xu L, Lam K, Cooper G (2003) The fat-derived hormone adiponectin alleviates alcoholic and nonalcoholic fatty liver diseases in mice. J Clin Investig 112:91–100

Yamakawa M, Liu LX, Date T, Belanger AJ, Vincent KA, Akita GY et al (2003) Hypoxia-inducible factor-1 mediates activation of cultured vascular endothelial cells by inducing multiple angiogenic factors. Circ Res 93:664–673

Yamauchi T, Hara K, Kubota N, Terauchi Y, Tobe K, Froguel P, Nagai R, Kadowaki T (2003) Dual roles of adiponectin/Acrp30 in vivo as an anti-diabetic and anti-atherogenic adipokine. Curr Drug Targets Immune Endocr Metabol Disord 3:243–254

Yamawaki H, Tsubaki N, Mukohda M, Okada M, Hara Y (2010) Omentin, a novel adipokine, induces vasodilation in rat isolated blood vessels. Biochem Biophys Res Commun 393:668–672

Yang R-Z, Lee M-J, Hu H, Pray J, H-B W, Hansen BC et al (2006) Identification of omentin as a novel depot-specific adipokine in human adipose tissue: possible role in modulating insulin action. Am J Physiol Endocrinol Metab 290:E1253–E1261

Ye J (2009) Emerging role of adipose tissue hypoxia in obesity and insulin resistance. Int J Obes 33:54–66

Ye J, Gao Z, Yin J, He Q (2007) Hypoxia is a potential risk factor for chronic inflammation and adiponectin reduction in adipose tissue of ob/ob and dietary obese mice. Am J Physiol Endocrinol Metab 293:E1118–E1128

Yin K, Liao D, Tang C (2010) ATP-binding membrane cassette transporter A1 (ABCA1): a possible link between inflammation and reverse cholesterol transport. Mol Med 16:438–449

Yki-Järvinen H (2005) Fat in the liver and insulin resistance. Ann Med 37:347–356

Yki-Järvinen H (2015) Nutritional modulation of non-alcoholic fatty liver disease and insulin resistance. Forum Nutr 7:9127–9138

Yki-Järvinen H, Westerbacka J (2005) The fatty liver and insulin resistance. Curr Mol Med 5:287–295

Zhang X, Lam K, Ye H, Chung S, Zhou M, Wang Y, Xu A (2010) Adipose tissue-specific inhibition of hypoxia-inducible factor 1α induces obesity and glucose intolerance by impeding energy expenditure in mice. J Biol Chem 285:32869–32877

Chapter 2
The Role of Brain in Energy Balance

Paulo Matafome and Raquel Seiça

Abstract Energy homeostasis is regulated by homeostatic and nonhomeostatic reward circuits which are closely integrated and interrelated. Before, during, and after meals, peripheral nutritional signals, through hormonal and neuronal pathways, are conveyed to selective brain areas, namely the hypothalamic nuclei and the brainstem, the main brain areas for energy balance regulation. These orexigenic and anorexigenic centers are held responsible for the integration of those signals and for an adequate output to peripheral organs involved in metabolism and energy homeostasis.

Feeding includes also a hedonic behavior defined as food intake for pleasure independently of energy requirement. This nonhomeostatic regulation of energy balance is based on food reward properties, unrelated to nutritional demands, and involves areas like mesolimbic reward system, such as the ventral tegmental area and the nucleus accumbens, and also opioid, endocannabinoid, and dopamine systems.

Herein, focus will be put on the brain circuits of homeostatic and nonhomeostatic regulation of food intake and energy expenditure.

Keywords Energy homeostasis • Energy expenditure • Brain • Food intake regulation • Hedonic eating

P. Matafome
Institute of Physiology, Institute for Biomedical Imaging and Life Sciences—IBILI,
Faculty of Medicine, University of Coimbra, Coimbra, Portugal

Department of Complementary Sciences, Coimbra Health School (ESTeSC),
Instituto Politécnico de Coimbra, Coimbra, Portugal

R. Seiça (✉)
Institute of Physiology, Institute for Biomedical Imaging and Life Sciences—IBILI,
Faculty of Medicine, University of Coimbra, Coimbra, Portugal
e-mail: rmfseica@gmail.com

© Springer International Publishing AG 2017
L. Letra, R. Seiça (eds.), *Obesity and Brain Function*, Advances in
Neurobiology 19, DOI 10.1007/978-3-319-63260-5_2

2.1 Introduction

Energy homeostasis and body weight are closely regulated by homeostatic and non-homeostatic reward systems. Homeostatic control uses a complex network that involves orexigenic and anorexigenic brain centers and peripheral organs such as gastrointestinal tract, pancreas, and adipose tissue. Such brain areas receive signals from these peripheral organs concerning the food intake and the energy status of the body, regulating short-term and long-term energy balance pathways. Peripheral signals through circulation and afferent nerves act mainly in those specific brain areas, inhibiting or exciting specific neurons in order to maintain energy homeostasis through neuronal and endocrine responses. The crosstalk between peripheral signals and hypothalamic and brainstem centers, the main brain areas for energy homeostasis regulation, typified the physiological system to control hungry/satiety, fat mass, and meal size and frequency (Wilson and Enriori 2015; Bauer et al. 2016).

Brain receives a vast number of nutritional signals before, during, and after food intake. Ingested nutrients are sensed in the gastrointestinal tract which induce gut-derived humoral and neuronal signals that are integrated by selective brain areas, in turn responsible for an adequate output to peripheral organs involved in metabolism and energy balance. The regulation of a meal comprises several factors that control meal initiation and size and inter-meal interval like ghrelin, glucagon like peptide-1 (GLP-1), and peptide tyrosine tyrosine (PYY). Most of the afferent gut signals are predominantly involved in the short-term circuitry regulating meal size and feeding frequency. Nevertheless, some of them may also act in concert with long-acting adiposity signals, which reflect the fat store levels specially related to leptin but also insulin (Bauer et al. 2016; Hagan and Niswender 2012).

Preprandial hunger signals are the first stimulus to food ingestion and are induced by sight and smell of food. At the beginning of the prandial phase the contact of food with mouth signals the brain in order to promote food intake. During the prandial period, the central nervous system (CNS) receives mechanical and chemical gut satiety signals. Furthermore, absorbed nutrients also reach CNS contributing to satiety (Harrold et al. 2012). The amount of energy stored as fat is also signaled to the CNS helping to maintain the balance between food intake and energy expenditure (Wilson and Enriori 2015; Abdalla 2017).

Apart from the homeostatic circuits, feeding includes hedonic behavior defined as food intake for pleasure independently of energy requirement. Palatable food can stimulate appetite beyond energy needs. This nonhomeostatic regulation of energy balance is based on food reward properties, unrelated to nutritional demands, and involves areas like mesolimbic reward circuitry, namely ventral tegmental area (VTA) and nucleus accumbens (NAc), and nuclei in the amygdale and hippocampus, interconnected to the hypothalamus and the brainstem (Sekar et al. 2017; Hagan and Niswender 2012; Yu et al. 2015).

This chapter introduces the major brain areas of homeostatic and nonhomeostatic regulation of food intake and energy expenditure, both crucial to maintain energy balance. In fact, a dysfunction at any level of this complex network including reward system malfunction may lead to an energy imbalance and thus to metabolic disorders.

2.2 Brain Circuits of Food Intake Regulation

2.2.1 Hypothalamus

The hypothalamus is a key organ in the regulation of food intake. Hypothalamic nuclei specifically arcuate nucleus (ARC), paraventricular nucleus (PVN), ventro-medial hypothalamus (VMH), dorsomedial hypothalamus (DMH), and lateral hypothalamic area (LHA) form an interconnected neuronal circuitry (Fig. 2.1) and are linked to brainstem and higher brain centers (Fig. 2.3). They respond to ingested food and changes in the energy status altering the expression of specific neuropeptides resulting in adjustments of food intake and energy expenditure to maintain

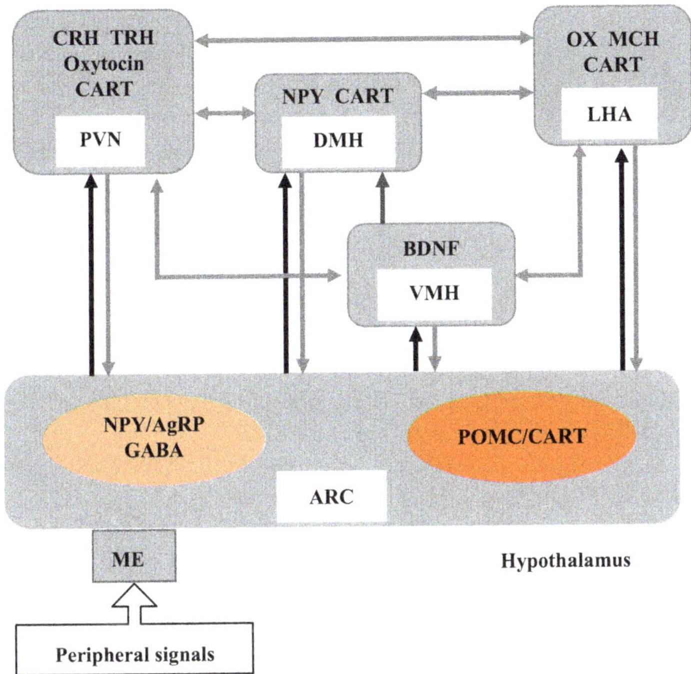

Fig. 2.1 Hypothalamic centers of energy balance. *ARC* arcuate nucleus, *PVN* paraventricular nucleus, *VMH* ventromedial hypothalamus, *DMH* dorsomedial hypothalamus, *LHA* lateral hypothalamic area, *ME* median eminence, *NPY* neuropeptide Y, *AgRP* agouti-related peptide, *GABA* γ-aminobutyric acid, *POMC* proopiomelanocortin, *CART* cocaine-and amphetamine-regulated transcript, *BDNF* brain-derived neurotrophic factor, *OX* orexin, *MCH* melanin-concentrated hormone, *CRH* corticotrophin-releasing hormone, *TRH* thyrotrophin-releasing hormone

energy balance (Jeong et al. 2014; Roh et al. 2016; López et al. 2007; Abdalla 2017; Harrold et al. 2012).

2.2.1.1 Arcuate Nucleus

The ARC expresses hormone receptors namely insulin, leptin and ghrelin receptors (Sutton et al. 2016; Jeong et al. 2014; Wilson and Enriori 2015) and nutrient sensors and lies immediately above the median eminence, a circumventricular organ with semipermeable blood–brain barrier (BBB) that allows peptides and proteins influx. It is considered the hypothalamic area that primarily senses circulating peripheral hormones, namely from gut and adipose tissue, and nutrient signals (Wilson and Enriori 2015; Roh et al. 2016; López et al. 2007; Sobrino Crespo et al. 2014).

The hypothalamic ARC (Fig. 2.1) comprises two neuronal populations with opposing effects on food ingestion: one group (NPY/AgRP neurons), found medially, produces the orexigenic neuropeptides, neuropeptide Y (NPY), the most potent stimulator of appetite, and agouti-related peptide (AgRP); the other group (POMC/CART neurons), in the lateral ARC, co-expresses the anorexigenic neuropeptides proopiomelanocortin (POMC) and cocaine- and amphetamine-regulated transcript (CART). It is in these first-order neurons that peripheral metabolic signals like leptin, insulin, and ghrelin primarily act (Roh et al. 2016; Jeong et al. 2014; Anderson et al. 2016; Wilson and Enriori 2015; López et al. 2007; Sobrino Crespo et al. 2014; Abdalla 2017).

Moreover, NPY/AgRP and POMC/CART neurons receive inputs from other brain areas, including from hypothalamic PVN, DMH, VMH, and LHA (Wang et al. 2015).

The NPY/AgRP and POMC/CART neurons project to second-order neurons located in other hypothalamic nuclei such as the VMH, PVN, DMH, and LHA and also to extrahypothalamic regions such as the autonomic preganglionic neurons in the brainstem and the spinal cord (Harrold et al. 2012; Abdalla 2017; Anderson et al. 2016; Roh et al. 2016; Roseberrya et al. 2015).

In the hypothalamus, the ARC is the major site of NPY expression (Sobrino Crespo et al. 2014). NPY, as a ligand of the receptors Y-1 (Y-1R) and Y-5 (Y-5R) in the PVN, stimulates appetite and reduces energy expenditure, albeit affects other hypothalamic nuclei like DMH and LHA as well as other brain areas (Sekar et al. 2017; Abdalla 2017). AgRP, an endogenous antagonist of α-melanocyte-stimulating hormone (α-MSH), co-localize with NPY exclusively in the ARC neurons. It competes with α-MSH for melanocortin receptors in the second-order neurons, namely in the PVN, antagonizing its effects, whereas NPY stimulates food intake binding to its receptors Y-1R or Y-5R (Roh et al. 2016; López et al. 2007; Abdalla 2017) (Fig. 2.2).

The NPY/AgRP neurons release, besides NPY and AgRP, the inhibitory neurotransmitter γ-aminobutyric acid (GABA) that seems to mediate most of the orexigenic effects of the NPY. Several anorexigenic neurons within the CNS receive the GABAergic inhibitory inputs from the NPY/AgRP neurons, including the POMC/

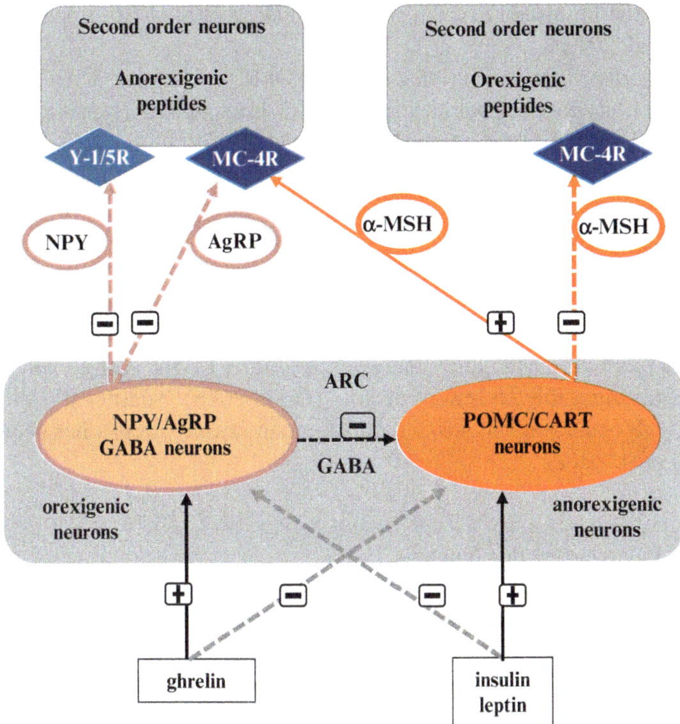

Fig. 2.2 Arcuate nucleus and anorexigenic and orexigenic neuropeptides connection. *ARC* arcuate nucleus, *NPY* neuropeptide Y, *AgRP* agouti-related peptide, *GABA* γ-aminobutyric acid, *POMC* proopiomelanocortin, *CART* cocaine-and amphetamine-regulated transcript, *Y-1/5R* Neuropeptide receptors 1 and 5, *α-MSH* α-melanocyte-stimulating hormone, *MC-4R* melanocyte receptor-4

CART neurons in the ARC (Fig. 2.2); this inhibitory signal associated with the direct inhibitory effect of NPY/AgRP on POMC/CART neurons form, within the hypothalamic ARC, an internal circuit of food intake regulation. In this way, it was suggested that GABA orexigenic pathways from the NPY/AgRP neurons increase food intake inhibiting the anorexigenic neurons within the CNS (Wilson and Enriori 2015; Waterson and Horvath 2015; Jeong et al. 2014).

Different effects have been assigned to NPY, GABA, and AgRP neuropeptides; NPY and GABA seem to be more important for acute feeding whereas AgRP has a major role in long-term regulation through its competition with melanocortin receptors (Wilson and Enriori 2015).

POMC is the precursor of α-MSH, the strongest brain anorexigenic neuropeptide that binds to the receptors MC-3R and MC-4R, particularly the MC-4R, on second-order neurons leading to reduced food intake and enhanced energy expenditure (Sekar et al. 2017; Jeong et al. 2014; Anderson et al. 2016; Roh et al. 2016; López et al. 2007) (Fig. 2.2). The POMC/CART neurons project to the second-order neurons, particularly in the PVN, VMH, and LHA, and also to autonomic preganglionic neurons in the brainstem and the spinal cord (Roh et al. 2016; López et al. 2007; Harrold et al. 2012).

The majority of the neurons that express POMC also express CART which putative receptor has not been yet identified. CART is expressed in the ARC and also in the PVN, LHA, and DMH neurons (Sekar et al. 2017; Gilon 2016). It is primarily an anorexigenic neuropeptide inhibiting food intake and increasing energy expenditure, but recent studies showed a possible orexigenic action. In fact, it was suggested that activation of CART neurons in the ARC and PVN induces anorexigenic effect while in the DMH and LHA neurons has opposite effects (Abdalla 2017). CART from ARC neurons controls the release of thyrotrophin-releasing hormone (TRH) in the PVN which, in turn, by regulating thyrotrophin-stimulating hormone (TSH) release in pituitary gland, alters energy expenditure. CART is also expressed in the NAc suggesting a role on reward activity (Gilon 2016). Additionally, an interconnection with endocannabinoids and dopamine in reward circuits has been reported (Harrold et al. 2012).

2.2.1.2 Paraventricular Nucleus

The hypothalamic PVN integrates signals from several brain areas, including those conveyed by POMC/CART and NPY/AgRP neurons from the ARC and orexin (OX) neurons from the LHA and also from the hypothalamic VMH and DMH and the brainstem (Harrold et al. 2012; Sutton et al. 2016; López et al. 2007). It also directly receives peripheral signals namely from leptin, ghrelin, and insulin (Jeong et al. 2014; Wilson and Enriori 2015) and sends signals to other hypothalamic nuclei (Wang et al. 2015) (Fig. 2.1). But the PVN primarily projects to extrahypothalamic regions, such as the brainstem, and controls sympathetic responses including those related to peripheral metabolism and energy expenditure (Fig. 2.3) (Roh et al. 2016; Sutton et al. 2016). Their neurons primarily produce anorexigenic neuropeptides such as corticotrophin-releasing hormone (CRH), TRH, and oxytocin (Roh et al. 2016; Wilson and Enriori 2015; Abdalla 2017). The PVN mediates the majority of the hypothalamic output to regulate food ingestion and energy expenditure. Furthermore, a recent study showed that, in rats, oxytocin receptors are expressed in the NAc suggesting its effect on reward circuitry (Abdalla 2017) (Fig. 2.3).

2.2.1.3 Ventromedial Hypothalamus and Dorsomedial Hypothalamus

The hypothalamic VMH expresses the anorexigenic neuropeptide, the brain-derived neurotrophic factor (BDNF), that may be activated by POMC/CART neurons from the ARC. It also receives NPY/AgRP neuronal projections from this nucleus (Abdalla 2017; Roh et al. 2016; López et al. 2007). The VMH projects, in turn, to the ARC, DMH, LHA, and PVN, as well as to the brainstem (López et al. 2007; Roh et al. 2016) (Fig. 2.1).

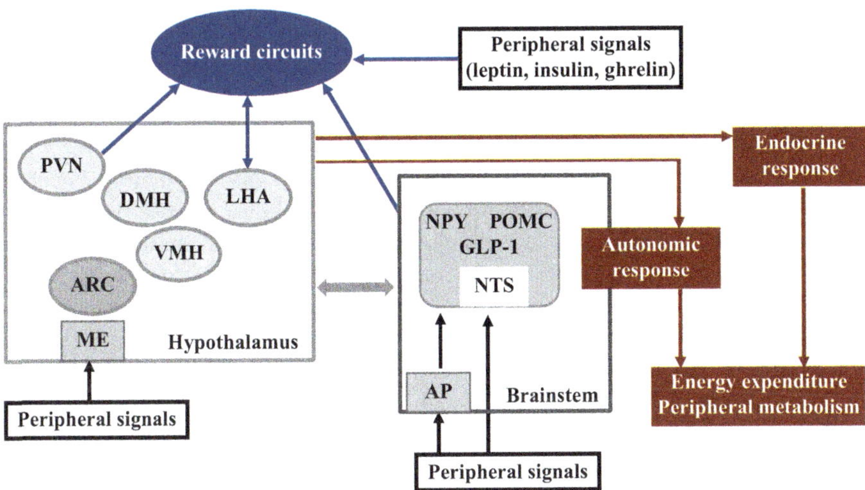

Fig. 2.3 Brain centers of energy balance and peripheral metabolism regulation. *ARC* arcuate nucleus, *PVN* paraventricular nucleus, *VMH* ventromedial hypothalamus, *DMH* dorsomedial hypothalamus, *LHA* lateral hypothalamic area, *ME* median eminence, *NTS* nucleus tractus solitarius, *AP* area postrema, *NPY* neuropeptide Y, *POMC* proopiomelanocortin, *GLP* glucagon-like peptide 1

The neuronal populations in the hypothalamic DMH are poor characterized (Abdalla 2017; Wilson and Enriori 2015) but it was reported that they express NPY and CART neuropeptides (Roh et al. 2016; López et al. 2007), receive NPY/AgRP and POMC/CART neuron projections from the ARC, and are also connected with other hypothalamic nuclei like PVN and LHA and the brainstem (Roh et al. 2016) (Fig. 2.1).

2.2.1.4   Lateral Hypothalamic Area

In contrast to the PVN, VMH, and DMH, the hypothalamic LHA is considered a feeding center. The LHA contains two distinct neuronal populations expressing the orexigenic neuropeptides, melanin-concentrating hormone (MCH) and orexin or hipocretin (OX-A and OX-B), receiving both the NPY/AgRP and POMC/CART neuronal projections from the ARC (Fig. 2.1) and project to extrahypothalamic areas such as the brainstem (Roh et al. 2016; Sobrino Crespo et al. 2014; López et al. 2007; Abdalla 2017). MCH neurons project to extrahypothalamic areas, including the NAc suggesting its role in hedonic feeding (Sobrino Crespo et al. 2014; Harrold et al. 2012) (Fig. 2.3). OX-A and OX-B neurons have extensive projections to the ARC, specially to the NPY neurons, to the VMH, DMH, and PVN, and also to the VTA, so its role in hedonic system was also proposed (Sobrino Crespo et al. 2014; Harrold et al. 2012; López et al. 2007) (Fig. 2.1).

2.2.1.5 Glucose and Lipid Sense Neurons

The brain areas associated to glucose metabolism regulation contain neurons that sense glucose concentrations in the extracellular fluid. These neurons are particularly located in the hypothalamic nuclei and the dorsal vagal complex in the brainstem and are divided into two types: glucose-excited and glucose-inhibited neurons that are activated by increased and decreased extracellular glucose, respectively. The former are mainly found in the hypothalamic VMH, ARC, and PVN, whereas the latter are located in the LHA, ARC, and PVN (Roh et al. 2016).

Considering the ARC neurons, some studies show that NPY/AgRP neurons are glucose-inhibited and POMC/CART are glucose-excited neurons. It was reported that glucose sensing neurons respond also to peripheral humoral signals related to fat stores such as leptin and insulin and are linked to hormonal and autonomic response that regulate glucose metabolism and energy balance (Levin et al. 2004).

The CNS, namely the hypothalamus, may also directly sense fatty acid concentrations. It was shown that intracerebroventricular administration of long-chain fatty acid oleic acid inhibits glucose production and food intake, associated to decreased hypothalamic NPY (Obici et al. 2002). Specialized neurons in the hypothalamic VMH and ARC and in other brain areas involved in the regulation of glucose metabolism and energy homeostasis seem to be sensitive to fatty acids (Moullé et al. 2014).

2.2.2 Brainstem

The brainstem is another key brain region in the regulation of food intake and energy expenditure and exist an extensive reciprocal nervous connection between brainstem and hypothalamus. The dorsal vagal complex (DVC) consists of the nucleus tractus solitarius (NTS), the dorsal motor nucleus of the vagus (DMV), and the area postrema (AP), a circumventricular organ with modified BBB similar to the hypothalamic median eminence. The DVC receives and integrates mechanical and chemical gut signals transmitted through gastrointestinal sensory nerves. Peripheral signals namely those assigned to the gastrointestinal tract, such as ghrelin and cholecystokinin, are sent to the NTS, not only via the sensory vagus nerve that extensively innervates gastrointestinal tract, but also via the systemic circulation through the area postrema. Thus, NTS integrates both humoral and neuronal peripheral signals and also neuronal projections from the hypothalamus, particularly from the PVN (Roh et al. 2016; Bauer et al. 2016; López et al. 2007; Wang et al. 2015). The NTS neurons project to the brainstem and other brain areas such as the hypothalamic ARC and PVN (Wang et al. 2015; Bauer et al. 2016). The NTS has the potential to regulate sympathetic outflow, namely to brown adipose tissue, being therefore involved in energy expenditure and peripheral metabolism (Sutton et al. 2016) (Fig. 2.3).

The NTS neurons express Y-1, Y-5, and MC-4 receptors and produce GLP-1 and the neuropeptides NPY and POMC-derived α-MSH (Roh et al. 2016; Sobrino Crespo et al. 2014; López et al. 2007). The POMC neurons in the NTS are crucial for the short-term regulation of food intake while POMC neurons from the hypothalamic ARC are particularly involved in the long-term control of energy balance (Jeong et al. 2014).

GLP-1 is synthesized in the NTS neurons and its receptor (GLP-1R) is expressed in the brainstem and in the hypothalamic nuclei ARC, PVN, and DMH (Sobrino Crespo et al. 2014; Sekar et al. 2017; Geloneze et al. 2017: López et al. 2007). The anorexigenic effect of central GLP-1 seems to be especially related to GLP-1R activation in the ARC leading to stimulation of POMC/CART and inhibition of the NPY/AgRP neurons. In the PVN, GLP-1 activates directly the release of the anorexigenic oxytocin and CRH. Some GLP-1 neurons project to the NAc and VTA, a reward centers, leading to reduction of palatable food intake via mesolimbic dopaminergic suppression (Sekar et al. 2017; Geloneze et al. 2017). It was also shown that GLP-1 at central level has a role in energy expenditure, inducing brown adipose tissue thermogenesis via sympathetic activation (Geloneze et al. 2017), and modulate circulating glucose utilization (Sekar et al. 2017).

2.2.3 Brain Reward Circuits

Brain homeostatic and hedonic systems have a role in hunger/satiety regulation. A negative energy balance leads to the physiological need to eat mainly mediated by hypothalamus and brainstem, but the desire to ingest palatable food may stimulate food intake independently of energy status. The meal beginning involves commonly nonhomeostatic mechanisms, whereas meal size and termination are mainly under homeostatic control particularly related to gut signals induced by food intake. Nevertheless, homeostatic and hedonic feeding are closely integrated and interrelated (Yu et al. 2015; Leigh and Morris 2016).

Reward/hedonic pathways include interactions between opioid, endocannabinoid, and dopamine systems, particularly the mesocorticolimbic dopaminergic system, and special brain areas like hippocampus, amygdala, prefrontal cortex, and midbrain and also hypothalamic LHA due to its linkage to the VTA and the NAc neural circuits (Abdalla 2017; López et al. 2007). Indeed, the LHA is an integrator center of both hedonic and homeostatic circuits; it receives homeostatic signals and modulates directly the VTA and the NAc. The OX and MCH neurons from the LHA project to the reward circuits and it was shown to change dopamine system (Leigh and Morris 2016; Sobrino Crespo et al. 2014; Harrold et al. 2012) (Fig. 2.3).

Moreover, hedonic behavior is modulated, specially through the modulation of mesolimbic dopaminergic pathways, by peripheral metabolic signals, mainly involved in the homeostatic control of energy balance like leptin, insulin, and ghrelin (Leigh and Morris 2016; Hagan and Niswender 2012; Harrold et al. 2012) (Fig. 2.3). In fact, insulin and leptin receptors are expressed in the mesolimbic system namely in the VTA, and negatively affect the dopaminergic reward neurons (Leigh and Morris 2016; Khanh et al. 2014; Davis et al. 2010). Furthermore, leptin inhibits orexin and MCH neurons in the hypothalamic LHA which, projecting to mesolimbic regions, control dopamine actions (Khanh et al. 2014). Ghrelin receptors are also expressed in the VTA and its activation impacts dopaminergic circuits leading to enhanced hedonic response (Harrold et al. 2012).

Food is rewarding and acts on the reward pathways that include dopaminergic system, a central component of the hedonic feeding. Dopamine, and its receptor D2, is implicated in the pleasure and emotions. After palatable food ingestion, dopamine is released from the neurons in the VTA of the midbrain which project to the NAc and the prefrontal cortex (Roseberrya et al. 2015; Abdalla 2017; Hagan and Niswender 2012). On the other hand, dopamine neurons receive afferent inputs from many different brain areas such as the hypothalamic LHA. Furthermore, evidences indicate that α-MSH and AgRP would affect food reward acting on the mesocorticolimbic and mesostriatal dopamine system, but the underlying mechanisms are unknown (Roseberrya et al. 2015).

Endocannabinoids are also involved in the hedonic control of food intake. The endocannabinoids, namely N-arachidonoyl-ethanolamine (anandamide) and 2-arachidonoylglycerol (2-AG), are produced from cell membrane phospholipids in the brain where they are widely expressed, including in the hypothalamus, the brainstem, and the corticolimbic system (Cristino et al. 2014). The endocannabinoids exert their effects through CB-1 and CB-2 receptors, being the former located in tissues associated to energy balance like hypothalamus, brainstem, mesolimbic regions, gastrointestinal tract, and adipose tissue. CB-2 receptor is related to immunologic processes but recently it was suggested that it may contribute to energy balance regulation (Abdalla 2017). Endocannabinoids and its receptor CB-1 act on the reward circuits such as on the NAc and the VTA and interact with both dopamine and opioid pathways leading to palatable food preferences (Cristino et al. 2014). Some peripheral hormones affect endocannabinoid activity like cholecystokinin, ghrelin, and leptin (Abdalla 2017; Cristino et al. 2014).

Endogenous opioids such as POMC-derived ß-endorphin and its μ-receptors, located in the VTA and the NAc, areas of the mesolimbic dopaminergic system, are also linked to the reward system. In fact, the activation of the μ-receptors is particularly implicated in the modulation of high palatable food intake. In addition, the interaction of opioids and the LHA orexin modulates both homeostatic and hedonic feeding (Abdalla 2017; Nogueiras et al. 2012).

The components of the reward system are associated with the pleasure (liking) that seems to be mediated by opioid and endocannabinoid systems, and with the motivation (wanting) to ingest food, a process that appears to be particularly related to mesolimbic dopaminergic neurons (Yu et al. 2015; Davis et al. 2010).

The pleasure we experience provided by food is important to decide about when, what, and how much to eat and is particularly linked to sweet and fat ingestion, in rats and in humans (Hagan and Niswender 2012). Malfunction of the reward circuits may explain, at least in part, the increased incidence of obesity. Indeed, the hedonic system may override homeostatic component of energy balance regulation leading to metabolic dysregulation (Yu et al. 2015).

2.3 Neuronal and Humoral Outputs to the Peripheral Organs

The brain is not only crucial to regulate food intake but also the energy expenditure, the other component of the energy balance, traditionally divided in energy expended in physical activity, resting energy expenditure and thermic effect of feeding.

Hypothalamus and brainstem receive humoral and neuronal peripheral signals and, therefore, modulate food intake, energy expenditure, and peripheral metabolism. They contain circuits that involve autonomic control of peripheral metabolic tissues such as gastrointestinal tract, pancreas, skeletal muscle, and white (WAT) and brown (BAT) adipose tissue. In fact, autonomic nervous system balances peripheral energy production and expenditure and peripheral metabolism (Seoane-Collazo et al. 2015). Pre-autonomic neurons in the hypothalamic nuclei particularly the PVN, the most important brain region for coordination of autonomic output, project to the lower autonomic areas in the spinal cord and brainstem mediating the autonomic efferent control of the peripheral organs (Yi and Tschöp 2012) (Fig. 2.3). Furthermore, the hypothalamic PVN response, besides the autonomic nervous system, involves endocrine outflow through adrenal and thyroid axis which also participate in heat production (Yi and Tschöp 2012).

The parasympathetic nervous system (PNS) and the sympathetic nervous system (SNS) regulate, among others, pancreatic endocrine function, like insulin and glucagon secretion, and hepatic and peripheral insulin sensitivity, being the first traditionally related to anabolism and the latter to catabolic response. Adipose tissue is particularly innervated by the SNS while liver, muscle, and endocrine pancreas receive both parasympathetic and sympathetic innervation (Seoane-Collazo et al. 2015). SNS activation induces WAT free fatty acid release and hepatic gluconeogenesis and glycogenolysis. In the muscle, it leads to free fatty acid oxidation, reduced glycogen synthesis and glucose oxidation and decreased glucose uptake (Yi and Tschöp 2012). Hypothalamic nuclei such as the LHA, VMH, and PVN send autonomic signals to the liver modulating hepatic glucose homeostasis. The LHA and VMH areas are involved, respectively, in hepatic parasympathetic and sympathetic innervation while the PVN receives and integrates different signals from other regions like the ARC and VMH and coordinates sympathetic and parasympathetic outputs to the liver (Seoane-Collazo et al. 2015).

Thus, a hypothalamic-autonomic nervous system axis might be considered as an important regulator of energy and metabolic homeostasis.

2.3.1 Central Nervous System and White Adipose Tissue

The CNS regulates WAT metabolism and secretory function via autonomic nervous system specifically the SNS. The connections between the hypothalamus and the autonomic nervous system are not well identified and may include different hypothalamic nuclei (Seoane-Collazo et al. 2015). Nevertheless, WAT lipolysis was linked to the hypothalamic VMH, LHA, PVN, and ARC. The VMH is related to SNS activation leading to enhanced WAT lipolysis. From the hypothalamic LHA, the OX-A neurons project to the brainstem and spinal cord regulating WAT lipolysis via SNS activation, while MCH neurons seem to induce lipid deposition and decreased lipid mobilization in the WAT (Yi and Tschöp 2012; Seoane-Collazo et al. 2015). The activation of melanocortin receptors in the hypothalamic PVN induced by ARC α-MSH increases energy expenditure. This effect appears to be mediated, at least in part, by the consequent activation of neuronal projections from the PVN to the brainstem NTS, leading to sympathetic stimulation. In fact, a decrease in α-MSH leads to increased adiposity and reduced lipolysis (Seoane-Collazo et al. 2015; Sutton et al. 2016). In turn, the NPY neurons from the ARC promote WAT energy accumulation and inhibit BAT activity through inhibition of the SNS outflow. Apart from this central effect of the NPY, this neuropeptide acts directly in the WAT as it is also expressed in the SNS neurons that innervate this adipose tissue (Seoane-Collazo et al. 2015).

2.3.2 Central Nervous System and Brown Adipose Tissue

Besides WAT lipolysis, lipid metabolism is regulated by modulation of BAT function and the CNS, mainly via the SNS, has a pivotal role (Roh et al. 2016; Seoane-Collazo et al. 2015; Contreras et al. 2015; Bartness et al. 2010; Roh and Kim 2016). BAT was considered relevant in rodents and hibernating mammals and also in newborn humans but negligible in adults. Evidences reported its presence in adult humans particularly located in cervical and supraclavicular areas and also in perirenal, intercostal, mediatinal, and periaortic regions. Recent imaging techniques showed that BAT, although limited to distinct depots, is a functionally active tissue in adult humans with important metabolic effects (Wankhade et al. 2016; Zhang and Bi 2015; Contreras et al. 2015; Seoane-Collazo et al. 2015).

BAT is an important regulator of energy expenditure, specialized for non-shivering thermogenesis, that dissipate energy in the form of heat, in order to maintain body temperature in response to cold exposition or to dissipate the excessive energy intake after hypercaloric food ingestion. In fact, BAT is composed of special adipocytes which generate heat instead of accumulating fat, via increased mitochondria number and expression and activity of mitochondrial uncoupling protein-1 (UCP-1) (Roh et al. 2016; Roh and Kim 2016; Wankhade et al. 2016; Contreras et al. 2015; Zhang and Bi 2015; Montanari et al. 2017).

In addition to white and brown adipocytes, a population of cells interdispersed in the WAT was recently identified, the brite/beige adipocytes or brown-like adipocytes. They are UCP-1 positive adipocytes that, expressing the characteristics of brown adipocytes, contribute to non-shivering thermogenesis and energy expenditure. Furthermore, the induction of WAT browning, that is to say the activation of the brite/beige adipocytes, may particularly occur in response to exercise, cold exposure, ß-adrenergic agonists, orexin, leptin, insulin, thyroid hormones, and nutrients (Contreras et al. 2015; Forest et al. 2016; Wankhade et al. 2016; Zhang and Bi 2015; Montanari et al. 2017; Roh et al. 2016). Recent studies suggest that adult human BAT appears to be composed by brown and brite/beige adipocytes (Contreras et al. 2015; Zhang and Bi 2015).

Non-shivering thermogenesis is particularly controlled by SNS and thyroid hormones. Indeed, the SNS, via norepinephrine and ß-adrenergic receptors, mainly β_3-receptors, is the principal regulator of BAT function, but paracrine and endocrine molecules are also involved. Some nutrient and hormonal signals such as leptin, insulin, and GLP-1, and also thyroid hormones which work in concert with norepinephrine, can interfere with sympathetic output to BAT (Roh et al. 2016; Contreras et al. 2015; Geloneze et al. 2017; Zhang and Bi 2015; Seoane-Collazo et al. 2015).

Activation of leptin receptors in the hypothalamus leads to SNS activation which increases BAT energy expenditure. A role of the hypothalamic melanocortin system has been suggested in BAT thermogenesis (Roh and Kim 2016; Seoane-Collazo et al. 2015; Bartness et al. 2010) and appears to mediate the action of leptin on SNS activation (van Swieten et al. 2014; Münzberg and Morrison 2015). It was proposed that leptin controls food intake acting in the hypothalamic ARC, inhibiting the orexigenic NPY/AgRP neurons and activating the anorexigenic POMC/CART neurons which, in turn, triggers the SNS activation and non-shivering thermogenesis (Seoane-Collazo et al. 2015; Dodd et al. 2015). In fact, leptin and insulin acting synergistically on distinct POMC/CART neurons in the hypothalamic ARC promote WAT browning and energy expenditure (Dodd et al. 2015; Forest et al. 2016; Roh and Kim 2016). The hypothalamic ARC is sensitive to insulin and sends signals to other hypothalamic nuclei, namely the PVN, to modulate glucose and lipid homeostasis. BAT and WAT are under the direct control of insulin but acting in brain centers, namely the hypothalamic ARC, insulin also acts indirectly through central SNS activation leading to BAT non-shivering thermogenesis and WAT increased lipogenesis and decreased lipolysis (Seoane-Collazo et al. 2015).

Several hypothalamic nuclei are associated with the regulation of brown and brite/beige thermogenesis such as the VMH, DMH, ARC, and PVN but the mechanisms underlying these circuits are incompletely understood (Seoane-Collazo et al. 2015; Roh et al. 2016; Zhang and Bi 2015). The VMH projects to other hypothalamic areas involved in sympathetic outflow, like the ARC and PVN, and also to other autonomic centers namely the NTS and the locus coeruleus. DMH neurons do not project directly to the spinal cord preganglionic neurons but seems to act indirectly via brainstem raphe pallidus (Seoane-Collazo et al. 2015; Zhang and Bi 2015). On the other hand, the hypothalamic PVN seems to be the mediator of the ARC regulation of energy expenditure.

Descending signals originated in hypothalamic nuclei control sympathetic enervation of BAT and WAT and modulate BAT thermogenesis and WAT browning (Zhang and Bi 2015).

2.4 Concluding Remarks

The CNS is crucial to integrate peripheral signals from gastrointestinal tract, pancreas, and adipose tissue and to orchestrate an adequate response in order to maintain energy homeostasis. Hypothalamus and brainstem are the most important brain centers of the homeostatic regulation of food intake and energy expenditure and are also interconnected with reward/non-homeostatic circuits. This homeostatic and hedonic feeding regulation involves a network of neurons and neuropeptides able to receive several inputs and to control food intake and energy expenditure.

Additionally, we eat in response to an array of additional stimuli such as emotion and cognitive cues. Recent evidences indicate that eating, in humans, is also regulated by emotion, memory, and cognitive control systems. Emotions modulate appetite that seems to involve the connection between the amygdala and the hypothalamus. Memory, mainly regulated by the hippocampus and parahippocampal formation, plays a role in eating disorders and thereby hippocampal dysfunction might induce increased food intake leading to obesity. Furthermore, it was theorized that impaired cognitive control may trigger increased reward responses to food and overeating but it remains unclear if obesity may give rise to this cognitive impairment or this causes obesity (Farr et al. 2016).

Thus, besides the traditional homeostatic and hedonic systems, a larger integrated energy homeostasis system involving also emotion, memory, and cognition has to be considered.

References

Abdalla MM (2017) Central and peripheral control of food intake. Endocr Regul 51:52–70

Anderson EJ, Çakir I, Carrington SJ, Cone RD, Ghamari-Langroudi M, Gillyard T, Gimenez LE, Litt MJ (2016) 60 YEARS OF POMC: regulation of feeding and energy homeostasis by α-MSH. J Mol Endocrinol 56:T157–T174

Bartness TJ, Vaughan CH, Song CK (2010) Sympathetic and sensory innervation of brown adipose tissue. Int J Obes 34:S36–S42

Bauer PV, Hamr SC, Duca FA (2016) Regulation of energy balance by a gut–brain axis and involvement of the gut microbiota. Cell Mol Life Sci 73:737–755

Contreras C, Gonzalez F, Fernø J, Diéguez C, Rahmouni K, Nogueiras R, López M (2015) The brain and brown fat. Ann Med 47:150–168

Cristino L, Becker T, Di Marzo V (2014) Endocannabinoids and energy homeostasis: an update. Biofactors 40:389–397

Davis JF, Choi DL, Benoit SC (2010) Insulin, leptin and reward. Trends Endocrinol Metab 21:68–74

Dodd GT, Decherf S, Loh K, Simonds SE, Wiede F, Balland E, Merry TL, Münzberg H, Zhang ZY, Kahn BB, Neel BG, Bence KK, Andrews ZB, Cowley MA, Tiganis T (2015) Leptin and insulin act on POMC neurons to promote the browning of white fat. Cell 160:88–104

Farr OM, Li CS, Mantzoros CS (2016) Central nervous system regulation of eating: insights from human brain imaging. Metabolism 65:699–713

Forest C, Joffin N, Jaubert AM, Noirez P (2016) What induces watts in WAT? Adipocytes 21(5):136–152

Geloneze B, Lima-Júnior JC, Velloso LA (2017) Glucagon-like peptide-1 receptor agonists (GLP-1RAs) in the brain–adipocyte axis. Drugs 77:493–503

Gilon P (2016) Cocaine- and amphetamine-regulated transcript: a novel regulator of energy homeostasis expressed in a subpopulation of pancreatic islet cells. Diabetologia 59:1855–1859

Hagan S, Niswender KD (2012) Neuroendocrine regulation of food intake. Pediatr Blood Cancer 58:149–153

Harrold JA, Dovey TM, Blundell JE, Halford JC (2012) CNS regulation of appetite. Neuropharmacology 63:3–17

Jeong JK, Kim JG, Lee BJ (2014) Participation of the central melanocortin system in metabolic regulation and energy homeostasis. Cell Mol Life Sci 71:3799–3809

Khanh DV, Choi YH, Moh SH, Kinyua AW, Kim KW (2014) Leptin and insulin signaling in dopaminergic neurons: relationship between energy balance and reward system. Front Psychol 5:846

Leigh SJ, Morris MJ (2016) The role of reward circuitry and food addiction in the obesity epidemic: an update. Biol Psychol. doi:10.1016/j.biopsycho.2016.12.013

Levin BE, Routh VH, Kang L, Sanders NM, Dunn-Meynell AA (2004) Neuronal glucosensing. What do we know after 50 Years? Diabetes 53:2521–2528

López M, Tovar S, Vázquez MJ, Williams LM, Diéguez C (2007) Peripheral tissue–brain interactions in the regulation of food intake. Proc Nutr Soc 66:131–155

Montanari T, Pošćić N, Colitti M (2017) Factors involved in white-to-brown adipose tissue conversion and in thermogenesis: a review. Obes Rev 18(5):495–513

Moullé VS, Picard A, Le Foll C, Levin BE, Magnan C (2014) Lipid sensing in the brain and regulation of energy balance. Diabete Metab 40:29–33

Münzberg H, Morrison CD (2015) Structure, production and signaling of leptin. Metabolism 64:13–23

Nogueiras R, Romero-Picó A, Vazquez MJ, Novelle MG, López M, Diéguez C (2012) The opioid system and food intake: homeostatic and hedonic mechanisms. Obes Facts 5:196–207

Obici S, Feng Z, Morgan K, Stein D, Karkanias G, Rossetti L (2002) Central administration of oleic acid inhibits glucose production and food intake. Diabetes 51:271–275

Roh E, Kim MS (2016) Brain regulation of energy metabolism. Endocrinol Metab 31:519–524

Roh E, Song Do K, Kim MS (2016) Emerging role of the brain in the homeostatic regulation of energy and glucose metabolism. Exp Mol Med 48:e216

Roseberrya AG, Stuhrmana K, Dunigana AI (2015) Regulation of the mesocorticolimbic and mesostriatal dopamine systems by α-melanocyte stimulating hormone and agouti-related protein. Neurosci Biobehav Rev 56:15–25

Sekar R, Wang L, Chow BK (2017) Central control of feeding behavior by the secretin, PACAP, and glucagon family of peptides. Front Endocrinol (Lausanne) 8:18

Seoane-Collazo P, Fernø J, Gonzalez F, Diéguez C, Leis R, Nogueiras R, López M (2015) Hypothalamic-autonomic control of energy homeostasis. Endocrine 50:276–291

Sobrino Crespo C, Perianes Cachero A, Puebla Jiménez L, Barrios V, Arilla Ferreiro E (2014) Peptides and food intake. Front Endocrinol (Lausanne) 5:58

Sutton AK, Myers MJ Jr, Olson DP (2016) The role of PVH circuits in leptin action and energy balance. Annu Rev Physiol 78:207–221

van Swieten MM, Pandit R, Adan RA, van der Plasse G (2014) The neuroanatomical function of leptin in the hypothalamus. J Chem Neuroanat 61-62:207–220

Wang D, He X, Zhao Z, Feng Q, Lin R, Sun Y, Ding T, Xu F, Luo M, Zhan C (2015) Whole-brain mapping of the direct inputs and axonal projections of POMC and AgRP neurons. Front Neuroanat 9:40

Wankhade UD, Shen M, Yadav H, Thakali KM (2016) Novel browning agents, mechanisms, and therapeutic potentials of brown adipose tissue. Biomed Res Int 2016:2365609

Waterson MJ, Horvath TL (2015) Neuronal regulation of energy homeostasis: beyond the hypothalamus and feeding. Cell Metab 22:962–970

Wilson JL, Enriori PJ (2015) A talk between fat tissue, gut, pancreas and brain to control body weight. Mol Cell Endocrinol 418:108–119

Yi CX, Tschöp MH (2012) Brain–gut–adipose-tissue communication pathways at a glance. Dis Model Mech 5:583–587

Yu YH, Vasselli JR, Zhang Y, Mechanick JI, Korner J, Peterli R (2015) Metabolic vs. hedonic obesity: a conceptual distinction and its clinical implications. Obes Rev 16:234–247

Zhang W, Bi S (2015) Hypothalamic regulation of brown adipose tissue thermogenesis and energy homeostasis. Front Endocrinol 31(6):136

Chapter 3
Neuroendocrinology of Adipose Tissue and Gut–Brain Axis

Paulo Matafome, Hans Eickhoff, Liliana Letra, and Raquel Seiça

Abstract Food intake and energy expenditure are closely regulated by several mechanisms which involve peripheral organs and nervous system, in order to maintain energy homeostasis.

Short-term and long-term signals express the size and composition of ingested nutrients and the amount of body fat, respectively. Ingested nutrients trigger mechanical forces and gastrointestinal peptide secretion which provide signals to the brain through neuronal and endocrine pathways. Pancreatic hormones also play a role in energy balance exerting a short-acting control regulating the start, end, and composition of a meal. In addition, insulin and leptin derived from adipose tissue are involved in long-acting adiposity signals and regulate body weigh as well as the amount of energy stored as fat over time.

P. Matafome
Institute of Physiology, Institute for Biomedical Imaging and Life Sciences—IBILI,
Faculty of Medicine, University of Coimbra, Coimbra, Portugal

Department of Complementary Sciences, Coimbra Health School (ESTeSC),
Instituto Politécnico de Coimbra, Coimbra, Portugal

H. Eickhoff
Institute of Physiology, Institute for Biomedical Imaging and Life Sciences—IBILI,
Faculty of Medicine, University of Coimbra, Coimbra, Portugal

Obesity Center, Hospital da Luz de Setúbal, Setúbal, Portugal

L. Letra
Institute of Physiology, Institute for Biomedical Imaging and Life Sciences—IBILI,
Faculty of Medicine, University of Coimbra, Coimbra, Portugal

Neurology Department, Centro Hospitalar do Baixo Vouga, Aveiro, Portugal

R. Seiça (✉)
Institute of Physiology, Institute for Biomedical Imaging and Life Sciences—IBILI,
Faculty of Medicine, University of Coimbra, Coimbra, Portugal
e-mail: rmfseica@gmail.com

© Springer International Publishing AG 2017
L. Letra, R. Seiça (eds.), *Obesity and Brain Function*, Advances in
Neurobiology 19, DOI 10.1007/978-3-319-63260-5_3

This chapter focuses on the gastrointestinal-, pancreatic-, and adipose tissue-derived signals which are integrated in selective orexigenic and anorexigenic brain areas that, in turn, regulate food intake, energy expenditure, and peripheral metabolism.

Keywords Energy homeostasis • Gastrointestinal peptides • Pancreatic hormones • Adipokines

3.1 Introduction

The gut–brain axis is a bidirectional communication pathway between the gut and the brain, particularly important regarding regulating energy balance which comprises neuronal and hormonal gut signals that are integrated in orexigenic and anorexigenic brain areas. It involves neuronal afferent signals to the brain via vagal and spinal (sympathetic) neurons, as well as gut hormones, while outputs from the brain to the gut are mediated via autonomic and endocrine pathways.

Mechanical and chemical signals induced by ingested nutrients stimulate gut peptide secretion that, in turn, informs the central nervous system (CNS) via paracrine action through the vagal and spinal route or via endocrine pathways. In a feedback loop, the CNS will then generate an appropriate response to maintain energy balance (Bauer et al. 2016; Wilson and Enriori 2015; López et al. 2007).

Long-term adiposity signals and short-term signals express the amount of body fat and the size and composition of ingested nutrients, respectively. Insulin and adipose tissue leptin are specifically involved in the long-acting adiposity signals and regulate body weight as well as the amount of energy stored as fat over time. On the other hand, mechanical forces, such as gastric distention, together with gut peptides like cholecystokinin (CCK), peptide tyrosine tyrosine (PYY), and glucagon-like peptide-1 (GLP-1), have a special role in the short-term signaling (Fig. 3.1). However, some evidences support the concept of a dual effect of some gut peptides and their involvement in both meal size and composition and long-term energy balance. Energy expenditure also contributes to energy homeostasis and body weight maintenance and gut and adiposity peptides can also influence energy output while acting on specific brain centers (Bauer et al. 2016; Wilson and Enriori 2015; Yi and Tschöp 2012).

This chapter has its focus on the peripheral signals from gastrointestinal tract, pancreas, and adipose tissue which are integrated in selective brain areas responsible for the maintenance of energy and metabolism homeostasis.

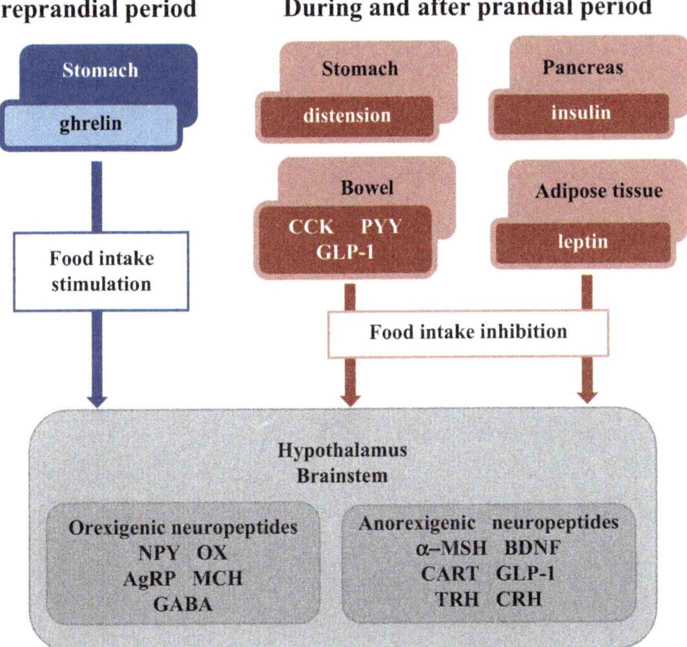

Fig. 3.1 Food intake regulation: principal preprandial and prandial signals and brain orexigenic and anorexigenic neuropeptides. *CCK* cholecystokinin, *GLP-1* glucagon-like peptide-1, *PYY* peptide tyrosine tyrosine, *NPY* neuropeptide Y, *AgRP* agouti-related peptide, *GABA* γ-aminobutyric acid, *α-MSH* α-melanocyte stimulating hormone, *CART* cocaine- and amphetamine-regulated transcript, *BDNF* brain-derived neurotrophic factor, *OX* orexin, *MCH* melanin-concentrated hormone, *CRH* corticotrophin-releasing hormone, *TRH* thyrotrophin-releasing hormone

3.2 Gastrointestinal Signals to the Brain

Ingested nutrients trigger the secretion of several peptides like CCK, GLP-1, and PYY that provide information to the brain about food intake to regulate appetite and energy expenditure. Gut and associated organs such as the endocrine pancreas play a pivotal role in the energy balance, namely in its short-term control, acting on the regulation of meal beginning, ending, and composition.

Intestinal vagal afferent neurons mediate the paracrine effects of gut peptides and mechanical stimuli and converge into the nucleus tractus solitarius (NTS) which, in turn, projects to other brain areas namely the hypothalamic nuclei. Thus, peripheral signals reach the hypothalamus indirectly via afferent neuronal pathways and the brainstem network. In the brainstem, the NTS is one of the major processors of afferent vagal signals. Spinal afferent nerves that represent an additional pathway, synapse on lamina I neurons of the dorsal horn of spinal cord which, in turn, discharge to the NTS. Thereby the NTS integrates vagal and spinal inputs (Bauer et al. 2016; López et al. 2007; Sobrino Crespo et al. 2014) (Fig. 3.2).

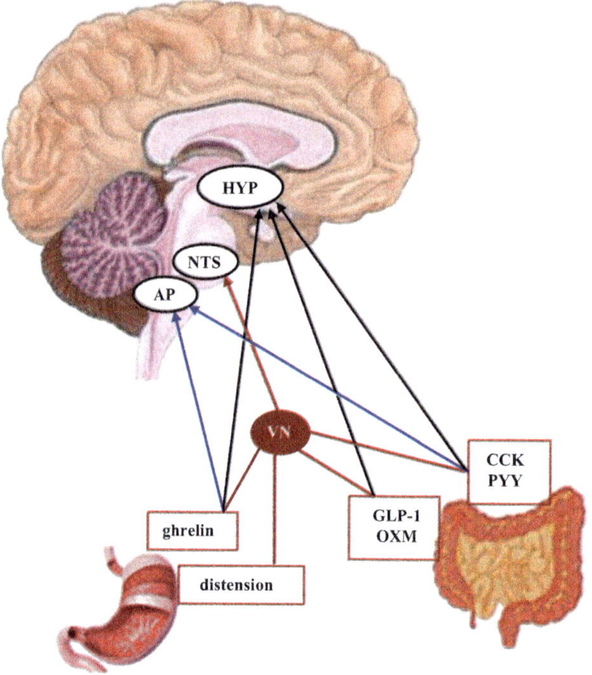

Fig. 3.2 Connecting pathways of the principal gastrointestinal signals to the brain centers of energy balance regulation. *HYP* hypothalamus, *NTS* nucleus tractus solitarius, *AP* area postrema, *VN* vagus nerve, *CCK* cholecystokinin, *OXM* oxyntomodulin, *GLP-1* glucagon-like peptide-1, *PYY* peptide tyrosine tyrosine

Gastrointestinal hormones can also influence specific satiety areas in the brain via the systemic circulation, reaching the NTS and/or the hypothalamic arcuate nucleus (ARC) through the area postrema (juxtaposed to the NTS) and the median eminence (adjacent to the ARC), which are semipermeable areas of the blood–brain barrier (BBB) (Bauer et al. 2016; Roh et al. 2016) (Fig. 3.2).

The hypothalamic ARC and brainstem NTS orexigenic and anorexigenic neurons project to second-order neurons located in part in the hypothalamic paraventricular nucleus (PVN) where anorexigenic thyrotrophin-releasing hormone (TRH) is released together with corticotrophin-releasing hormone (CRH) and oxytocin, and also in the lateral hypothalamic area (LHA) where orexigenic melanin-concentrated hormone (MCH) and orexin (OX) are secreted (Sobrino Crespo et al. 2014; Bauer et al. 2016).

Thus after reaching the brain centers, a hypothalamic and brainstem neural network integrates and synergizes short- and long-term signals and generates proper responses regarding fasting and feeding in order to maintain the balance of energy intake and expenditure.

After food ingestion, gastrointestinal signals are generated and sent to the brain centers involved in the central control of hunger and satiety, to restrain food intake. Initially, a mechanical signal following gastric distension is generated and reaches the CNS through gastric vagal and splanchnic afferent neurons (Bauer et al. 2016). With the arrival of nutrients to the small intestine, gastric emptying is slowed, leading to the maintenance of stomach distension, and gut neurohumoral signals are induced and sent to the central nucleus, to promote meal ending (Fig. 3.1). Enteroendocrine cells sense the ingested nutrients and produce a set of hormones that via autocrine, paracrine, or endocrine pathways control gastrointestinal functions and energy homeostasis (Monteiro and Batterham 2017; Bauer et al. 2016). In fact, synthesis and secretion of gut peptides are induced by ingested nutrients through the chemosensory machinery on the enteroendocrine luminal membrane. Gut peptides enter the systemic circulation or signal to the brain via paracrine actions on specific receptors located in afferent neurons in the digestive tract wall, namely the vagus nerve, which expresses receptors for gut hormones like CCK, GLP-1, PYY, and ghrelin. In addition, gut hormones may indirectly activate vagal and spinal afferent neurons via enteric nervous system stimulation which also appears to express gut hormone receptors (Bauer et al. 2016; López et al. 2007). The action of gut hormones on visceral afferent neurons is one of the first steps in the regulation of gastric motility and food ingestion. Several gut peptides use this pathway to increase or to reduce food intake such as ghrelin, CCK, GLP-1, and PYY. Vagal afferent neurons mediate orexigenic and anorexigenic pathways stimulating or inhibiting food intake and gastric emptying in response to fasting and feeding conditions. This neuronal switch is crucial to starting and finishing meals (Bauer et al. 2016; Wilson and Enriori 2015; Yi and Tschöp 2012).

3.2.1 Orexigenic Factors

3.2.1.1 Ghrelin

Ghrelin, which in its active form is an acylated peptide, acts as an endogenous ligand of the growth hormone secretagogue receptor (GHS-R1a). It is secreted mainly from the gastric oxyntic gland cells, the X/A-like-type cells in rodents or P/D1-type cells in humans and, to a far lesser extent, in the small intestine, particularly in the duodenum (Mishra et al. 2016; Monteiro and Batterham 2017; Sobrino Crespo et al. 2014).

Ghrelin was also found in the hypothalamic nuclei, specifically in the ARC, PVN, and ventromedial hypothalamus (VMH) (Stoyanova 2014). GHS-R1a is expressed in brain areas of homeostatic feeding, such as hypothalamus and brainstem, and in hedonic feeding areas. In the latter areas, ghrelin acts on the mesolimbic dopaminergic circuitry being thus involved in the hedonic reward system (Monteiro and Batterham 2017; Leigh and Morris 2016; Harrold et al. 2012; Stoyanova 2014).

Plasma ghrelin levels increase during fasting and before the onset of meals and decrease after nutrient ingestion proportionally to energy intake (Mishra et al. 2016; Monteiro and Batterham 2017; Stoyanova 2014; Sobrino Crespo et al. 2014). Ghrelin release increases in response to epinephrine, norepinephrine, secretin, and endothelin and decreases in response to hyperglycemia, insulin, leptin, insulin-like growth factor (IGF-1), PYY, oxyntomodulin (OXM), GLP-1, CCK, gastrointestinal polypeptide (GIP), somatostatin, high-fat diet, and obesity (Stoyanova 2014; Monteiro and Batterham 2017).

Ghrelin is the only known orexigenic hormone that induces an increase in food intake and adiposity in rodents and humans. Such effects occur through activation of the GHS-R1a in the neuropeptide Y/agouti-related peptide (NPY/AgRP) neurons in the hypothalamic ARC and in the brainstem (Fig. 3.1) via afferent vagus nerve (Stoyanova 2014; Wilson and Enriori 2015) (Fig. 3.2).

More than 90% of NPY/AgRP neurons of the ARC express the GHS-R1a which is the main pathway that communicates the ghrelin feeding effect, whereas less than 8% of proopiomelanocortin/cocaine- and amphetamine-regulated transcript (POMC/CART) neurons express ghrelin receptors. However, exogenous ghrelin administration induces a decrease in POMC/CART neurons activity and thus in the POMC-derived α-melanocyte-stimulating hormone (α-MSH), resulting in higher energy intake, which seems to be induced by increased GABA-mediated inhibitory inputs from NPY/AgRP neurons to POMC/CART neurons (Wilson and Enriori 2015; Stoyanova 2014). Ghrelin receptors are also expressed in the LHA OX neurons and in the brainstem NTS neurons as well as in the area postrema, suggesting the involvement of these brain areas in the orexigenic effect of ghrelin (López et al. 2007) (Fig. 3.2).

Alongside its effects on the central regulation of satiety and hunger, ghrelin indirectly stimulates gastric acid secretion via vagus nerve stimulation (Date et al. 2001; Stoyanova 2014) and accelerates gastric emptying. It increases growth hormone (GH), prolactin, corticotrophin (ACTH), and cortisol levels and inhibits insulin and pancreatic polypeptide secretion. It may be involved in the sleep-wake state, learning and memory, and shows neuroprotective effects (Stoyanova 2014).

Ghrelin takes part in glucose and lipid metabolism and regulation of insulin sensitivity. Apparently, ghrelin acts on glucose metabolism inducing hyperglycemia and a decrease in serum insulin levels in healthy lean volunteers (Broglio et al. 2001). The importance of ghrelin in glucose homeostasis is supported by the discovery of the islet ghrelin cell (Wierup et al. 2002) and reinforced by the description of GHS-R1a expression in pancreatic beta-cells (Dezaki et al. 2008). Besides the role of circulating ghrelin, a paracrine effect on insulin secretion was proposed. In isolated mouse islets, ghrelin in very high concentrations increased glucose-stimulated insulin release, whereas low concentrations inhibited insulin secretion (Salehi et al. 2004). However, most studies report an inhibitory effect of ghrelin on insulin release that, in a physiological context, acts in a reciprocal association with insulin (Dezaki et al. 2008; Peng et al. 2012; Tong et al. 2010; Wierup et al. 2014). Thus, during fasting high plasma ghrelin levels are linked to low insulin levels, and the postprandial increase in insulin secretion is accompanied by a decrease in circulating ghrelin.

Apparently, inhibitory effects on insulin secretion are additionally mediated by suppression of GLP-1-induced insulin release by exogenous and endogenous islet-derived ghrelin (Damdindorj et al. 2012).

Recent evidence showed that the unacylated ghrelin or des-acylghrelin, the major form of ghrelin in systemic circulation, so far considered a nonfunctional peptide, might have an opposite effect with the acyl/des-acyl ghrelin ratio being particularly important in physiologic and pathologic conditions (Monteiro and Batterham 2017; Stoyanova 2014). It was suggested that acute effects of des-acylghrelin are partially mediated by the hypothalamus reducing circulating acylghrelin and increasing postprandial insulin levels while improving insulin sensitivity and decreasing fat mass (Stoyanova 2014).

3.2.1.2 Endocannabinoids

The endocannabinoids, namely *N*-arachidonoyl-ethanolamine (anandamide) and 2-arachidonoylglycerol (2-AG), are produced from cell membrane phospholipids and act in an autocrine or paracrine manner on their receptors, CB-1R and CB-2R. The brain is the major organ involved in endocannabinoid-induced regulation of food intake, but activation of cannabinoid receptors within the gut may also play a role. The endocannabinoid system is expressed in the gut and takes part in the gastrointestinal functions namely peristalsis and gastric acid secretion inhibition (DiPatrizio 2016).

Endocannabinoid signals have been proposed to be involved in the gut–brain energy balance circuit through the vagus nerve pathway (DiPatrizio 2016). It was shown that increased circulating endocannabinoids are associated with palatable food but the source is not known (Mennella et al. 2015), although it was hypothesized that the release of endocannabinoids in the small intestine in response to hedonic food reaches homeostatic and reward brain regions via the systemic circulation (DiPatrizio 2016). An increase of 2-AG levels in rat's jejunum mucosa after fasting was also reported (DiPatrizio et al. 2015). Consequently, vagal CB-1R would be activated or CCK production in enteroendocrine I-cells would be inhibited (Sykaras et al. 2012), leading to decreased vagal afferent signals and enhanced appetite.

3.2.2 Anorexigenic Factors

3.2.2.1 Cholecystokinin

Cholecystokinin (CCK) is a gastrointestinal peptide secreted from duodenal and jejunal I-cells in response to fat and proteins, as well as in the enteric nervous system and the CNS. CCK is responsible for stimulation of exocrine pancreatic secretion and enzyme synthesis, gallbladder contraction and gastric emptying delay, and is also involved in short-term control of food intake, reducing meal size and duration (López et al. 2007; Bauer et al. 2016; Monteiro and Batterham 2017). The main form of CCK in the body is CCK-8 which binds to CCK-1 receptor (CCK-1R) which are mainly expressed in the gastrointestinal tract and to CCK-2 receptor (CCK-2R) predominantly located in the brain (Monteiro and Batterham 2017). CCK-8 anorectic effects are mediated by CCK-1R activation on gut vagal afferent nerve fibers through paracrine pathways which, in turn, project to the NTS and then to the hypothalamus namely to the PVN (López et al. 2007; Bauer et al. 2016; Monteiro and Batterham 2017). Besides the neuronal signal, that seems to be crucial for its effect on satiation, CCK crosses the BBB and may directly, via systemic circulation, reach the brainstem NTS and the NPY/AgRP neurons in the dorsomedial hypothalamus (DMH) (López et al. 2007; Sobrino Crespo et al. 2014; Bauer et al. 2016) (Fig. 3.2).

Moreover, the role of the CCK-2R in long-term regulation of hunger and satiety was questioned as receptor-deficient mice develop hyperphagia, obesity, and impairment of glucose homeostasis (Clerc et al. 2007). Additionally, CCK-mediated actions on GLP-1 secretion (Beglinger et al. 2010) and on the secretion of ghrelin and PYY have been suggested (Degen et al. 2007).

3.2.2.2 Glucagon-like-Peptide-1

Preproglucagon is processed by convertases producing glucagon in α-pancreatic cells, glucagon-like peptide-2 (GLP-2), oxyntomodulin (OXM), and glucagon-like peptide-1 (GLP-1) in intestinal enteroendocrine L-cells and the latter in the brainstem NTS as well (Monteiro and Batterham 2017; López et al. 2007; Sobrino Crespo et al. 2014).

Thus, GLP-1 is expressed in intestinal L-cells, mainly found in the distal jejunum, ileum, and colon, as well as in brainstem NTS neurons and pancreatic α-cells (D'Alessio 2016; López et al. 2007). It is released in response to carbohydrates, lipids, and proteins. Once in circulation, GLP-1$_{7-36,}$ the major secretory product that exerts its insulinotropic and anorexigenic effects binding to GLP-1 receptor (GLP-1R), is rapidly cleaved by the enzyme DPP-IV to metabolites long considered to be biologically inactive. However, recent evidence shows that these by-products of GLP-1 cleavage have cytoprotective effects on pancreatic ß-cells, inhibit hepatic glucose production, and have cardiovascular and neuronal protective actions (Guglielmi and Sbraccia 2016).

GLP-1 is primarily an incretin hormone that increases glucose-stimulated insulin secretion, an effect particularly manifest after meals, and is responsible for about 60% of the postprandial release of insulin, which is involved in the long-term regulation of energy homeostasis (Mikulaskova et al. 2016). Recently, it was shown that GLP-1 from α-cells plays a role on ß-cell function via paracrine effects (D'Alessio 2016). Besides effects on insulin secretion and biosynthesis, GLP-1 also inhibits glucagon secretion, increases ß-cell mass in rats, reduces gastrointestinal motility and secretion, and has cardio and neuroprotective and, possibly, insulinomimetic effects (Guglielmi and Sbraccia 2016).

Peripheral GLP-1 acts on the central nervous system in specific areas of the hypothalamus inhibiting food intake, in rats and humans, particularly through inhibition of NPY/AgRP neurons (Mikulaskova et al. 2016) (Fig. 3.1). GLP-1R is widely expressed in the CNS, namely in the hypothalamic nuclei like ARC, VMH, and PVN and in the brainstem NTS neurons as well as in peripheral tissues such as the pancreas and gastrointestinal tract. The expression of GLP-1R in intestinal vagal afferent neurons suggests that a paracrine activation of these receptors followed by sensory transmission to the NTS neurons in the brainstem is responsible for the short-term effect of GLP-1 on food intake (Sobrino Crespo et al. 2014; Monteiro and Batterham 2017; Sekar et al. 2017) (Fig. 3.2). This paracrine effect of GLP-1 also contributes to pancreatic insulin secretion (Sekar et al. 2017). Central GLP-1R activation in hypothalamic VMH can control the energy expenditure inducing brown adipose tissue thermogenesis through sympathetic stimulation (Geloneze et al. 2017).

GLP-1 neurons in brainstem NTS send their projections to the hypothalamic ARC and PVN inducing an anorexigenic response and to the ventral tegmental area (VTA) and nucleus accumbens (NAc) which also suggests a role in hedonic feeding circuits (Harrold et al. 2012; Geloneze et al. 2017).

3.2.2.3 Glucagon-like Peptide-2

Glucagon-like peptide-2 (GLP-2) also derives from the cleavage of proglucagon in intestinal L-cells, together with GLP-1. GLP-2 is released following food ingestion in response to luminal carbohydrates and lipids, particularly short-chain fatty acids; it is generally thought that GLP-2 secretion is regulated by the same factors that influence GLP-1 release. GLP-2 effects are mediated by interaction with a specific GLP-2 receptor (GLP-2R) localized mainly in the gut and in the brain regions involved in energy homeostasis such as the hypothalamus and brainstem (Baldassano et al. 2016).

GLP-2 is primarily an intestinotrophic factor; it is related to increased nutrient absorption and inhibition of gastrointestinal motility and maintenance of intestinal mucosa morphology and integrity. However, it was suggested that GLP-2, besides inhibition of gastric emptying, is also an anorexigenic peptide acting at a central and/or peripheral level in various animal species but its effects in humans still need to be explored (Baldassano et al. 2016).

3.2.2.4 Peptide Tyrosine Tyrosine

Peptide tyrosine tyrosine (PYY) belongs to the family of NPY and pancreatic polypeptide (PP). PYY and PP are expressed in the gut and pancreas, whereas NPY is widely distributed throughout the gut–brain axis, namely in enteric afferent and sympathetic nerves and neurons of CNS. PYY is produced in L-cells of the distal small bowel and, in increasing concentrations, in the colon and rectum, where it co-localizes with proglucagon products. It is also expressed in gastric enteric neurons and pancreatic endocrine cells (Adrian et al. 1985; Lundberg et al. 1982; Holzer et al. 2012). The expression of PYY in the stomach, duodenum, and jejunum has recently been demonstrated as well as its co-secretion with other gastrointestinal hormones like CCK and secretin (Monteiro and Batterham 2017; Persaud and Bewick 2014). PYY is secreted in response to food intake, particularly nutrients with a high lipid and protein content, with its release being affected by intestinal microbiota (Adrian et al. 1985; Holzer et al. 2012).

In the CNS the subtype PYY_{3-36}, the major form of circulating hormone, interacts as a preferential agonist of the Y2 receptor (Y-2R) on ARC neurons. In fact, PYY_{3-36} induces satiety through activation of presynaptic Y-2R on NPY/AgRP neurons in the hypothalamic ARC leading to NPY inhibition; suppressing the inhibition of POMC/CART neurons, α-MSH is released and may exert its anorexigenic effects (Fig. 3.1). PYY_{3-36} also affects feeding control by reaching the brainstem directly via the area postrema and indirectly via the vagal afferent pathway. While the key effect is mediated directly in brain areas involved in the control of appetite, some studies showed that PYY_{3-36} also acts through the stimulation of Y-2R on intestinal vagal afferent neurons (Bauer et al. 2016; Mishra et al. 2016; López et al. 2007; Persaud and Bewick 2014; Holzer et al. 2012; Sobrino Crespo et al. 2014) (Fig. 3.2). It has been demonstrated that PYY_{3-36}, besides its effects on homeostatic food intake, contributes to hedonic feeding (Harrold et al. 2012).

After a meal, PYY_{3-36} levels remain elevated for several hours suggesting an endocrine long-term effect on satiety, in contrast to CCK and GLP-1 which are more important in short-term control (Bauer et al. 2016; Mishra et al. 2016). It also delays gastric emptying, contributing to reduction of food intake, reduces intestinal motility and gastric secretion, and increases ileal absorption (López et al. 2007; Holzer et al. 2012; Persaud and Bewick 2014).

PYY_{3-36} regulates glucose homeostasis by improvement of insulin sensitivity but does not affect glucose-induced insulin secretion (Persaud and Bewick 2014). Recent studies in rodents suggested that PYY_{3-36} acts on GLP-1 secretion from intestinal L-cells, through activation of peripheral Y-2R, leading indirectly to insulin release but these findings were not sustained in humans (Chandarana et al. 2013; Persaud and Bewick 2014). However, a role of PYY_{3-36} in improvement of glucose tolerance is supported through its effects on insulin sensitivity and GLP-1 induced insulin secretion (Persaud and Bewick 2014).

3.2.2.5 Oxyntomodulin

Oxyntomodulin (OXM) is another enterohormone derived from proglucagon and secreted from L-cells in the distal ileum in response to a meal. Actions of OXM include inhibition of basal and postprandial gastric acid secretion, gastroduodenal motility, and gastric emptying (Schjoldager et al. 1989). Furthermore, OXM reduces food intake in rodents (Dakin et al. 2001) and humans (Cohen et al. 2003). In obese nondiabetic volunteers, administration of OXM resulted in significant weight loss coupled with reduction in energy intake and increased energy expenditure (Monteiro and Batterham 2017). Recent research suggested that dual activation of GLP-1R and glucagon receptor (GCGR) by OXM is involved in weight loss and maintenance of glucose homeostasis in mice (Du et al. 2012; Kosinski et al. 2012; Pocai 2014; Monteiro and Batterham 2017). Apparently OXM promotes glucose-dependent insulin secretion and improves glucose tolerance in type 2 diabetes through GCGR activation. However, the anorectic effect of OXM may require the GLP-1R and includes paracrine vagal stimulation and direct CNS activation (Fig. 3.2). Despite GLP-1R-mediated effects, hypothalamic mechanisms seem to be different from those of GLP-1 (Monteiro and Batterham 2017).

3.2.2.6 Gastric Leptin

Although leptin is mostly produced in white adipose tissue, it is also secreted by gastric endocrine cells into the systemic circulation and by gastric chief cells into the gastric lumen, independently to adipose-derived hormone (Cammisotto and Bendayan 2012; Monteiro and Batterham 2017). Gastric leptin secretion occurs in response to secretory factors like food ingestion, insulin, secretin, and CCK through a mechanism mediated by the vagus nerve. Chief cells release leptin bound to a soluble receptor which allows it to resist the hydrolytic conditions of gastric juice and proteolysis. After its influx into the duodenum, leptin interacts with its membrane receptor and exerts intestinal effects. In addition, acting either in a paracrine fashion or crossing the intestinal mucosa to reach the systemic circulation, gastric leptin signals brain centers to regulate food intake (Cammisotto and Bendayan 2012; Monteiro and Batterham 2017). Gastric leptin takes part in short-term control of food intake while leptin derived from adipose tissue plays a pivotal role in long-term regulation (Cammisotto and Bendayan 2012).

3.2.2.7 Apolipoprotein-IV

Intestinal apolipoprotein-IV (ApoA-IV) is a chylomicron-derived lipoprotein synthesized and released from small intestine enterocyte following absorption and uptake of long-chain fatty acids into chylomicrons. ApoA-IV is a short-term anorexigenic signal that reaches brain centers via the systemic circulation or through the stimulation of CCK from local enteroendocrine cells with the consequent activation of vagal afferent neurons (Bauer et al. 2016; Monteiro and Batterham 2017). An incretin effect was also assigned to ApoA-IV (Monteiro and Batterham 2017).

3.2.2.8 Neurotensin

Neurotensin (NT) is a peptide expressed in the CNS and in enteroendocrine cells that acts via NT receptors (NT-1R, NT-2R, and NT-3R). Central and peripheral administration of NT, biding to its receptor NT-1R, reduces food intake through increased POMC in the hypothalamic ARC or via afferent vagal stimulation. In fact, NT acts through endocrine or paracrine mechanisms in response to its local concentrations. It also affects hedonic circuits interacting with leptin and the dopaminergic system (Monteiro and Batterham 2017).

It was recently reported that NT, PYY, and GLP-1 are co-expressed and co-secreted and may act synergistically (Monteiro and Batterham 2017). Luminal nutrients, mainly fat, induce gastrointestinal NT release. NT controls gut motility, pancreatic and biliary secretion and may have incretin effects (Monteiro and Batterham 2017).

3.2.2.9 Secretin

Secretin, a peptide secreted from the S-cells of the duodenum in response to intestinal luminal acid, has primarily the function of stimulating bicarbonate secretion, particularly by the pancreas, to neutralize luminal acid. It is also expressed in the brain, namely in the hypothalamic PVN and ARC and in the brainstem NTS. Secretin receptors (SCT-R) are expressed in the PVN and ARC among other brain regions (Sekar et al. 2017).

An anorectic effect of peripheral and central secretin was recently shown in mice. Although secretin is able to cross the BBB, its anorectic effects appear to be mediated by paracrine activation of SCT-R in intestinal vagal afferents fibers (Sekar et al. 2017). An increase in POMC and a reduction in AgRP levels were observed in the hypothalamic ARC after intracerebroventricular and intraperitoneal secretin administration. In addition, central and peripheral secretin seem to increase melanocortin-4 receptor (MC-4R) in the hypothalamic PVN (Sekar et al. 2017).

3.2.2.10 GUT Microbiota

Recent research established a relationship between gut microbiology and obesity. Microorganisms colonize the human gut since early life and the composition of gut microbiota depends, among others, on dietary patterns, ethnicity, and genetic factors (Mishra et al. 2016).

Altered gut microbiota and/or a reduced bacterial diversity were observed in obese animal and human subjects in comparison with leaner counterparts (Bauer et al. 2016; Mishra et al. 2016). These microorganisms colonize and survive within the host gut, in a symbiotic relationship that provides an advantageous environment for the microbiota, and structural and functional beneficial effects for the host

(Bauer et al. 2016). The role of symbiotic gut microflora is not only related to digestive functions but also to metabolism and energy homeostasis.

An important activity of colonic microbes is the breakdown of dietary substrates such as fiber and resistant starch into short-chain fatty acids (SCFA) while also releasing acetate, propionate, and butyrate (Bauer et al. 2016; Mishra et al. 2016; Monteiro and Batterham 2017). Butyrate is the principal energy source for the colonic epithelium while propionate enters the portal circulation to be used in gluconeogenesis and acetate reaches peripheral tissues through the systemic circulation and is used to form acetyl-CoA. SCFA are thought to interfere with host energy balance acting on gut peptide release in both rodents and humans (Bauer et al. 2016).

Free fatty acid receptors FFA-3 and FFA-2 play an important role in the interaction of SCFA with the host and are expressed in many tissues including enteroendocrine cells particularly L-cells and P/D1-cells. FFA-3 and FFA-2 receptor activation was shown to induce GLP-1 and PYY secretion and to reduce ghrelin release (Monteiro and Batterham 2017; Holzer et al. 2012; Bauer et al. 2016).

FFA-3 and FFA-2 receptors are also localized in adipose tissue highlighting its role as a major target for gut microbiota metabolites (Mishra et al. 2016). It is plausible that changes in gut microbiota can modify gut peptide signaling and consequently the gut–brain axis. Recently, it was suggested that gut microbiota directly impacts the CNS centers of energy homeostasis regulation (Bauer et al. 2016).

Taken together, gut microbiota can interfere with local and central signaling of energy homeostasis regulation.

3.3 Endocrine Pancreas-Derived Signals to the Brain

Neuroendocrine signals from the pancreas contribute to energy balance and are related to gastrointestinal peptides, like incretin hormones, and to adipose tissue, the other player in the energy homeostasis. Pancreatic hormones are involved in energy homeostasis, regulation of satiety and hunger, and body weight control. They exert these effects in concert with gastrointestinal and adipose tissue signals and brain centers.

3.3.1 Insulin

Insulin is secreted by the pancreatic ß-cells in accordance with glycemia levels through short-term feedback regulation. On the other hand, it also plays a role in long-term control of satiety as it is secreted in proportion to the amount of stored fat (Roh et al. 2016; López et al. 2007). However, albeit implicated in the meal size regulation (Lutz 2012), insulin is mainly an adiposity signal with the hypothalamus being the target for insulin-induced appetite regulation (Fig. 3.3). Insulin receptors are expressed in the hypothalamic ARC, DMH, and PVN but insulin effects on food ingestion are specifically mediated by NPY/AgRP and POMC/CART neurons in the ARC. In fact, insulin promotes food intake reduction through inhibition of orexigenic NPY/AgRP and stimulation of anorexigenic POMC/CART neurons (López et al. 2007) (Fig. 3.1). Insulin receptors are also expressed in the brainstem but the direct effect of insulin in the brainstem has not yet been established (Filippi et al. 2013). Insulin receptors are particularly expressed in the cerebral cortex, olfactory bulbs, hippocampus, hypothalamus, cerebellum, and reward circuits and account for insulin effects such as central and peripheral metabolism, homeostatic and hedonic food intake and energy expenditure and modulation of memory and cognitive processes and neuronal development (Csajbók and Tamás 2016; Filippi et al. 2013).

Insulin crosses the BBB but its synthesis in the brain, primary in the cerebral cortex, has also been suggested (Csajbók and Tamás 2016).

3.3.2 Glucagon

Glucagon, a hormone secreted by pancreatic α-cells in response to low glycemia, antagonizes insulin action stimulating hepatic glucose production and lipolysis. Glucagon crosses the BBB and might exert part of its effects on peripheral homeostasis acting in the hypothalamus and the brainstem (Filippi et al. 2013). The mechanisms underlying the central effects of glucagon on appetite reduction remain unknown and further studies are required (Filippi et al. 2013; Sekar et al. 2017). However, the involvement of hypothalamic CRH neurons and the glucose-sensitive neurons in the hypothalamic LHA has been suggested (Filippi et al. 2013). Moreover, it was shown that glucagon affects meal size rather than meal interval, and hepatic vagal afferents seem to mediate this effect (Sekar et al. 2017) (Fig. 3.3).

Insulin and glucagon do not seem to have opposite metabolic effects in the CNS; furthermore, both cross the BBB and both seem to reduce food intake through complementary effects (Filippi et al. 2013).

3.3.3 Amylin

Amylin or islet amyloid polypeptide is produced in pancreatic ß-cells. It is co-stored with insulin and co-secreted with this hormone in response to meals and to insulinotropic factors. Amylin promotes meal ending via receptors in the area postrema, the primary target for this peptide, and in the brainstem NTS and lateral parabrachial nucleus which transmit the signal to the LHA and other hypothalamic areas like the VMH (Mikulaskova et al. 2016; Sobrino Crespo et al. 2014; Lutz 2012; Lutz 2013). The anorexigenic effects of amylin, including delay of gastric emptying, seem to be mediated by direct humoral action on area postrema neurons (Sobrino Crespo et al.

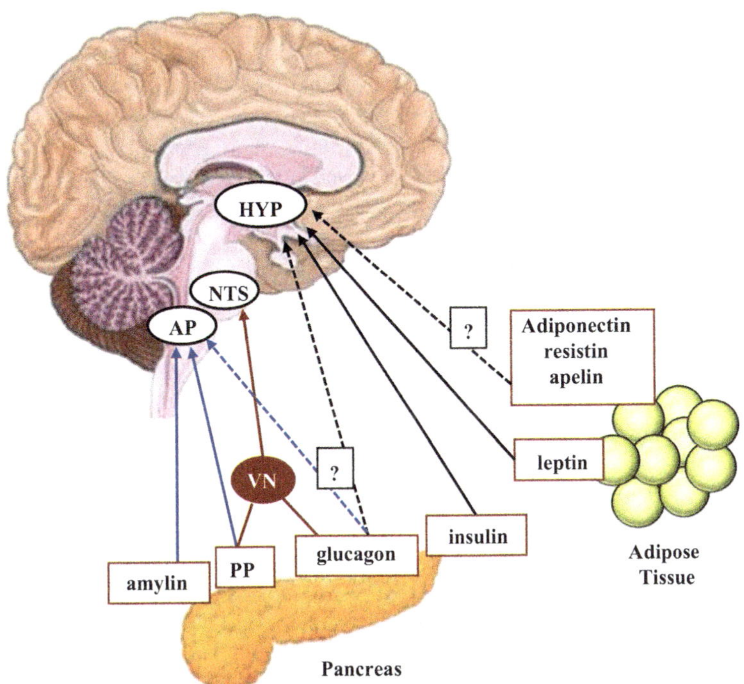

Fig. 3.3 Connecting pathways of the pancreatic and adipose tissue signals to the brain centers of energy balance regulation. *HYP* hypothalamus, *NTS* nucleus tractus solitarius, *AP* area postrema, *VN* vagus nerve, *PP* pancreatic polypeptide

2014; Lutz 2012) (Fig. 3.3). Amylin appears to increase energy expenditure but the underlying mechanism is unknown (Lutz 2012). It also seems to act as an adiposity signal because amylin levels are well correlated with body fat mass (Lutz 2012).

Some studies show a synergistic effect of amylin and other peptides on meal size and adiposity. This synergism was reported for leptin, insulin, and gastrointestinal peptides like CCK, PYY_{3-36}, and perhaps GLP-1 which, in turn, induces an enhancement of amylin release, through its incretin effect (Lutz 2013).

3.3.4 Pancreatic Polypeptide

Pancreatic polypeptide (PP) is released under cholinergic control during the post-prandial state by PP (or F) cells located in the pancreatic islets. To a lesser extent, it is expressed in the distal gut. PP delays gastric emptying and food intake by activating its preferential Y receptor, Y-4R subtype, in the vagus nerves. It can also reach the brain directly through the area postrema (Holzer et al. 2012; Sobrino Crespo et al. 2014; Mishra et al. 2016) (Fig. 3.3). The anorectic effects of PP, apart from the reduction of gastric emptying, are most likely mediated by the hypothalamic VMH and PVN and the brainstem. It also increases energy expenditure and has digestive actions such as the inhibition of exocrine pancreatic secretion and gallbladder motility (Sobrino Crespo et al. 2014; Holzer et al. 2012).

As circulatory PP levels are sustained after a meal, a possible role in the long-term regulation of food intake has been proposed. However, the physiologic role of PP on energy balance needs to be further clarified (Mishra et al. 2016).

3.4 Adipose Tissue-Derived Signals to the Brain

Long-term signaling of energy homeostasis reflects the levels of fat mass and regulates body weight as well as the amount of fat stores, and adipose tissue plays a critical role in this mechanism. White adipose tissue possesses an active secretory function producing endocrine and paracrine factors commonly known as adipokines. They act on the brain and on peripheral organs such as the liver, pancreas, and skeletal muscle to regulate several processes such as food intake, energy expenditure, and glucose and lipid metabolism (Harwood 2012).

3.4.1 Leptin

The predominant signal derived from adipose tissue is leptin which is mainly secreted from adipocytes in proportion to fat stores, both in rodents and humans (van Swieten et al. 2014). Although leptin levels are proportional to fat mass, its secretion is stimulated in times of nutrient availability and decreased in times of energy demand (Harwood 2012; van Swieten et al. 2014; Münzberg and Morrison 2015).

After a meal, leptin levels increase to suppress appetite centrally and to increment energy expenditure. During fasting, leptin levels are lower and, in concert with insulin, appetite is stimulated and sympathetic nerves activity, thyroid hormone actions, and thermogenesis are inhibited (Harwood 2012).

Recent evidence suggests that hypothalamic and extra-hypothalamic sites contribute to the effects of leptin on energy balance (Münzberg and Morrison 2015; Wilson and Enriori 2015). Leptin receptors b (LRb) are highly expressed in the brain namely in the hypothalamic nuclei, especially the ARC, DMH, and VMH (Harwood 2012; van Swieten et al. 2014; Münzberg and Morrison 2015; Parimisetty et al. 2016), while their expression in the PVN and LHA is lower (Harwood 2012). Activation of hypothalamic LRb decreases appetite and increases the activity of the sympathetic nervous system (SNS), which stimulates non-shivering thermogenesis in the brown adipose tissue (Wilson and Enriori 2015). Leptin acts on both hypothalamic orexigenic and anorexigenic neurons to control the energy balance (Fig. 3.1). The proximity of the hypothalamic ARC to the semipermeable BBB makes this nucleus a particularly sensitive brain area for leptin action (Fig. 3.3). In fact, the anorexigenic POMC/CART and the orexigenic NPY/AgRP neurons are direct targets of leptin. Leptin activates POMC/CART neurons and stimulates α-MSH secretion; conversely, it inhibits AgRP/NPY neurons and reduces NPY, AgRP, and GABA release. It was shown that the reduction in the inhibitory input to POMC/CART neurons is mediated by LRbs on presynaptic GABAergic neurons (Wilson and Enriori 2015; van Swieten et al. 2014; Münzberg and Morrison 2015; Parimisetty et al. 2016). The effects of leptin on the hypothalamic PVN and LHA neurons appear to be primarily indirect, from ARC neuron projections. In the PVN leptin leads to activation of the anorexigenic TRH, CRH, and oxytocin neurons while it inhibits orexigenic MCH and OX neurons in the hypothalamic LHA (Harwood 2012; van Swieten et al. 2014; Münzberg and Morrison 2015).

The development of leptin resistance in the hypothalamus changes the central sensing of nutrient reserves and inhibits appetite suppression after meals. Such mechanisms are currently believed to contribute to hyperphagia in obese subjects.

Leptin is also involved in food reward through the mesolimbic dopaminergic system in the VTA and NAc (Münzberg and Morrison 2015). A synergy with CCK (Sobrino Crespo et al. 2014; Harrold et al. 2012) and a crosstalk with insulin (Harwood 2012) was suggested as well as its expression in the brain, namely in the hypothalamus (Parimisetty et al. 2016).

Leptin action in the hypothalamic neurons inhibits food intake and increases energy expenditure and improves, independently of its anorectic effects, glucose and lipid metabolism (Wilson and Enriori 2015; Roh et al. 2016). As such, apart from its peripheral effects in the liver, pancreas, muscle, adipose tissue, and in the immune and cardiovascular systems, leptin has an important role in the central control of glucose homeostasis and brown adipose tissue thermogenesis.

3.4.2 Adiponectin

Adiponectin is another adipose tissue-derived hormone mainly secreted by mature adipocytes and its plasma concentrations are negatively correlated with obesity. It modulates several metabolic processes such as glucose and lipid metabolism and insulin sensitivity; it also has effects on the immune and cardiovascular system (Harwood 2012; Parimisetty et al. 2016).

The central effects of adiponectin in energy homeostasis are not yet clarified. However, adiponectin receptors, particularly Adipo-R1/R2, are expressed in the hypothalamus, brainstem, and pituitary gland (López et al. 2007; Harwood 2012). An effect on energy expenditure mediated by the hypothalamic melanocortin system was suggested (López et al. 2007; Harwood 2012).

It was shown that adiponectin and leptin synergistically activate the ARC neurons affecting energy balance, particularly thermogenesis, and glucose metabolism, in rodents (Sun et al. 2016). Two opposite signaling pathways in POMC/CART neurons in relation to brain glucose concentrations were recently reported. In rodents, intracerebroventricular injection of adiponectin increased or decreased food intake under high or low brain glucose levels, respectively (Suyama et al. 2016).

3.4.3 Resistin

Resistin is an adipokine expressed and secreted in humans by adipose tissue macrophages in relation to adiposity (Parimisetty et al. 2016). The mechanism underlying the central effects of resistin and its receptor were not yet identified (Parimisetty et al. 2016) but resistin expression was reported in the hypothalamus and cerebral cortex. It was shown in rodents that resistin reduces sympathetic outflow to brown adipose tissue and decreases thermogenesis (Kosarı et al. 2013) while inducing a modest and transient reduction of energy intake (Tovar et al. 2005).

3.4.4 Apelin

Apelin is a ubiquitous peptide considered as an adipokine as it is produced and secreted by adipocytes. It has beneficial glucose-lowering properties and is upregulated in obese and insulin-resistant models. Both apelin and its receptor APJ are expressed in several peripheral tissues and in the CNS, particularly in the hypothalamus. Apelin mRNA is present in different hypothalamic nuclei including the supraoptic, PVN, and ARC which implies the existence of an apelinergic neuronal system and thus a dual action of apelin as circulating peptide and neurotransmitter (Reaux-Le Goazigo et al. 2011; Castan-Laurell et al. 2011). It is not clear though whether peripheral apelin can reach the hypothalamus and modulate its apelin levels (Castan-Laurell et al. 2011). The central effects of apelin on energy metabolism are complex and controversial, as well as dependent on nutritional status and on the amount of apelin in the hypothalamus. Some authors hypothesize that a rise in hypothalamic apelin levels could be involved in the transition from the euglycemic to the diabetic status, since acute and chronic icv administration of apelin induces fasting hyperglycemia, hyperinsulinemia, and insulin resistance (Drougard et al. 2016; Castan-Laurell et al. 2011). In fact, higher hypothalamic apelin levels have been described in obese/diabetic mice (Reaux-Le Goazigo et al. 2011), favoring the hypothesis that elevated cerebral apelin has deleterious effects on energy metabolism, as opposed to its recognized beneficial peripheral actions. Several mechanisms have been pointed out and include a proinflammatory role of high amounts of apelin in the hypothalamus along with an increase in circulating proinflammatory cytokines (e.g., interleukin-1ß), decreased energy expenditure, and impaired thermogenesis (Drougard et al. 2016).

3.5 Concluding Remarks

The interplay between gut-derived short-term and adipose tissue-derived long-term signaling controls meal duration and composition as well as body nutrient reserves to maintain energy homeostasis. Thereby gastrointestinal, pancreatic, and adipose tissue outputs play an important role in the regulation of energy homeostasis acting via neuronal and endocrine pathways in selective brain areas, among which the hypothalamus is undoubtedly the key integrator. In recent years, important advances have been made in the understanding of the neuronal circuits involved in regulation of appetite and overall body metabolism. Interestingly, nonneuronal cells, such as astrocytes and tanycytes, seem to be involved in sensing and integrating metabolic signals (Freire-Regatillo et al. 2017; Yi and Tschöp 2012) which, in turn, makes this adipose tissue–gut–brain axis an even more complex and intricate network.

References

Adrian TE, Ferri GL, Bacarese-Hamilton AJ, Fuessl HS, Polak JM, Bloom SR (1985) Human distribution and release of a putative new gut hormone, peptide YY. Gastroenterology 89:1070–1077

Baldassano S, Amato A, Mulè F (2016) Influence of glucagon-like peptide 2 on energy homeostasis. Peptides 86:1–5

Bauer PV, Hamr SC, Duca FA (2016) Regulation of energy balance by a gut–brain axis and involvement of the gut microbiota. Cell Mol Life Sci 73:737–755

Beglinger S, Drewe J, Schirra J, Göke B, D'Amato M, Beglinger C (2010) Role of fat hydrolysis in regulating glucagon-like peptide-1 secretion. J Clin Endocrinol Metab 95:879–886

Broglio F, Arvat E, Benso A, Gottero C, Muccioli G, Papotti M, van der Lely AJ, Deghenghi R, Ghigo E (2001) Ghrelin, a natural GH secretagogue produced by the stomach, induces hyperglycemia and reduces insulin secretion in humans. J Clin Endocrinol Metab 86:5083–5086

Cammisotto P, Bendayan M (2012) A review on gastric leptin: the exocrine secretion of a gastric hormone. Anat Cell Biol 45(1):16

Castan-Laurell I, Dray C, Attané C, Duparc T, Knauf C, Valet P (2011) Apelin, diabetes, and obesity. Endocrine 40:1–9

Chandarana K, Gelegen C, Irvine EE, Choudhury AI, Amouyal C, Andreelli F, Withers DJ, Batterham RL (2013) Peripheral activation of the Y2-receptor promotes secretion of GLP-1 and improves glucose tolerance. Mol Metab 2:142–152

Clerc P, Coll Constans MG, Lulka H, Broussaud S, Guigné C, Leung-Theung-Long S, Perrin C, Knauf C, Carpéné C, Pénicaud L, Seva C, Burcelin R, Valet P, Fourmy D, Dufresne M (2007) Involvement of cholecystokinin 2 receptor in food intake regulation: hyperphagia and increased fat deposition in cholecystokinin 2 receptor-deficient mice. Endocrinology 148:1039–1049

Cohen MA, Ellis SM, Le Roux CW, Batterham RL, Park A, Patterson M, Frost GS, Ghatei MA, Bloom SR (2003) Oxyntomodulin suppresses appetite and reduces food intake in humans. J Clin Endocrinol Metab 88:4696–4701

Csajbók ÉA, Tamás G (2016) Cerebral cortex: a target and source of insulin? Diabetologia 59:1609–1615

D'Alessio D (2016) Is GLP-1 a hormone: whether and when? J Diabetes Investig 7:50–55

Dakin CL, Gunn I, Small CJ, Edwards CM, Hay DL, Smith DM, Ghatei M, Bloom SR (2001) Oxyntomodulin inhibits food intake in the rat. Endocrinology 142:4244–4250

Damdindorj B, Dezaki K, Kurashina T, Sone H, Rita R, Kakei M, Yada T (2012) Exogenous and endogenous ghrelin counteracts GLP-1 action to stimulate cAMP signaling and insulin secretion in islet β-cells. FEBS Lett 586:2555–2562

Date Y, Nakazato M, Murakami N, Kojima M, Kangawa K, Matsukura S (2001) Ghrelin acts in the central nervous system to stimulate gastric acid secretion. Biochem Biophys Res Commun 280:904–907

Degen L, Drewe J, Piccoli F, Gräni K, Oesch S, Bunea R, D'Amato M, Beglinger C (2007) Effect of CCK-1 receptor blockade on ghrelin and PYY secretion in men. Am J Phys Regul Integr Comp Phys 292:R1391–R1399

Dezaki K, Sone H, Yada T (2008) Ghrelin is a physiological regulator of insulin release in pancreatic islets and glucose homeostasis. Pharmacol Ther 118:239–249

DiPatrizio NV (2016) Endocannabinoids in the gut. Cannabis Cannabinoid Res 1:67–77

DiPatrizio NV, Igarashi M, Narayanaswami V, Murray C, Gancayco J, Russell A, Jung KM, Piomelli D (2015) Fasting stimulates 2-AG biosynthesis in the small intestine: role of cholinergic pathways. Am J Phys Regul Integr Comp Phys 309:R805–R813

Drougard A, Fournel A, Marlin A, Meunier E, Abot A, Bautzova T, Duparc T, Louche K, Batut A, Lucas A, Le-Gonidec S, Lesage J, Fioramonti X, Moro C, Valet P, Cani PD, Knauf C (2016) Central chronic apelin infusion decreases energy expenditure and thermogenesis in mice. Sci Rep 6:31849

Du X, Kosinski JR, Lao J, Shen X, Petrov A, Chicchi GG, Eiermann GJ, Pocai A (2012) Differential effects of oxyntomodulin and GLP-1 on glucose metabolism. Am J Physiol Endocrinol Metab 303:E265–E271

Filippi BM, Abraham MA, Yue JT, Lam TK (2013) Insulin and glucagon signaling in the central nervous system. Rev Endocr Metab Disord 14:365–375

Freire-Regatillo A, Argente-Arizón P, Argente J, García-Segura LM, Chowen JA (2017) Non-neuronal cells in the hypothalamic adaptation to metabolic signals. Front Endocrinol (Lausanne) 8:51

Geloneze B, Lima-Júnior JC, Velloso LA (2017) Glucagon-like peptide-1 receptor agonists (GLP-1RAs) in the brain–adipocyte axis. Drugs 77:493–503

Guglielmi V, Sbraccia P (2016) GLP-1 receptor independent pathways: emerging beneficial effects of GLP-1 breakdown products. Eat Weight Disord 22(2):231–240

Harrold JA, Dovey TM, Blundell JE, Halford JC (2012) CNS regulation of appetite. Neuropharmacology 63:3–17

Harwood HJ (2012) The adipocyte as an endocrine organ in the regulation of metabolic homeostasis. Neuropharmacology 63:57–75

Holzer P, Reichmann F, Farzi A (2012) Neuropeptide Y, peptide YY and pancreatic polypeptide in the gut-brain axis. Neuropeptides 46:261–274

Kosari S, Camera DM, Hawley JA, Stebbing M, Badoer E (2013) ERK1/2 in the brain mediates the effects of central resistin on reducing thermogenesis in brown adipose tissue. Int J Physiol Pathophysiol Pharmacol 5:184–189

Kosinski JR, Hubert J, Carrington PE, Chicchi GG, Mu J, Miller C, Cao J, Bianchi E, Pessi A, Sinharoy R, Marsh DJ, Pocai A (2012) The glucagon receptor is involved in mediating the body weight-lowering effects of oxyntomodulin. Obesity 20:1566–1571

Leigh SJ, Morris MJ (2016) The role of reward circuitry and food addiction in the obesity epidemic:an update. Biol Psychol. doi: 10.1016/j.biopsycho.2016.12.013

López M, Tovar S, Vázquez MJ, Williams LM, Diéguez C (2007) Peripheral tissue–brain interactions in the regulation of food intake. Proc Nutr Soc 66:131–155

Lundberg JM, Tatemoto K, Terenius L, Hellström PM, Mutt V, Hökfelt T, Hamberger B (1982) Localization of peptide YY (PYY) in gastrointestinal endocrine cells and effects on intestinal blood flow and motility. Proc Natl Acad Sci 79:4471–4475

Lutz TA (2012) Control of energy homeostasis by amylin. Cell Mol Life Sci 69:1947–1965

Lutz TA (2013) The interaction of amylin with other hormones in the control of eating. Diabetes Obes Metab 15:99–111

Mennella I, Ferracane R, Zucco F, Fogliano V, Vitaglion P (2015) Food liking enhances the plasma response of 2-arachidonoylglycerol and of pancreatic polypeptide upon modified sham feeding in humans. J Nutr 145:2169–2175

Mikulaskova B, Maletínská L, Zicha J, Kunes J (2016) The role of food intake regulating peptides in cardiovascular regulation. Mol Cell Endocrinol 436:78–92

Mishra AK, Dubey V, Ghosh AR (2016) Obesity: an overview of possible role(s) of gut hormones, lipid sensing and gut microbiota. Metabolism 65:48–65

Monteiro MP, Batterham RL (2017) The importance of the gastrointestinal tract in controlling food intake and regulating energy balance. Gastroenterology 152(7):1707–1717.e2

Münzberg H, Morrison CD (2015) Structure, production and signaling of leptin. Metabolism 64:13–23

Parimisetty A, Dorsemans AC, Awada R, Ravanan P, Diotel N, Lefebvre d'Hellencour C (2016) Secret talk between adipose tissue and central nervous system via secreted factors—an emerging frontier in the neurodegenerative research. J Neuroinflammation 13(1):67

Peng Z, Xiaolei Z, Al-Sanaban H, Chengrui X, Shengyi Y (2012) Ghrelin inhibits insulin release by regulating the expression of inwardly rectifying potassium channel 6.2 in islets. Am J Med Sci 343:215–219

Persaud SJ, Bewick GA (2014) Peptide YY: more than just an appetite regulator. Diabetologia 57:1762–1769

Pocai A (2014) Action and therapeutic potential of oxyntomodulin. Mol Metab 3:241–251

Reaux-Le Goazigo A, Bodineau L, De Mota N, Jeandel L, Chartrel N, Knauf C, Raad C, Valet P, Llorens-Cortes C (2011) Apelin and the proopiomelanocortin system: a new regulatory pathway of hypothalamic α-MSH release. Am J Physiol Endocrinol Metab 301:E955–E966

Roh E, Song Do K, Kim MS (2016) Emerging role of the brain in the homeostatic regulation of energy and glucose metabolism. Exp Mol Med 48:e216

Salehi A, Dornonville de la Cour C, Håkanson R, Lundquist I (2004) Effects of ghrelin on insulin and glucagon secretion: a study of isolated pancreatic islets and intact mice. Regul Pept 118:143–150

Schjoldager B, Mortensen PE, Myhre J, Christiansen J, Holst JJ (1989) Oxyntomodulin from distal gut. Role in regulation of gastric and pancreatic functions. Dig Dis Sci 34:1411–1419

Sekar R, Wang L, Chow BK (2017) Central control of feeding behavior by the secretin, PACAP, and glucagon family of peptides. Front Endocrinol (Lausanne) 8:18

Sobrino Crespo C, Perianes Cachero A, Puebla Jiménez L, Barrios V, Arilla Ferreiro E (2014) Peptides and food intake. Front Endocrinol (Lausanne) 5:58

Stoyanova II (2014) Ghrelin: a link between ageing, metabolism and neurodegenerative disorders. Neurobiol Dis 72:72–83

Sun J, Gao Y, Yao T, Huang Y, He Z, Kong X, KJ Y, Wang RT, Guo H, Yan J, Chang Y, Chen H, Scherer PE, Liu T, Williams KW (2016) Adiponectin potentiates the acute effects of leptin in arcuate Pomc neurons. Mol Metab 5:882–891

Suyama S, Maekawa F, Maejima Y, Kubota N, Kadowaki T, Yada T (2016) Glucose level determines excitatory or inhibitory effects of adiponectin on arcuate POMC neuron activity and feeding. Sci Rep 6:30796

van Swieten MM, Pandit R, Adan RA, van der Plasse G (2014) The neuroanatomical function of leptin in the hypothalamus. J Chem Neuroanat 61-62:207–220

Sykaras AG, Demenis C, Case RM, McLaughlin JT, Smith CP (2012) Duodenal enteroendocrine i-cells contain mRNA transcripts encoding key endocannabinoid and fatty acid receptor. PLoS One 7(8):e42373

Tong J, Prigeon R, Davis HW, Bidlingmaier M, Kahn SE, Cummings DE, Tschöp MH, D'Alessio D (2010) Ghrelin suppresses glucose-stimulated insulin secretion and deteriorates glucose tolerance in healthy humans. Diabetes 59:2145–2151

Tovar S, Nogueiras R, Tung LY, Castaneda TR, Vazquez MJ, Morris A, Williams LM, Dickson SL, Carlos Diéguez C (2005) Central administration of resistin promotes short-term satiety in rats. Eur J Endocrinol 153:R1–R5

Wierup N, Svensson H, Mulder H, Sundler F (2002) The ghrelin cell: a novel developmentally regulated islet cell in the human pancreas. Regul Pept 107:63–69

Wierup N, Sundler F, Heller RS (2014) The islet ghrelin cell. J Mol Endocrinol 52:R35–R49

Wilson JL, Enriori PJ (2015) A talk between fat tissue, gut, pancreas and brain to control body weight. Mol Cell Endocrinol 418:108–119

Yi CX, Tschöp MH (2012) Brain–gut–adipose-tissue communication pathways at a glance. Dis Model Mech 5:583–587

Part II
Obesity as a Risk Factor for Neurological Disease

Chapter 4
Hypothalamic Dysfunction in Obesity and Metabolic Disorders

Sara Carmo-Silva and Cláudia Cavadas

Abstract The hypothalamus is the brain region responsible for the maintenance of energetic homeostasis. The regulation of this process arises from the ability of the hypothalamus to orchestrate complex physiological responses such as food intake and energy expenditure, circadian rhythm, stress response, and fertility. Metabolic alterations such as obesity can compromise these hypothalamic regulatory functions. Alterations in circadian rhythm, stress response, and fertility further contribute to aggravate the metabolic dysfunction of obesity and contribute to the development of chronic disorders such as depression and infertility.

At cellular level, obesity caused by overnutrition can damage the hypothalamus promoting inflammation and impairing hypothalamic neurogenesis. Furthermore, hypothalamic neurons suffer apoptosis and impairment in synaptic plasticity that can compromise the proper functioning of the hypothalamus. Several factors contribute to these phenomena such as ER stress, oxidative stress, and impairments in autophagy. All these observations occur at the same time and it is still difficult to discern whether inflammatory processes are the main drivers of these cellular dysfunctions or if the hypothalamic hormone resistance (insulin, leptin, and ghrelin) can be pinpointed as the source of several of these events.

Understanding the mechanisms that underlie the pathophysiology of obesity in the hypothalamus is crucial for the development of strategies that can prevent or attenuate the deleterious effects of obesity.

Keywords Hypothalamus • Obesity • Energy expenditure • Food intake • Circadian rhythm • Stress response • Fertility

S. Carmo-Silva • C. Cavadas (✉)
CNC—Center for Neuroscience and Cell Biology, University of Coimbra, Coimbra, Portugal

Faculty of Pharmacy, University of Coimbra, Coimbra, Portugal
e-mail: ccavadas@ci.uc.pt

© Springer International Publishing AG 2017
L. Letra, R. Seiça (eds.), *Obesity and Brain Function*, Advances in Neurobiology 19, DOI 10.1007/978-3-319-63260-5_4

4.1 Introduction

The hypothalamus is the brain region responsible for the maintenance of body homeostasis and acts as a control center for endocrine functions. The hypothalamus has three main regions: periventricular, medial, and lateral (Elizondo-Vega et al. 2015). Each of these regions is composed by several nuclei that contain different neuronal cells. These neurons communicate within and between nuclei to regulate physiological functions (Machluf et al. 2011; Pearson and Placzek 2013). Although each nuclei is specifically related to certain physiological functions, the communication between nuclei is necessary for complex physiological functions coordinated by this brain region. Moreover, the neuronal connections between nuclei support the interactions between different physiological functions (Gao and Sun 2016).

The hypothalamus has relevant physiological functions such as feeding, body temperature, stress response, sexual behavior and reproduction, and circadian rhythm (Ganong 2000; Clifford et al. 2015). And it has been long considered a key region in the pathogenesis of obesity and diabetes (Gerozissis 2008; Cavadas et al. 2016). Obesity pathophysiology includes the disruption of hypothalamic functions where changes in circadian rhythm, stress response, and energy balance occur. Moreover, the hypothalamic regulation of the overall homeostasis that is achieved by connections between different hypothalamic nuclei and major metabolic organs is also compromised upon obesity and metabolic dysfunction (Seoane-Collazo et al. 2015).

In this chapter we will present and discuss the major physiological functions of the hypothalamus, highlighting the contribution of hypothalamic regulation of peripheral organs and the impact of obesity on hypothalamic physiology.

4.2 Changes of Hypothalamic Food Intake Regulation Controlled by the Hypothalamus

Feeding is one of the most basic physiological functions of the organism, required for the maintenance of the physiological homeostasis. A state of energy homeostasis reflects the balance between energy intake and energy expenditure. This equilibrium is assured and modulated by peripheral signals that reflect body energetic status, but also by environmental and behavioral aspects. The hypothalamus is the sensor integrating peripheral signals from organs such as the stomach, gut, pancreas, and adipose tissue, and orchestrating a response regarding the energetic needs of the organism (Blouet and Schwartz 2010; Ueno and Nakazato 2016). Several hypothalamic nuclei are directly involved in the regulation of food intake and energy expenditure, such as the hypothalamic arcuate nucleus (ARC), ventromedial and dorsomedial hypothalamus (VMH and DMH), paraventricular nucleus (PVN), and lateral hypothalamic area (LHA) (Schwartz et al. 2000; Gonnissen et al. 2013). In the ARC, the orexigenic molecules as neuropeptide Y (NPY) and agouti-related peptide (AgRP) send signals that promote the feeling of hunger and the consequent food ingestion. On the opposite, the ARC also presents anorexigenic molecules, proopiomelanocortins (POMC) and cocaine-and amphetamine-regulated transcript (CART) that produce signals of satiety (Schwartz et al. 2000).

Other brain areas, such as the nucleus tractus solitarius (NTS) of the brainstem, the paraventricular thalamus (PVT), the parabrachial nucleus (PBN), the ventrolateral periaqueductal gray and dorsal raphe complex (PAGvl/DR), and the ventral tegmental area (VTA), also participate on food intake and energy expenditure regulation (Waterson and Horvath 2015). However, the main energetic center is located at the hypothalamus involving several nuclei, namely ARC, PVN, VMH, DMH, and the LHA (Williams and Elmquist 2012; Waterson and Horvath 2015). Although the different hypothalamic nuclei interact to form a response and lesions to specific nuclei can elicit different outcomes in feeding behavior (Hetherington and Ranson 1940; Anand and Brobeck 1951a, b), the ARC is considered the critical energy balance sensor.

The ARC has a privileged location within the hypothalamus that allow this nucleus to receive signals from the periphery, given the fact that the blood brain barrier (BBB) is fenestrated at the median eminence that is in close contact with the ARC (Schwartz et al. 2000; Myers et al. 2009; Langlet 2014). The integration of peripheral signals is achieved by the two main populations of neurons in ARC: the orexigenic neurons that produce and release NPY and AgRP, and the anorexigenic neurons that produce and release POMC and CART. NPY/AgRP neurons stimulate food intake while POMC/CART decrease food intake, promoting satiety and energy expenditure (Hahn et al. 1998; Schwartz et al. 2000). Briefly, upon stomach emptying cells in the gastric epithelium produce and release ghrelin that reaches ARC and activates ghrelin receptors located in the NPY/AgRP neurons, thus releasing the AgRP and NPY neuropeptides that promote food intake (Kamegai et al. 2000; Betley et al. 2015). Upon food intake, ghrelin levels decrease and leptin is secreted by the adipose tissue and insulin by the pancreatic beta cells. Leptin and insulin inhibit NPY/AgRP neurons and activate POMC/CART neurons to decrease food intake and promote energy expenditure (Barsh and Schwartz 2002; Friedman 2002).

Beyond peripheral hormones, hypothalamic neuropeptides play a role in the regulation of energy balance. However, the modulation of neuropeptides upon high-fat diet (HFD) feeding is still a controversial subject. Some studies report an increase in orexigenic input with an increase in the expression of NPY and AgRP and the concomitant decrease in anorexigenic input with a decrease in POMC and CART. On the opposite, some studies report the exact opposite observations (Gao et al. 2002; Torri et al. 2002; Briggs et al. 2010). These differences in observations might be related with the duration and the type of diets used in the different studies.

4.2.1 Obesity Induces Insulin, Leptin, and Ghrelin Resistance in the Hypothalamus

Circulating insulin and leptin levels are proportional to body adiposity and in obesity phenomena of hyperinsulinemia and hyperleptinemia are associated with the pathophysiology of obesity-related metabolic consequences. Ghrelin deregulation can also occur and mediate some of the pathophysiological phenomena observed in obesity. Ghrelin, insulin, and leptin increase can desensitize the respective receptors in the hypothalamus. Thus, hormone resistance occurs in consequence of the increased weight gain and the organism inability to counteract the excessive increase in this circulating hormones (Könner and Brüning 2012). Hypothalamic insulin resistance is one of the most well described consequences of obesity. However, similarly to ghrelin and leptin resistance it is hard to define the line where insulin resistance starts being the cause for obesity rather than a consequence. The better understanding of the mechanisms involved in insulin resistance might shed new light in new targets for the potential treatment of obesity.

4.2.1.1 Hypothalamic Insulin Resistance in Obesity

Insulin was discovered in 1922 (Banting et al. 1922) and since then has been primarily studied in the context of type 1 and 2 diabetes. Insulin is secreted by the β-cells in the islets of Langerhans of the endocrine pancreas and participates in several cellular mechanisms namely in the context of body weight and glucose homeostasis. Insulin levels are adjusted in response to the nutritional status, namely in response to increased blood glucose. Moreover, insulin levels correlate with adiposity (Baskin et al. 1999; Schwartz et al. 2000). The increase in insulin acts on peripheral organs to promote glucose uptake in adipose tissue and muscle, lipogenesis in adipose tissue and liver, and glycolysis and glycogen synthesis in muscle (Czech et al. 2013). In the brain, insulin signaling is required for body weight and metabolism but also for CNS-mediated regulation of fat and glucose metabolism (Obici et al. 2002; Konner et al. 2007; Belgardt and Bruning 2010; Scherer et al. 2011).

Insulin can cross the BBB through receptor-mediated transcytosis (Banks et al. 1997) and through an astrocyte-mediated process (García-Cáceres et al. 2016). In the brain, insulin reaches the hypothalamus and exerts its anorexigenic effect by activating insulin receptors (IR) in POMC/CART neurons and increasing α-melanocyte-stimulating hormone (α-MSH) and CART, while inhibiting NPY/ AgRP neurons, thus promoting energy expenditure (Schwartz et al. 1992; Sipols et al. 1995; Benoit et al. 2002). The anorexigenic effect of insulin depends on PI3K pathway activation; insulin-mediated signaling in both NPY/AgRP and POMC neurons is determinant for the global insulin sensitivity (Konner et al. 2007; Steculorum et al. 2016). Upon insulin binding to its receptor, there is intrinsic autophosphorylation of the receptor and the recruitment and phosphorylation of the insulin receptor substrate (IRS) 1 and 2. This action potentiates insulin action. At the same time, insulin also activates the PI3K signaling cascade that activates Akt and other downstream effectors such as mTOR that negatively phosphorylate IRS and inhibit insulin activity. All these signaling events ultimately lead to the translocation of glucose transporter 4 (GLUT4) to the membrane to allow glucose uptake (Biddinger and Kahn 2006).

Insulin resistance can result from insulin signaling pathways impairment. Insulin resistance occurs in obesity and is determinant in the development of detrimental consequences of obesity such as type 2 diabetes mellitus and cardiovascular diseases. There are several phenomena that occur upon HFD feeding and obesity that might help to consider insulin resistance as both result and cause for the pathophysiology of obesity. Hypothalamic insulin resistance occurs possibly through many mechanisms (Könner and Brüning 2012). The initial step of insulin transport into the brain is impaired in obesity (Banks et al. 1997; Kaiyala et al. 2000; Urayama and Banks 2008). Obesity is characterized by hyperinsulinemia that results either from hypersecretion or from decreased insulin clearance, which can lead to the desensitization of the IR (Michael et al. 2000; Farris et al. 2003, 2004). Hyperinsulinemia first occurs as a compensatory mechanism towards the increase in glucose derived from the diet. This compensatory hyperinsulinemia is the main driver of insulin resistance; however, it can still promote some of insulin-mediated actions while leading to the overactivation of other insulin-induced responses. This defines insulin resistance as a state of both failure and overactivation of insulin signaling that is specific to certain cell types (Muoio and Newgard 2008; Könner and Brüning 2012). Therefore, chronic high insulin levels can promote desensitization of IR or dysregulation in insulin signaling pathways in the hypothalamus.

Obese Zucker rats show decreased insulin-stimulated PI3K activation in the hypothalamus (Carvalheira et al. 2003). Moreover, intracerebroventricular (ICV) insulin infusion had lower anorexigenic action in obese rats relative to lean controls (Clegg et al. 2011). Selective PI3K signaling impairment in the ARC neurons can contribute to hyperphagia and hyperglycemia observed upon obesity. Moreover, SF-1 neurons of the VMH show increased insulin signaling upon HFD while POMC neurons of the ARC under the same condition become insulin resistant. SF-1 neurons project to POMC neurons, an innervation that is blunted upon HFD. The deletion of IR in SF-1 neurons restored insulin signaling in POMC neurons even in HFD (Klöckener et al. 2011). These observations show that hypothalamic insulin sensitivity depends on the specific insulin signaling in the ARC and the VMH.

The cellular mechanisms that mediate leptin resistance are similar to those observed in insulin resistance. SOCS3 and PTP1B increase in obesity show suppressive effects over insulin signaling, thus contributing for insulin resistance. Moreover, insulin and leptin signaling both induce STAT3 activation; some authors believe that this is the connecting link between both anorexigenic hormones signaling pathways and the reason why leptin resistance is also a major contributor for insulin resistance (Thon et al. 2016). ER stress and hypothalamic inflammations are also main effectors in insulin resistance. Endoplasmatic reticulum (ER) stress can, on the one hand, induce β-cell death and compromise insulin release and impair hypothalamic IR signaling (Ozcan et al. 2004; Liang et al. 2015). Hyperinsulinemia can promote hypothalamic inflammation and central insulin resistance, and on the other side tumor necrosis factor-α (TNF-α) blockade protected mice from diet-induced insulin resistance (Hotamisligil et al. 1993; Uysal et al. 1997; Fishel et al. 2005). These reports strengthen the idea that the molecular mechanisms involved in the pathophysiology of obesity cannot be completely dissociated from insulin resistance.

Obesity can promote post-translational modifications in several effectors of the insulin pathways (Tanti and Janger 2009; Belgardt et al. 2010). The effect of maternal obesity and metabolic reprogramming effect on insulin resistance has also been a hot topic of research. Maternal obesity and excessive enriched feeding in early life is also correlated with predisposition for insulin resistance (Nicholas et al. 2016; Benite-Ribeiro et al. 2016; Hur et al. 2017).

4.2.1.2 Hypothalamic Leptin Resistance in Obesity

Leptin was identified in 1994 as the product of the gene *Ob* and was the first adipose-derived cytokine to be related with energy balance (Zhang et al. 1994; Lehr et al. 2012). Leptin has been implicated in several physiological processes such as bone formation, reproduction, and cardiovascular regulation; however the main interest around this hormone revolves around its anorexigenic action. Leptin is mainly produced by the adipose tissue and its levels can be directly correlated with energy reserves. An increase in fat depots, associated with a positive energy balance, increases leptin production and secretion to the bloodstream (Friedman and Halaas 1998). After crossing the BBB (Golden et al. 1997), leptin binds to its receptor (LepRb or ObR) in the neurons of the ARC. Leptin binding to LepRb can have two opposite effects on the neuronal populations of the ARC. On one hand, leptin excites POMC/CART neurons promoting the release of anorexigenic neuropeptides that inhibit food intake and promote energy expenditure, and, at the same time, leptin can inhibit the orexigenic NPY/AgRP neurons, thus ensuring the arrest on food intake (Korner et al. 1999; Banks et al. 2000; Lee et al. 2013; Mercer et al. 2014). Leptin administration to humans with congenital leptin deficiency decreases body weight and food intake (Farooqi et al. 1999). However, unexpectedly, leptin administration does not decrease body weight and food intake in nongenetic obese humans (Halaas et al. 1997). This arises from the fact that obesity promotes hyperleptinemia and leptin resistance. Additionally, obese patients with higher circulating levels of leptin after diet present an increased propensity for weight regain (Crujeiras et al. 2010, 2014). Interestingly, leptin resistance in obesity appears to be selective to some specific leptin actions. In obese and hyperleptinemic mouse models, exogenous leptin administration was unable to promote an anoretic response; however it promoted renal sympathetic activation and increased blood pressure (Correia et al. 2002; Rahmouni et al. 2005).

Leptin resistance is a multifactorial condition; the selective characteristic turns the hypothalamus one of the major affected areas. The hypothalamic dysfunction induced by HFD can promote leptin resistance that in turn can aggravate the diet-induced obesity (DIO) phenotype.

The knowledge of leptin resistance consequences for the development of obesity and its related metabolic consequences are not still completely known. In the one hand, LepRb antagonists administered by ICV had no effect on food intake or body weight in both lean and obese and hyperleptinemic mice (Ottaway et al. 2015). On the opposite, hypothalamic leptin resistance, namely in the neurons of the ARC, is well described as a contributor to obesity development (Myers et al. 2010; Könner and Brüning 2012). These observations suggest that leptin resistance is specific to certain brain regions and cell types, which might be related with differential signaling cascade activation upon LepRb binding (Cui et al. 2017). This specific leptin resistance might impact body homeostasis differently.

There are several potential mechanisms that underlie the selective leptin resistance, more specifically regarding hypothalamic leptin resistance (Könner and Brüning 2012). The main two hypotheses regarding the cause for leptin resistance in obesity are: (a) inability of leptin to cross the BBB and (b) the inhibition of the leptin signaling pathways resulting from defects/desensitization of the receptor or downstream signaling molecules (Crujeiras et al. 2015; Cui et al. 2017). Leptin can circulate in a free form or bound to an inactive form of its receptor, LepRe. In normal lean individuals most leptin is found bound to LepRe; however, in obese individuals, leptin can be found in its free form. This results in leptin increase within the brain of obese individuals which might desensitize LepRb and increase leptin resistance (Sinha et al. 1996; Mark 2013; Simonds et al. 2014). These observations might explain the ineffectiveness of exogenous leptin administration in DIO. Moreover, leptin passage through the BBB is achieved by transporters that become saturated due to the constant high levels of leptin in obesity, impairing the correct leptin transport to the brain (Banks 2001; Banks and Farrell 2003). Triglycerides decrease leptin transport at the BBB; considering that obesity is characterized by an increase in triglycerides, this phenomenon can further promote leptin resistance (Banks et al. 2004). Leptin binding to LepRb activates JAK2 that phosphorylates tyrosine residues within the receptor and the recruitment and phosphorylation of STAT3. STAT3 translocates to the nucleus where it can regulate the expression of certain genes, such as the suppressor of cytokine signaling 3 (SOCS3), an inhibitor of leptin signaling (Bjørbaek et al. 1999; Banks et al. 2000). Alterations in this signaling pathway can contribute to leptin resistance.

Obesity leads to the increase of protein tyrosine phosphatases (PTPs) in the hypothalamus that can prevent the necessary phosphorylation for leptin downstream signaling, thus blunting leptin sensitivity (Cui et al. 2017). Moreover, HFD can increase SOCS3 expression in the ARC neurons, a negative regulator of STAT3 activation and a promotor of leptin resistance (Münzberg et al. 2004; Gamber et al. 2012; Olofsson et al. 2013). In fact, SOCS3 silencing in the hypothalamus can protect against DIO and leptin resistance (Liu et al. 2011). Hypothalamic inflammation and ER stress are hallmarks of obesity (Cavadas et al. 2016) and also play a determinant role in the development of leptin resistance. Decrease of ER stress specifically in the hypothalamus, using different experimental approaches, protects mice from obesity and leptin resistance. Moreover, blockade of TNF-α signaling in the hypothalamus also improves leptin sensitivity (Cui et al. 2017).

More recently, others showed that leptin resistance is linked to epigenetic and neonatal metabolic programming (Crujeiras et al. 2015; Cui et al. 2017). Diet and environmental habits induce epigenetic alterations such as DNA methylation. Studies showed that a methylation of the proximal region of *LEP* promoter is associated with decreased *LEP* expression (Marchi et al. 2011). *LEP* methylation has also been correlated with success in weight loss programs, where individuals that had a better response to the program with consequent higher weight loss percentage have lower levels of leptin methylation (Cordero et al. 2011). More indirectly, hypermethylation of *POMC* gene blocks the effect of hyperleptinemia in obesity (Marco et al. 2013; Shi et al. 2013). Excess nutrition during prenatal and postnatal life contributes to a higher predisposition for obesity and metabolic disorders later in life (Hur et al. 2017). Rats raised in small litters are more susceptible to postnatal overfeeding. These rats tend to be leptin resistant, with decreased leptin effect on neurons of the ARC and a decrease in LepRb expression in the hypothalamus. Moreover, rats raised in small litters present increased expression of AgRP and NPY in the ARC, despite higher circulating leptin levels what supports the presence of leptin resistance (Duque-Guimarães and Ozanne 2013; Cui et al. 2017).

Moreover, in obese rats, insulin-stimulated PI_3K activation in the hypothalamus is compromised, leading to selective hypothalamic insulin resistance (Carvalheira et al. 2003; Clegg et al. 2011). Similarly, obese rats have reduced expression of LepRb in the hypothalamus and a decrease in hypothalamic STAT3 activation that mediates leptin signaling (Maes et al. 1997; Gao et al. 2004). The loss of LepRb expression in the hypothalamus per se can promote obesity (Bingham et al. 2008). It is now accepted that insulin and leptin resistance in the hypothalamus can contribute to the development of hyperphagia and glucose metabolism impairment, which promotes the development of obesity and the aggravation of the obesity-induced metabolic dysfunction (Myers et al. 2010; Könner and Brüning 2012).

4.2.1.3 Hypothalamic Ghrelin Resistance in Obesity

Ghrelin was first identified as an endogenous ligand of the growth hormone secretagogue receptor (GHSR1) (Kojima et al. 1999; Nakazato et al. 2001). Although ghrelin is involved in a plethora of physiological functions, this hormone gained interest due to its orexigenic effects within the hypothalamic appetite-regulatory mechanisms. Thus, ghrelin is a potent orexigenic peptide mostly derived from the stomach (Kojima et al. 1999) and the peripheral signal that stimulates food intake. When energetic storage is low, ghrelin is released from the stomach and activates the NPY/AgRP neurons in the ARC thus promoting the secretion and release of neuropeptides (NPY and AgRP) that promote food intake (Cowley et al. 2003; Egecioglu et al. 2008). After food intake, ghrelin levels in the plasma decrease (Tschöp et al. 2001a; Cummings 2006).

Given the potent orexigenic effect of ghrelin, this hormone was studied as a possible therapeutic target for metabolic disorders such as obesity. However, the observations regarding ghrelin signaling upon the positive energy balance in obesity, revealed some difficulty in considering this hormone a good therapeutic target. Ghrelin plasma levels of obese individuals are decreased and the meal-related fluctuations of ghrelin are also impaired in these individuals (Tschöp et al. 2001b). In obese individuals, ghrelin levels do not drop after a meal (English et al. 2002). These observations gave rise to the hypothesis that obesity and fat-enriched diets could desensitize ghrelin-producing cells and promote ghrelin resistance. Studies in mice further corroborated these observations. In DIO mice (after 12 weeks of HFD) ghrelin is decreased in the plasma as well as the GHSR mRNA levels in the hypothalamus (Briggs et al. 2011). Moreover, in DIO mice ghrelin does not increase upon fasting (Perreault et al. 2004) and ghrelin transport across the BBB is impaired (Banks et al. 2008). Regarding hypothalamic neurons activation, in DIO mice, the injection of ghrelin had no effect on NPY/AgRP firing rate, neuropeptide release, and consequently on food intake (Briggs et al. 2011). All these observations point to ghrelin ineffectiveness upon fat-enriched diet consumption; however, the mechanisms behind a possible ghrelin resistance in obesity are still not quite understood.

Interestingly, *ob/ob* mice that are obese and glucose intolerant can maintain ghrelin sensitivity. Leptin administration in these mice can promote the decrease in ghrelin sensitivity, which suggests that the hyperleptinemia observed in obesity is responsible for ghrelin resistance (Briggs et al. 2011). Hypothalamic inflammation was also pointed as a possible inductor of ghrelin resistance; however, the phenotype of the *ob/ob* mice also includes hypothalamic inflammation which discards this hypothesis (Zigman et al. 2016). Regarding cellular mechanisms, ghrelin exerts its orexigenic action through the activation of AMPK and mTOR pathways. Chronic activation of AMPK can promote ghrelin-dependent hyperphagia and obesity, while the inhibition of mTOR can decrease the orexigenic potential of ghrelin (Cui et al. 2017). The deregulation of this pathway in the hypothalamus in obesity can be either an effector to or a cause of ghrelin resistance; more studies are necessary to better understand the mechanisms behind this phenomenon.

It is accepted that ghrelin resistance may protect against the unmeasured weight gain facing high fat diet feeding (Zigman et al. 2016; Cui et al. 2017). Weight loss can resensitize the brain to ghrelin (Briggs et al. 2013), which supports this hypothesis. At this point, ghrelin resistance appears to have a conservative/protective role in obesity in opposition to insulin and leptin resistance. However, given the plethora of actions achieved by ghrelin, it is necessary to understand if it is possible to surpass the decrease in ghrelin as regulatory effect, to assure the rest of ghrelin actions in other mechanisms beyond food intake regulation.

4.2.2 Obesity Changes Hypothalamic Synaptic Plasticity

Hypothalamic synaptic plasticity is required for the rapid adjustment of neurons involved in energy homeostasis to nutritional changes. When exposed to HFD, NPY/AgRP and POMC/CART neurons present a decrease in the number of synapses on their perikarya (Horvath 2006; Chun and Jo 2010). This observation supports the idea that both neuronal populations are affected during obesity which might not allow for significant differences between orexigenic and anorexigenic neuropeptide expression.

Although reports regarding energy balance regulation by the hypothalamus and obesity are still controversial, obesity impairs the balance between the orexigenic and anorexigenic input thus disrupting energy homeostasis. Hypothalamic insulin and leptin resistance might play a more determinant role in the dysregulation of hypothalamic physiology, specially in energy balance regulation.

4.3 Circadian Rhythm Controlled by the Hypothalamus and Obesity

Circadian rhythms dictate the impact of the earth rotation around its own axis in the physiology of the organism. Hence, the light-dark cycles determine the adaptation of the organism in a rhythmic pace to maintain homeostasis. Circadian clocks (circadian deriving from the Latin circa *diem*, "about a day") enable organisms to prepare for environmental changes such as light and temperature and adapt its behavior accordingly, throughout the 24 h of the day. Zeitgebers (the German word for "time givers") are signals that control body circadian clock synchronization (Sahar and Sassone-Corsi 2012; Gerber et al. 2015). Several physiological functions have circadian rhythmicity, such as body temperature, activity, sleep, heart rate, blood pressure, and hormone and neurotransmitters secretion (Hastings et al. 2008). The disruption of the circadian clock is directly correlated with chronic debilitating disorders such as obesity and cardiovascular diseases, diabetes and cancer (Nagai et al. 1994; Achermann and Borbély 2003; Meier-Ewert et al. 2004; Froy 2010; Sahar and Sassone-Corsi 2012; Eckel-Mahan and Sassone-Corsi 2013; Challet 2015; Coomans et al. 2015).

The main Zeitgeber is the solar time, which gives the information of the beginning of the light period. Light is received by the ganglion cells in the retina and transmitted via the optic tract to the hypothalamus (Albrecht 2012). It has been known for nearly half a century that lesions of the mediobasal hypothalamus cause loss of circadian rhythmicity of locomotor activity, feeding, and drinking (Richter et al. 2014). In 1972, the localization of biological master clock was described within the mammalian suprachiasmatic nucleus (SCN) of the hypothalamus (Moore and Eichler 1972; Stephan and Zucker 1972). All cells in different tissues of the body present internal clocks that are mainly synchronized by the SCN. The SCN drives rest activity cycles and feeding-fasting rhythms that are Zeitgebers to peripheral tissues (Dibner et al. 2010; O'Neill and Reddy 2012; Rosenwasser and Turek 2015).

The relationship between circadian rhythm deregulation and obesity was already observed in a number of models, human and rodent. People that work in shifts or work nights have higher risk for the development of obesity and metabolic diseases (Peplonska et al. 2015). Moreover, in the night eating syndrome, people binge on food in the night period but with the choice of highly caloric palatable foods (Gallant et al. 2012; Blancas-Velazquez et al. 2017). In rodents, mice fed with chow diet during the inactive period present metabolic abnormalities (Kohsaka et al. 2007). Moreover, mice fed a high fat diet solely during the day have worse metabolic parameters when compared to mice fed a high fat diet just at night (Hatori et al. 2012). In addition, mice fed with a high fat diet ad libitum display abnormal eating patterns eating during the resting phase (Kuroda et al. 2012). It is now accepted that feeding at metabolic inactive phases is directly related with obesity and metabolic disturbances.

Given that food intake regulation displays circadian rhythmicity and functions itself as a Zeitgeber, is easy to conceive that obesity and the deregulation of the internal circadian rhythms can deeply affect the hypothalamic physiology controlling food intake. In the hypothalamic nuclei SCN, as well as in other hypothalamic nuclei, neurons present rhythmic activity. However, the ARC is the only nucleus that maintains clear circadian firing rate in the absence of the SCN input (Guilding et al. 2009). This observation further supports the conserved role of food intake regulation for survival.

When mice are fed a high palatable diet there is over-ingestion of calories but also a fragmented feeding pattern. Upon high fat diet feeding, rodents eat more frequently smaller amounts of food rather than large meals. This "snacking" behavior is similar to what is displayed by obese humans. Moreover, the ingestion of small amounts of food takes part throughout the 24 h of the day instead of the preferential nighttime feeding (Kohsaka et al. 2007; Pendergast et al. 2013; Branecky et al. 2015; Mifune et al. 2015). Given the fact that caloric intake is rhythmic and mostly constricted to the active period, this alteration in feeding patterns might reflect a disruption in hypothalamic physiology, namely in circadian rhythm regulation.

At molecular level, AgRP, NPY, orexin and melanocyte-stimulating hormone (MSH) and leptin receptor show rhythmic hypothalamic expression throughout the day. The same is not observed for POMC and CART expression (Stütz et al. 2007; Opperhuizen et al. 2016). Ablation of NPY receptors signaling can promote feeding during the light period (Wiater et al. 2011), while NPY overexpression in LHA reduces locomotor activity and restores the disruption of feeding patterns (Tiesjema et al. 2007). These observations support a role for circadian regulation of hypothalamic neuropeptides. Upon high fat diet, leptin and NPY lose its rhythmicity in the hypothalamus (Cano et al. 2008; Gumbs et al. 2016). Although few studies investigated the effect of high fat diet feeding on neuropeptide levels rhythmicity in the different hypothalamic nuclei, it is possible that disruption of feeding patterns in obesity results from the loss of rhythmicity within the hypothalamic expression of its main effectors.

Time keeping and circadian rhythmicity is ensured by a transcription-translation feedback system (Reppert and Weaver 2002). Briefly, two-transcription factors circadian locomotor output cycles kaput (*CLOCK*) and aryl hydrocarbon receptor nuclear translocator like (*ARNTL1 or BMAL1*) heterodimerize and bind to the target gene promoter driving the rhythmic expression of *Period* (*Per1, Per2,* and *Per3*) and *cryptochrome* (*Cry1* and *Cry2*). PER and CRY then complex and translocate to the nucleus to inhibit CLOCK:BMAL1-induced gene expression. This loop takes around 24 h to be completed and happens in all types of cells, driving all different clocks (Dardente and Cermakian 2007; Brown et al. 2012). The rhythmic expression of the clock genes varies between the different hypothalamic nuclei. Moreover, mutations in the clock genes are associated with metabolic alterations and obesity (Rudic et al. 2004; Turek et al. 2005; Blancas-Velazquez et al. 2017). However, upon high fat diet feeding there is no alteration in the expression of these genes within rodent whole-hypothalamus (Kohsaka et al. 2007). When the expression of these clock genes was assessed in the different nuclei of the hypothalamus of mice fed a high fat diet, the only alteration observed was an increase in *Bmal1* in the SCN (Cunningham et al. 2016). These results suggest that although there are significant changes in the circadian rhythm upon obesity and that circadian rhythm alterations directly promote metabolic alterations, the hypothalamus seems to be partly resistant to changes in rhythmicity. These observations might result from an evolutionary method to ensure the hypothalamic physiology during circadian alterations, thus maintaining energetic homeostasis. Moreover, the changes observed in food intake rhythmic patterns might derive from alterations in the brain reward system upon high fat diet feeding, rather than disruptions in hypothalamic signaling (Blancas-Velazquez et al. 2017).

In conclusion, obesity may change circadian regulation and this will have relevant impact on several physiological functions regulated by the hypothalamus such as sleep and metabolic homeostasis. The disruption of those functions is directly correlated to chronic disorders such as cardiovascular diseases, obesity, diabetes, and cancer (Eckel-Mahan and Sassone-Corsi 2013; Challet 2015; Coomans et al. 2015).

4.4 Hypothalamic Regulation of Stress Adaptation in Obesity

The physiological response to stress involves the activation of the autonomous nervous system and the hypothalamic–pituitary–adrenocortical (HPA) axis. Focusing on the HPA axis response, this process begins at the PVN where neurosecretory neurons release corticotrophin-releasing hormone (CRH) and arginine-vasopressin (AVP) into portal circulation. CRH stimulates the pituitary gland to release adrenocorticotrophin (ACTH), which in turn acts on the inner adrenal cortex to initiate the synthesis and release of glucocorticoid hormones, namely cortisol (mammals) or corticosteroids (rats and mice) (Armario et al. 2006; Ulrich-Lai and Herman 2009). Glucocorticoids release promotes several central and peripheral adaptations, on the one hand to prevent excessive reaction to stress inhibiting the ongoing activation of the HPA axis, and on the other, mobilizing energy and potentiating autonomous nervous system activation to prepare for further stress (Sapolsky et al. 2000; Frank et al. 2013; Belda et al. 2015). The lack of an adaptive response to stress can result in psychiatric disorders such as depression (Jurena 2014). Reciprocally, depression is correlated with the development of metabolic syndrome (Kyrou and Tsigos 2009; Marazziti et al. 2014), and chronic stress can promote obesity, type 2 diabetes, and related cardiometabolic complications (Harrell et al. 2016).

It is now accepted that obesity and metabolic deregulation impairs HPA axis whether by overactivation, by decreased sensitivity to negative feedback, or by loss of sensitivity of peripheral tissues to glucocorticoid activity (Seimon et al. 2013). However, the effects of obesity and fat/sugar-enriched diets on the HPA axis and the hypothalamus are still controversial. In the one hand, HPA activation can be correlated with weight loss and reduced food intake; in the other hand, HPA activation can also increase the intake of palatable food and obesity (Dallman et al. 2003; Adam and Epel 2007). Some studies show that rats fed with a high fat diet have reduced plasma ACTH and corticosterone, suggesting a dampened HPA axis activation (Pecoraro et al. 2004), while others suggest that obesity promotes an anxiety-like state in rats (de Noronha et al. 2017) which could be accounted as an overactivation of the HPA axis. These observations have also to be discussed considering the duration of the high fat diet, given that a longer exposure to the metabolic insult can exacerbate response of the HPA axis through the dampening of the negative feedback over the HPA (Duong et al. 2012; Michopoulos et al. 2012; Balsevich et al. 2014).

At molecular level, the melanocortin system is tightly related with food intake and HPA activation. The activation of MC-3 and MC-4 receptors induces an anorexigenic stimuli and MC-2 receptor binds ACTH. Mice knockout for MC-4R are obese and have impaired adaptation to stress and lower activation of PVN neurons (Ryan et al. 2014). Hypothalamic neuropeptide levels upon high fat diet are still a controversial issue. In fact, some studies report that rodents under HFD have an increase in orexigenic inputs and a decrease in anorexigenic signaling, while others report the exact opposite (Gao et al. 2002; Torri et al. 2002; Briggs et al. 2010) However, NPY is known to promote HPA activity, having an antidepressive effect. Injections of NPY in the PVN stimulate ACTH and corticosterone release (Wahlestedt et al. 1987). In the context of obesity, where there are alterations in NPY levels in the hypothalamus (Levin and Dunn-Meynell 1997; Lin et al. 2000), this might be the cause for a compromise in HPA axis activation. Leptin is also a major known player in the pathogenesis of depressive disorders, characterized by impaired HPA functioning. And depressed patients show higher circulating leptin levels, similar to obese patients (Esel et al. 2005; Morris et al. 2012). On the opposite, mice subjected to stress have lower levels of leptin receptor (LepRb) expression in the hypothalamus (Yang et al. 2016). The high circulating leptin levels with the decrease in LepRb expression might result in the deficient leptin signaling which might directly impact on the HPA activation.

The pathophysiology of obesity affects directly the hypothalamus, namely with an increase in proinflammatory markers that might trigger HPA axis dysfunction and, in turn, promote stress-related disorders (Lamers et al. 2013). However, further studies are needed to dissect the HPA axis changes in obesity and the underlying hypothalamic mechanisms.

4.5 Contribution of the Hypothalamus to Fertility and Reproductive Dysfunctions Induced by Obesity

The hypothalamic–pituitary–gonadal (HPG) axis is responsible for the integration of hormonal signals and the regulation of reproduction. The HPG axis response includes the gonadotrophin-releasing hormone (GnRH) neurons, which are mainly located at preoptic area (POA) and project to the median eminence (Prevot et al. 2010; Rudolph et al. 2016). Upon GnRH release to the pituitary occurs the production of follicle stimulating (FSH) and luteinizing (LH) hormones that are released into the bloodstream and act on the gonads. LH and FSH at the ovary promote estradiol and progesterone secretion. At the testis, LH acts primarily on Leydig cells to promote testosterone production while FSH acts preferentially in Sertoli cells; the concerted action of both hormones at the testis regulates male gonadal function (Schwanzel-Fukuda et al. 1992; Simoni et al. 1997; Chrousos et al. 1998; Dufau 1998; Rudolph et al. 2016).

Several reports show a direct link between the nutritional status and energetic balance and fertility, both in males and females. In the one hand, undernutrition and caloric restriction affect reproduction and fertility reducing GnRH secretion (I'Anson et al. 2000); in the other hand, obesity induces reproductive dysfunctions and infertility by the disruption in GnRH secretion (reviewed in Clarke and Arbabi 2016; Oliveira et al. 2017). GnRH neurons release GnRH and receive projections from POMC neurons from the ARC (Tilbrook et al. 2002). This observation suggests a link between energy balance and fertility-reproduction. In general, orexigenic neuropeptides inhibit reproduction while anorexigenic neuropeptides stimulate reproduction (reviewed in Clarke and Arbabi 2016; Muroi and Ishii 2016). NPY, through Y-2 receptor activation, reduces LH secretion (Barker-Gibb et al. 1995; Clarke et al. 2005) and can inhibit GnRH production (Catzeflis et al. 1993; Kalra et al. 1992). AgRP also prevents LH surge in rats and has a determinant role for puberty in mice (Schioth et al. 2001; Sheffer-Babila et al. 2013). On the opposite, the melanocortin system stimulates the HPG axis promoting LH secretion (Han et al. 2005). Moreover, NPY/AgRP and POMC/CART neurons indirectly regulate the activity of GnRH neurons via the KnDY (Kisspeptin/neurokin B/dynorphin) neurons in the ARC. The kisspeptin neuronal system has been studied extensively as a link between energy balance and reproduction (De Bond and Smith 2014; Wahab et al. 2013).

Kisspeptin directly activates GnRH neurons (Han et al. 2005); moreover, infusion of kisspeptin in the third ventricle increases NPY levels while reduces POMC in the ARC (Backholer et al. 2010). Although it is still not clear whether NPY regulates kisspeptin neurons, POMC and CART are able to activate kisspeptin neurons thus promoting GnRH neurons activity (De Bond and Smith 2014; Muroi and Ishii 2016). During food restriction it can be observed a decrease in hypothalamic kisspeptin levels which might be responsible for the decline in the activation of the HPG axis (Zhou et al. 2014). On the opposite, in leptin deficient mice, *ob/ob* mice show a decrease in hypothalamic kisspeptin, which also may account for the decreased fertility (Swerdloff et al. 1976). Exogenous leptin infusion increases kisspeptin levels and revert hypothalamic hypogonadism, thus reverting infertility (Cleary et al. 2001; Tsatsanis et al. 2015; Clarke and Arbabi 2016). Moreover, in Ob/Ob mice, the ablation of AgRP neurons of the hypothalamus restores body weight and fertility, supporting a role for hypothalamic ARC neurons on fertility (Wu et al. 2013; Dietrich and Horvath 2012). Moreover, changes of peripheral hormones, such as ghrelin and glucagon-like peptide 1 (GLP-1), are also known to act on the HPG axis and interfere directly with GnRH secretion (Alves et al. 2016; Clarke and Arbabi 2016; Muroi and Ishii 2016).

Although there is not much consensus about the mechanisms relating obesity and infertility, a possible mechanism is through the increase of NPY in the hypothalamus, the concomitant increase of leptin and the consequent leptin resistance (Tortiello et al. 2004; Alves et al. 2016). Leptin resistance in the hypothalamus, together with the increase in NPY and decrease in kisspeptin, will culminate in the suppression of the activation of the GnRH neurons and the inability of the HPG axis to respond, hence promoting the dysfunctions in fertility observed in obesity.

Furthermore, obesity is characterized by a systemic inflammation that affects primarily the hypothalamus (Jais and Brüning 2017) but also it has been shown to delay puberty onset mainly through the decline in leptin signaling (Ballinger et al. 2003; DeBoer et al. 2010) and to decrease male fertility, reducing testosterone levels (Tengstrand et al. 2009). The increase of inflammatory markers in the hypothalamus can disrupt the signaling of anorexigenic hormones such as leptin and insulin, dysregulating NPY/AgRP and POMC/CART neurons and compromising their interactions with kisspeptin neurons and with the GnRH neurons in the POA.

Given that fertility and reproduction is a result from the orchestrated inputs between several types of neurons, culminating in the activation of the GnRH neurons, it is easy to envisage that alterations might occur from the disturbed neuronal communication in obesity.

4.6 Obesity Compromises the Hypothalamic Control Over Periphery

The central nervous system receives and integrates signals relative to energetic needs from the periphery and orchestrates a response through the modulation of food intake and energy expenditure. The autonomic nervous system (ANS) plays the determinant role in this response through the innervation of peripheral metabolic organs such as the liver, white adipose tissue (WAT), brown adipose tissue (BAT), pancreas, and skeletal muscle. The ANS is constituted by the sympathetic nervous system (SNS) and the parasympathetic nervous system (PNS). Although the SNS is associated with catabolic responses and the PNS with anabolic responses, these two systems work together and might be activated at the same time under some circumstances. The liver, pancreas, and skeletal muscle receive innervation from both systems while the adipose tissue is mainly innervated by the SNS. The action of the ANS in these organs is mediated by different brain regions, including the hypothalamus (Buijs 2013; Harlan et al. 2013).

4.6.1 Obesity Changes Hypothalamic Autonomic Control of Liver

Liver is a key organ involved in energy storage and glucose homeostasis. Liver receives fatty acids from the gut resulting from the hydrolysis of triglycerides and from lipolysis of WAT. The liver can also synthetize fatty acids from excess glucose, de novo lipogenesis. Fatty acids are then used by the liver and assembled into triglycerides and stored or secreted into the plasma in very-low-density lipoprotein (VLDL) particles (Gibbons 1990; Nguyen et al. 2008; Bruinstroop et al. 2014). The liver is also involved in insulin sensitivity responding to glucose and insulin levels by the modulation of hepatic glucose production (HGP) (Ramnanan et al. 2011).

The ARC is involved in VLDL-TG secretion, whereas NPY stimulates its secretion. On the opposite, blunted melanocortin system in hypothalamic neurons promotes lipogenesis and hepatic steatosis (reviewed in Bruinstroop et al. 2014). Several hypothalamic nuclei have been implicated in the regulation of hepatic glucose production (HGP). In the one hand, VMH stimulation promotes glucose output and decreases glycogen in the liver (Shimazu 1979), whereas stimulation of LHA increases hepatic glycogen synthesis (Shimazu and Ogasawara 1975). In the ARC, the absence of LepRb in POMC neurons and the increase in NPY promote HGP (Coppari et al. 2005; Parton et al. 2007; van den Hoek et al. 2008; Bisschop et al. 2014), while insulin signaling can inhibit HGP by the inhibition of NPY (Schwartz et al. 1992; Obici et al. 2002).

As discussed previously, obesity and metabolic dysfunction can promote insulin and leptin resistance that might directly serve to increase the HGP, hence causing hyperglycemia and type 2 diabetes. Moreover, obesity can promote chronic sympathetic activation, leading not only to the desensitization of regulatory processes but also promoting directly an increase in HGP (Valenzano et al. 2016; Arrieta-Cruz and Gutiérrez-Juárez 2016). Hypothalamic inflammation induced by a HFD feeding can also deregulate the regulatory effect of the hypothalamus over the liver. In fact, inhibiting hypothalamic inflammation in diet-induced obese mice improves hepatic insulin action and reduces HGP and hepatic steatosis (Milanski et al. 2012). Hypothalamic amino-acid (AA) sensing and metabolism is determinant for insulin action and glucose metabolism. The BBB increased permeability in obesity allows circulating AAs to reach the hypothalamic neurons in higher concentrations. The AAs increase, such as leucine and proline, in the hypothalamus can inhibit HGP. However, when the metabolism of the AAs in the hypothalamus is inhibited, this effect is blunted (Arrieta-Cruz and Gutiérrez-Juárez 2016). The elevation of AAs in the hypothalamus beyond a certain threshold might lead to an overload of the hypothalamic response, contributing for insulin resistance and hyperglycemia. Liver functioning, namely in respect to glucose metabolism, is directly regulated by the hypothalamus. Given the major physiological alterations in this brain region upon obesity and/or metabolic dysfunction, the autonomous regulation of the hypothalamus over the liver is compromised and further aggravates the already compromised metabolic status.

4.6.2 Obesity Changes Hypothalamic Autonomic Control of Brown Adipose Tissue

Brown adipose tissue (BAT) is responsible for energy expenditure, mostly through non-shivering thermogenesis (NST). BAT is highly vascularized to allow energy dissipation. Brown adipocytes present a high number of mitochondria (responsible for the brown coloration of this tissue) and multilocular lipid droplets. BAT is highly sensitive to certain diets and physical exercise, conditions that can change its physiology and functioning. BAT function is regulated by the central and peripheral nervous system, and SNS activation is determinant for BAT thermogenesis. The activation of SNS promotes norepinephrine release and the activation of β-adrenergic receptors and the promotion of thermogenesis. Moreover, BAT also promotes clearance of plasma triglyceride and improved glucose homeostasis, uptake of lipids and glucose from the circulation to use as substrate for thermogenesis (Whittle et al. 2011; Contreras et al. 2015; Morrison et al. 2014).

As responsible for the regulation of food intake and energy expenditure, the hypothalamus was early related to BAT function (Fuller et al. 1975; Imai-Matsumura et al. 1984). Within the hypothalamus, the POA, the VMH, the DMH, and the ARC are directly connected to BAT activity. Upon activation of neurons in these nuclei there is an increase in thermogenesis regulated by BAT (reviewed in Contreras et al. 2015). In the case of the VMH, this effect is abolished upon β-adrenergic receptors blockade (Holt et al. 1987). DMH neurons do not project directly to BAT (Cao et al. 2004; Dimicco and Zaretsky 2007), which means that the interactions between the different hypothalamic nuclei might lay beyond the effect of this nucleus in BAT activation. Some evidences point for a role of the PVN in BAT regulation; however, the observations are controversial and might result from an orchestrated response with other nuclei, namely with the ARC (Contreras et al. 2015). The main implication of the ARC in BAT thermogenesis is mediated with the melanocortin system since this nucleus is responsible for the integration of leptin signaling (Butler and Cone 2002; Farooqi and O'Rahilly 2006). Leptin sensing and signaling pathways promote AMPK (5′ AMP-activated protein kinase) activation. AMPK activation in the VMH is a negative regulator of BAT activation (López et al. 2010; Whittle et al. 2011; Tanida et al. 2013), while leptin in the DMH promotes thermogenesis. Moreover, in the ARC leptin activates POMC/CART neurons responsible for the promotion of thermogenesis, while inhibiting NPY/AgRP neurons (Harlan et al. 2011; Morrison et al. 2014). Moreover, NPY knockdown in the DMH also promotes BAT thermogenesis (Chao et al. 2011).

Diet and exercise are main influencers of BAT activity. Initial studies correlating obesity and BAT functioning observed that rats fed a fat-enriched diet, exhibiting hyperphagia, had increased thermogenesis (Rothwell and Stock 1981, 1997). However, this was later accepted as a regulatory mechanism to counteract the effects of the positive energy balance promoted by the diet (Mercer and Trayhurn 1987). Studies in *ob/ob* mice showed an increase in energy storage even when pair-fed with ad libitum wild type mice, resulting from a decrease in NST (Thurlby and Trayhura 1979). Moreover, obese mice resulting from a lesion to the VMH showed increased lipid deposition in BAT and decreased mitochondrial content, although no differences in thermogenesis (Seydoux et al. 1981; Hogan et al. 1982).

The effect of obesity in BAT function is controversial, specially regarding hypothalamic regulation over BAT. However, more recent studies show that obesity is associated with decreased BAT activity in humans (van Marken Lichtenbelt et al. 2009) and the overactivation of the SNS that occurs in obesity (Valenzano et al. 2016) can lead to the desensitization of hypothalamus-BAT interaction. Moreover, the hypothalamic inflammation typical of obesity reduces BAT thermogenesis. In fact, ICV injection of TNF-α decreases the expression of genes usually upregulated under conditions that promote thermogenesis, such as cold and fasting, and TNF-R1 KO mice have increased BAT activation and are resistant to diet-induced obesity (Romanatto et al. 2009; Arruda et al. 2010; Arruda et al. 2011). As mentioned, leptin action in the ARC is one of the main activators of the process of NST (Morrison et al. 2014; Contreras et al. 2015), hence the hypothalamic leptin resistance installed in obesity (Könner and Brüning 2012; Cui et al. 2017) might contribute to BAT deregulation.

4.6.3 Obesity Changes Hypothalamic Control of White Adipose Tissue (WAT)

WAT function as a reservoir of chemical energy in the form of triglycerides, but also functions as an endocrine "gland." Adipokines participate in glucose and lipid metabolism and are involved in obesity-related disorders (Ahima and Flier 2000; Cancello et al. 2004). The SNS regulates WAT functioning, namely lipolysis, adipocyte number, and WAT proteins regulation (Nogueiras et al. 2010). Metabolic regulation of WAT can determine the storage of energy in fat depots, the release of triglycerides into circulation or its accumulation in other tissues such as the liver (Guilherme et al. 2008).

The hypothalamus, as a regulator of energy balance, is tightly involved in WAT regulation. The SNS-WAT axis includes several hypothalamic nuclei such as the ARC, DMH, LHA, PVN, and SCN (Bamshad et al. 1998; Bartness et al. 2005). Electrical stimulation of the VMH and LHA can promote lipolysis (Ruffin and Nicolaidis 1999; Shen et al. 2008; Perez-Leighton et al. 2014). In other hypothalamic nuclei the neuropeptide system involved in food intake and energy expenditure takes a major part in WAT functioning. MC-3/4R agonists in the brain reduce adiposity (Raposinho et al. 2003), while the blockade of melanocortin system in CNS can stimulate lipid storage and triglyceride synthesis (Nogueiras et al. 2007). On the opposite, NPY and orexin can inhibit lipolysis and promote adiposity (Baran et al. 2002; Shen et al. 2008; Sousa-Ferreira et al. 2011). Moreover, NPY can also supress cathecolamine release and consequent lipolysis (Kaushik et al. 2012). Insulin and leptin receptors colocalize in hypothalamic neurons that project to WAT (Adler et al. 2012). Insulin infusion in the medial basal hypothalamus (MBH) supresses lipolysis, while insulin receptor knockout in mice can decrease lipogenesis and increase lipolysis (Scherer et al. 2011). Leptin infusion in the MBH can inhibit lipogenesis (Buettner and Camacho 2008); however, this infusion in the third ventricle increases lipolysis (Tajima et al. 2005).

Upon obesity, WAT is characterized by an increase in adipocyte size resulting from the increased food intake and the consequent accumulation of triglycerides. Moreover, inflammatory processes can also be observed in WAT concomitant with adipokine imbalance (Ouchi et al. 2001; Shimizu and Walsh 2015).

It is now accepted that browning of WAT, i.e., the process where white adipocytes acquire brown-like characteristics, occurs in a normal energetic status and promotes energy expenditure. In fact, browning of WAT can attenuate diet-induced obesity and it is looked upon as a possible therapeutic strategy for obesity (Wu et al. 2013). However, blockade of WAT browning can promote obesity and, in turn, the alterations in WAT-regulatory circuits (SNS-Hypothalamus-WAT) in obesity can impair WAT browning, thus exacerbating the pathology (Wu et al. 2013). Furthermore, hypothalamic AMPK mediates the browning of WAT. AMPK deregulation is well known to occur in obesity and metabolic dysfunction, thus contributing for the impairment of this process (López et al. 2016). Moreover, hypothalamic inflammation that occurs in obesity is responsible per se for the increase in white adipocytes and WAT inflammation (Guilherme et al. 2008; Williams 2012; Valdearcos et al. 2015). As mentioned, insulin and leptin exert their regulatory effects over the adipose tissue directly and through receptors in hypothalamic neurons, the hypothalamic resistance to these hormones (Könner and Brüning 2012) turns this regulation obsolete. AMPK signaling also mediates insulin and leptin signaling in the hypothalamus (López et al. 2016), hence its deregulation in obesity compromises these hormone effects on the hypothalamus and consequently in WAT physiology.

4.6.4 Obesity Changes Hypothalamic Control of Pancreas

The endocrine pancreas is constituted by the islets of Langerhans that secrete insulin and glucagon. The islets are composed mainly by α- and β-cells that control glucagon and insulin secretion, respectively. The orchestrated activity of these cells regulate glucose homeostasis; however, this is a process tightly regulated and composed by several organs beyond the pancreas (Rodriguez-Diaz and Caicedo 2014). In contrast to most organs, islets of Langerhans are self-sufficient, within these cells are all the necessary mechanisms to sense glucose changes and to orchestrate the hormone secretion. For this reason, it is accepted that ANS regulation over the pancreas serves mostly as an adaptive response, to adjust glucose homeostasis to food intake and/or stress (Teff and Townsend 2004).

Central regulation of the pancreas is mainly achieved by the presence of glucose-sensing neurons in the hypothalamus, that serve to influence glucose metabolism and HPG in the liver, but also to regulate islet function (Thorens 2011). Studies in dogs, rats, and mice support the role of different hypothalamic nuclei in the glucose counterregulatory response (Ogunnowo-Bada et al. 2014). Within the hypothalamus, the ARC, PVN, VMH, and LHA are the main nuclei presenting glucose-sensing neurons (Silver and Erecińska 1998). The VMH plays a major role in islet regulation; lower levels of glucose in this nucleus promote glucagon secretion whereas increased glucose suppresses it (Borg et al. 1995, 1997). On the opposite, insulin injection in the VMH inhibits glucagon secretion and the reduction of insulin receptors in this nucleus impairs pancreatic cells function (Paranjape et al. 2010, 2011). AgRP neurons in the ARC also appear to be involved in pancreatic regulation. Mice lacking AgRP become obese and hyperinsulinemic even in chow diet, whereas on HFD show reduced body weight gain and paradoxical improved glucose tolerance (Joly-Amado et al. 2012).

Obesity is correlated with secondary metabolic dysfunctions, namely insulin resistance and type 2 diabetes. Obesity is frequently associated with hyperglycemia that has a direct impact on islet functioning; however, it is still unclear how this can impact hypothalamic regulation over the pancreas. The ectopic accumulation of fat in the pancreas can lead to nonalcoholic fatty pancreas disease (NAFPD) that can override the central regulation of the ANS over insulin and glucagon secretion (Della Corte et al. 2015; Alempijevic et al. 2017). Moreover, patients with diabetes develop impairments in the counterregulatory response to hypoglycemia, suggesting alterations in glucose sensing circuitry on the hypothalamus (Ogunnowo-Bada et al. 2014). Insulin signaling in the hypothalamus also appears to be an important regulator of pancreatic function (Paranjape et al. 2010, 2011), thus hypothalamic insulin resistance compromises the regulatory feedback mechanisms orchestrated through the hypothalamus. Moreover, AMPK is determinant for glucose homeostasis in pancreas, hyperglycemia inhibits AMPK in the PVN, VMH, LHA, and ARC. AMPK deregulation in these hypothalamic nuclei might be responsible for the pancreatic function disruption (López et al. 2016). Hypothalamic inflammation can impair the release of insulin by the pancreas, by a mechanism independent of body weight (Calegari et al. 2016; Purkayastha et al. 2011).

4.6.5 Obesity Changes Hypothalamic Control of Skeletal Muscle

The skeletal muscle is one of the major sites of insulin-stimulated glucose uptake. Glucose and lipids are used as a thermogenic substrate by the skeletal muscle. The decrease in glucose uptake by the skeletal muscle can promote hyperglycemia and improper insulin-mediated responses (Marino et al. 2011).

The regulation of glucose uptake by the skeletal muscle through hypothalamic circuitry is related with the hypothalamic regulation of energy expenditure. The VMH, LHA, and ARC neurons have a direct effect on glucose uptake (Minokoshi 2017). Electrical stimulation of the VMH and leptin injection in this nucleus promote glucose uptake by the skeletal muscle independently of circulating insulin levels (Minokoshi et al. 1988; Sudo et al. 1991). Furthermore, leptin can also improve insulin sensitivity in skeletal muscle through the melanocortin system in the VMH (Roman et al. 2010; Koch et al. 2008). AMPK regulates glucose uptake and fatty acid oxidation in skeletal muscle, leptin injection in the VMH and ARC increases AMPK signaling (Minokoshi et al. 2002). This effect is mediated by the melanocortin system in the VMH neurons and the POMC neurons of the ARC (Gavini et al. 2016; Minokoshi 2017). Moreover, orexin injection in the VMH promotes skeletal muscle glucose uptake and enhances insulin sensitivity; however, ICV injection of NPY and/or AgRP have no effect on skeletal muscle glucose homeostasis (Shiuchi et al. 2009; Minokoshi 2017).

Not much is known regarding the effects of obesity on the hypothalamic regulation of skeletal muscle glucose homeostasis. Obesity can impair skeletal muscle insulin sensitivity and the glucose uptake (Kahn and Flier 2000). Obesity-induced hypothalamic inflammation can impair the central regulatory processes over the muscle (Valdearcos et al. 2015). As mentioned, hypothalamic AMPK signaling mediates leptin-induced glucose uptake (Minokoshi et al. 2002); thus, this deranged signaling pathway in obesity (López et al. 2016) can dampen the hypothalamus-muscle interaction. The same effect can be exacerbated by the leptin resistance in obesity, given that leptin signaling (either through the melanocortin system or the direct binding to the Ob-Rb) is determinant for skeletal muscle glucose uptake (Minokoshi 2017). Moreover, considering that POMC neurons are required for the leptin-induced glucose uptake (Gavini et al. 2016), the selective apoptosis of POMC neurons in obesity (Thaler et al. 2012) can impair the hypothalamus-skeletal muscle regulatory effects contributing for the deregulations observed in obesity.

4.7 Hypothalamic Inflammation in Obesity

The search for the mechanisms underlying obesity and its metabolic consequences such as insulin resistance and diabetes has highlighted the impact of inflammation in the mediation of obesity-induced dysfunctions. It is acccpted that obesity is characterized by a low-grade systemic inflammation (Hotamisligil 2006; Kolb and Mandrup-Poulsen 2010; Odegaard and Chawla 2013). However, obesity-induced inflammation is different from generalized inflammation. The inflammatory processes occur in the brain but also in peripheral organs such as the liver, adipose tissue, pancreas, skeletal muscle, and heart. In the one hand, inflammation occurs to regulate energy homeostasis, and, in the other hand, inflammation deregulates adaptive responses that further compromise the metabolic status of the organism (Saltiel and Olefsky 2017).

Given the fact that the hypothalamus is the brain region responsible for integration of signals that reflect the metabolic status of the organism and the regulation of energy homeostasis, it is easy to understand why this area is so susceptible to HFD effects. In fact, hypothalamic inflammation has been correlated to the development and progression of obesity. Moreover, hypothalamic inflammation has also been considered the main driver for the deregulation of cellular mechanisms that mediate the pathophysiology of obesity such as ER stress, autophagy, and oxidative stress (Williams 2012; Jais and Brüning 2017).

In this section we discuss the effects of obesity on hypothalamic inflammation and the contribution of this mechanism for the development of obesity pathophysiology.

Consumption of fat-enriched diets is associated with hypothalamic deregulation in the human brain. Obesity-induced systemic inflammation can alter the integrity of the brain structures directly involved in reward and feeding behaviors (Cazettes et al. 2011). Studies in the human brain also showed increased inflammatory markers and gliosis in the hypothalamus of obese individuals (Thaler et al. 2012; Puig et al. 2015).

The privileged location of the hypothalamus within the brain makes it susceptible to inflammation. In fact, hypothalamic inflammation arises even before significant body weight gain. The increase in proinflammatory markers in the hypothalamus can be observed just one day after HFD ingestion (Thaler et al. 2012; Waise et al. 2015). The BBB acts therefore as an intermediate between the brain and the periphery and it is deeply affected by HFD. The median eminence is fenestrated allowing the hypothalamus to receive signals from the periphery (Elizondo-Vega et al. 2015). Overnutrition and peripheral inflammatory markers such as the interleukin IL-1β decrease the expression of tight junctions contributing for BBB dysfunction (Nitta et al. 2003; Beard et al. 2014). Vascular endothelial growth factor (VEGF) signaling at the BBB is determinant for permeability, long-term HFD exposure can increase VEGF expression in hypothalamic tanycytes, increasing permeability and promoting BBB integrity disruption (Lee et al. 2007; Langlet et al. 2013).

Within the hypothalamus, all cell types appear to be affected by the inflammatory processes and contribute for its maintenance. Microglia activation is associated with dietary content and mediates inflammatory processes (Milanski et al. 2009; Thaler et al. 2012). Activated hypothalamic microglia produces proinflammatory cytokines such as TNF-α and the interleukins IL-1β and IL-6, which can lead to the expression of fractalkine (CX3CL1) in neurons (Nakanishi et al. 2007; Lambertsen et al. 2009; Ropelle et al. 2010). This chemokine will further recruit peripheral immune cells to the hypothalamus, exacerbating the inflammation (Jones et al. 2010). Hypothalamic astrocytes are also activated upon HFD (Thaler et al. 2012; Buckman et al. 2015), and can produce several proinflammatory cytokines (Gupta et al. 2012). Astrocytosis can also trigger the activation of the NF-κB pathway through toll-like receptors (TLRs) and through the transforming growth factor-β1 (TGF-β1)-induced cellular stress mechanisms responses (stress granules activation) (Gorina et al. 2011; Yan et al. 2014).

The inflammatory process in the hypothalamus can be divided in two different phases, a transient inflammatory phase upon HFD initial exposure and a prolonged phase that arises from prolonged exposure to HFD and includes the severe deregulation of cellular mechanisms. The hypothalamic inflammation can persist even after weight loss and increases the predisposition to further weight regain (Wang et al. 2012). This observation suggests that the prolonged hypothalamic damage induced by a HFD could be irreversible (Moraes et al. 2009; van de Sande-Lee et al. 2011; Thaler et al. 2012).

Overnutrition can promote the release of cytokines and the activation of inflammatory pathways in the hypothalamus. Short-term HFD feeding is sufficient to promote inflammation (Thaler et al. 2012; Waise et al. 2015) and to reduce hypothalamic sensitivity to insulin (Clegg et al. 2011). Decreased insulin sensitivity can mediate hypothalamic dysfunction.

Lipids from the diet such as long-chain saturated fatty acids (SFAs) can cross the BBB and accumulate in the hypothalamus (Posey et al. 2009; Borg et al. 2012). SFAs can blunt leptin and insulin signaling, thus contributing to a positive energy balance. This effect of SFAs is partly mediated by the activation of proinflammatory signaling cascades through binding and activation of TLR4 (Lee et al. 2001; Lee et al. 2003; Shi et al. 2006; Kleinridders et al. 2009). Pharmacological inhibition of neuronal TLR4 can prevent insulin and leptin resistance (Posey et al. 2009; Milanski et al. 2009) and, in fact, mice lacking this receptor are protected against obesity (Tsukumo et al. 2007). TLR4 activation promotes apoptotic mechanisms within the hypothalamus (Moraes et al. 2009). TLR4 activates IKKβ that activates NF-κB inflammatory pathway (Hayden and Ghosh 2008). Stimulation of the IKKβ/NF-κB pathway induces the SOCS3 expression in the hypothalamus, leading to the inhibition of both leptin and insulin signaling (Zhang et al. 2008). TLR4 activation can also promote AMPK activation through JNK-mediated signaling. JNK signaling is also associated with increased inflammation and tightly correlated with insulin sensitivity (Solinas and Karin 2010). JNK activation on AgRP neurons is sufficient to promote hyperphagia and weight gain (Tsaousidou et al. 2014) and the deletion of JNK1 protects mice from obesity and insulin resistance (Hirosumi et al. 2002). TNF-α is a known activator of JNK, the central administration of TNF-α blunted leptin and insulin action (Romanatto et al. 2007). Moreover, TNF-α can also increase the hypothalamic levels of PTP1B via the NF-κB pathway, negatively regulating insulin and leptin sensitivity (Zabolotny et al. 2002; Ito et al. 2012). SFAs can also promote the increase in the synthesis of ceramides and sphingolipids that impair insulin and leptin signaling (Summers and Nelson 2005; Holland et al. 2007, Holland et al. 2011).

However, POMC neurons appear to be more sensitive to the effect of obesity in the synaptic organization (Horvath et al. 2010). Furthermore, prolonged inflammation can promote POMC neurons apoptosis (Moraes et al. 2009; Thaler et al. 2012). There are other cellular mechanisms underlying obesity that can be a cause of the hypothalamic inflammation caused by overnutrition or just a concomitant effect of the overall compromise of the hypothalamic physiology in obesity.

4.8 Obesity Changes Neurogenesis in the Hypothalamus

Neurogenesis is the process that results in the production of new neurons, where new cells are generated from the proliferation of neuroprogenitor cells (NPCs). In the adult brain, neurogenesis occurs mostly at the subventrical zone (SVZ) of the lateral ventricles and at subgranular zone (SGZ) of the hippocampus (Götz and Huttner 2005; Lledo et al. 2006). Neurogenesis in the hypothalamus resides in the tanycytes of the median eminence (Lee et al. 2012), the process of neurogenesis plays an important role in energy balance (Kokoeva et al. 2005; Sousa-Ferreira et al. 2014).

Several evidences support that an inadequate metabolic environment impair the formation of NPC population and consequently the neuronal network (Sousa-Ferreira et al. 2014). Maternal high fat diet feeding in rats promotes the increase of orexigenic neurons in the PVN of newborns, thus increasing the predisposition for obesity (Chang et al. 2008). On the opposite, nutritional restriction during the gestational period decreased NPC proliferative capacity, resulting on decreased body weight of the offspring and metabolic deficits (Desai et al. 2011).

Adult hypothalamic neurogenesis is regarded as a protective mechanism to maintain homeostasis facing nutritional challenge. HFD increases the number of neurons adopting an anorexigenic phenotype, possibly to prevent the increase in food intake and body weight (Gouaze et al. 2013). Moreover, in mice NPY/AgRP neurons degeneration can promote an increase in hypothalamic cell proliferation to maintain energetic homeostasis (Pierce and Xu 2010). On the other hand, HFD decreases the neurogenic capacity of hypothalamic neural stem cells (htNSC) through apoptosis of newly divided cells and decreased survival of proliferating progenitor glial cells (McNay et al. 2012).

These observations suggest that obesity can alter hypothalamic neuronal population thus contributing to the failure of hypothalamic physiology.

4.9 Obesity Induces Hypothalamic ER Stress

The crosstalk between mechanisms that maintain cellular homeostasis can also reflect a deep compromise in all these protective mechanisms upon insult. HFD can elicit several exacerbated cellular responses that become deleterious at long term and can in fact contribute to the development and progression of obesity (Cavadas et al. 2016).

ER stress can be induced by metabolic stressors such as dietary fats, leading to the activation of the unfolded protein response (UPR) as a consequence of NF-κB activation (Zhang et al. 2008). ER stress can also activate JNK pathway and increase PTP1B, thus inhibiting insulin and leptin signaling (Hosoi et al. 2008). Central administration of chemical chaperones that reverse UPR improves hypothalamic leptin action (Zhang et al. 2008). Interestingly, the deletion of X-box binding protein 1 (Xbp1), an ER stress activator, promotes leptin resistance and obesity (Ozcan et al. 2009) while the Xbp1 expression in POMC neurons protects against HFD-induced obesity (Williams et al. 2014). These observations suggest that ER stress might work as a protective mechanism; however, the continuous stress resulting from overnutrition and the compromise of other cellular mechanisms can lead to deleterious overactivation of the UPR.

4.10 Obesity Induces Oxidative Stress in the Hypothalamus

Oxidative stress is a key feature of several disorders, namely neurodegenerative disorders. Indeed, high fat diet promotes oxidative stress that precedes other pathological observations such as insulin resistance (Matsuzawa-Nagata et al. 2008). The increase in triglycerides in rats was also correlated with an increase in mitochondrial respiration and ROS formation in the hypothalamus (Benani et al. 2007). Increased reactive oxygen species (ROS) formation in the hypothalamus can impair glucose and lipid sensing, thus compromising the metabolic sensing ability of hypothalamic neurons (Leloup et al. 2006; Colombani et al. 2009). In obese mice, the increased size of mitochondria in NPY/AgRP neurons as well as the decrease of proteins involved in mitochondrial fusion can protect from DIO (Dietrich et al. 2013). On the opposite, the same decrease in POMC/CART neurons promotes obesity (Schneeberger et al. 2013). ROS presence appears to be related with activation of POMC/CART neurons while ROS formation suppression activates NPY/AgRP neurons, thus promoting obesity (Schrader and Fahimi 2006; Diano et al. 2011). It is still not completely understood if inflammation can also promote mitochondrial dysfunction (Valdearcos et al. 2015); thus, it is possible that the hypothalamic inflammation observed in obesity might serve as a driver for mitochondrial dysfunction and oxidative stress.

4.11 Obesity Deregulates Hypothalamic Autophagy

Autophagy is a "self-cleaning" process that maintains cellular homeostasis by the degradation of damaged cytosolic organelles and proteins (Singh and Cuervo 2011). Autophagy is involved in several stress-induced cellular mechanisms such as ER stress and oxidative stress (Williams 2012).

In the hypothalamic neurons, autophagy has been involved in food intake and energy balance regulation (Singh and Cuervo 2011). While nutrient deprivation and caloric restriction can promote autophagy in hypothalamic neurons, obesity can disrupt autophagic processes (Kaushik et al. 2011; Meng and Cai 2011; Coupé et al. 2012; Aveleira et al. 2015). Autophagy is induced both by ER stress and oxidative stress upon HFD (Butler and Bahr 2006; Yorimitsu et al. 2006). Mice with *Atg7* deletion in the MBH showed increased weight gain and the activation of the inflammatory pathway, NF-κB. This observation suggested not only a role for autophagy in the global regulation of body weight but also an interplay between autophagy and inflammation (Meng and Cai 2011). Furthermore, autophagy plays a differential role in NPY/AgRP and POMC/CART neurons. The disruption of autophagy (by deleting *Atg7*) in AgRP neurons promoted a decrease in body weight (Kaushik et al. 2011) while this in POMC neurons promoted increased food intake and obesity (Kaushik et al. 2012). The disruption in hypothalamic autophagy might result from hypothalamic inflammation; however, the compromise of autophagy per se appears to be enough to induce severe disturbances in weight and it can for certain mediate the pathophysiology of obesity.

4.12 Conclusion

The hypothalamus as the center regulator of energetic homeostasis plays a determinant role in the pathogenesis of obesity. Obesogenic diets compromise the structure and circuitry in the hypothalamus. Diet-induced hypothalamic dysfunction affects all the physiological functions regulated by this brain structure, not only food intake and energy balance. Alterations in circadian rhythm, reproductive behavior, and stress response also affect directly energetic homeostasis and can further contribute to obesity. Moreover, overnutrition also impairs the autonomic regulation of the hypothalamus over peripheral organs that contribute to energy balance (Fig. 4.1).

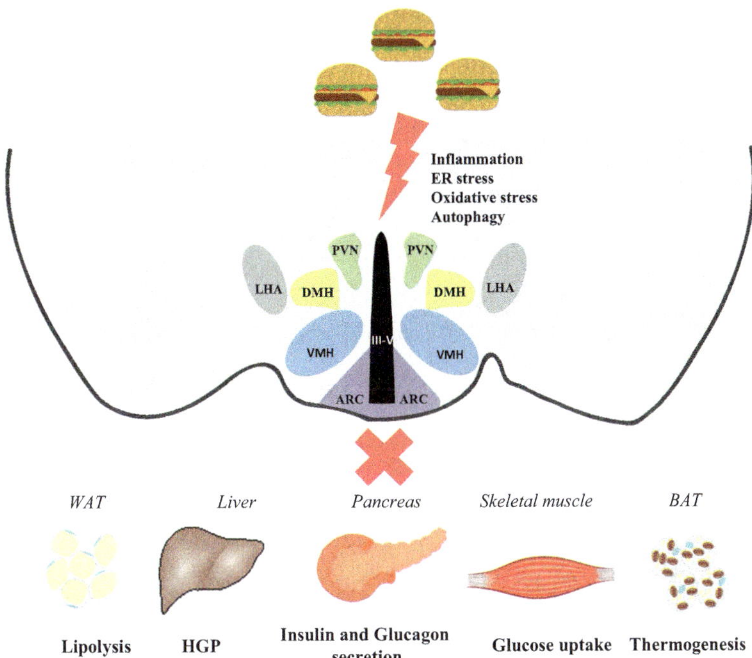

Fig. 4.1 Effect of an obesogenic diet in the hypothalamus. A fat-enriched diet promotes hypothalamic injury through the impairment or deleterious activation of several cellular mechanisms such as inflammation, ER stress, oxidative stress, and autophagy. The damage to the hypothalamus affects different nuclei that mediate metabolic homeostasis controlled by the central nervous system (CNS). The hypothalamus can directly regulate energy balance through the regulation of food intake through the ARC and VMH, but also regulates WAT lipolysis through the LHA, VMH, and ARC. Glucose homeostasis is regulated through sympathetic and parasympathetic innervation of the liver, pancreas, and skeletal muscle. The LHA, VMH, PVN, and ARC are involved in the hepatic glucose metabolism, specially in the regulation of hepatic glucose production (HPG). Hypothalamic modulation of pancreatic secretion of insulin and glucagon is achieved mainly through the VMH, DMH, and ARC. The VMH, DMH, and PVN are directly related with the stimulation of thermogenesis in BAT. Hypothalamic dysfunction observed in obesity can impair these regulatory mechanisms that further potentiate metabolic unbalance and potentiate the development and progression of obesity and its related pathologies such as type 2 diabetes and cardiovascular diseases. *ARC* arcuate nucleus, *BAT* brown adipose tissue, *DMH* dorsomedial hypothalamus, *HGP* hepatic glucose production, *LHA* lateral hypothalamic area, *PVN* paraventricular nucleus, *VMH* ventromedial hypothalamus, *WAT* white adipose tissue, *III-V* third ventricle

Hypothalamic inflammation might precede the impairment of response cellular mechanisms such as autophagy, oxidative stress, and ER stress, which suggests that inflammatory mediators in the hypothalamus might drive the pathogenesis of obesity. Understanding the mechanisms involved in the effects of obesogenic diets in the hypothalamus might help to uncover new targets for possible therapeutic approaches.

Acknowledgments This work was funded by the European Regional Development Fund (ERDF), through the Centro 2020 Regional Operational Programme: project CENTRO-01-0145-FEDER-000012-HealthyAging2020, the COMPETE 2020—Operational Programme for Competitiveness and Internationalisation, and the Portuguese national funds via FCT—Fundação para a Ciência e a Tecnologia, I.P.: project POCI-01-0145-FEDER-007440, Strategic Project UID/NEU/04539/201, and fellowship SFRH/BD/89035/2012.

References

Achermann P, Borbély AA (2003) Mathematical models in sleep deprivation. Front Biosci 8:s683–s693

Adam TC, Epel ES (2007) Stress, eating and the reward system. Physiol Behav 91:449–458

Adler ES, Hollis JH, Clarke IJ, Grattan DR, Oldfield BJ (2012) Neurochemical characterization and sexual dimorphism of projections from the brain to abdominal and subcutaneous white adipose tissue in the rat. J Neurosci 32:15913–15921

Ahima RS, Flier JS (2000) Adipose tissue as an endocrine organ. Trends Endocrinol Metab 11(8):327–332

Albrecht U (2012) Timing to perfection: the biology of central and peripheral circadian clocks. Neuron 74:246–260

Alempijevic T, Dragasevic S, Zec S, Popovic D, Milosavljevic T (2017) Non-alcoholic fatty pancreas disease. Postgrad Med J 93(1098):226–230

Alves MG, Jesus TT, Sousa M, Goldberg E, Silva BM, Oliveira PF (2016) Male fertility and obesity: are ghrelin, leptin and glucagon-like peptide-1 pharmacologically relevant? Curr Pharm Des 22:783–791

Anand BK, Brobeck JR (1951a) Localization of a "feeding center" in the hypothalamus of the rat. Proceedings of the Society for Experimental Biology and Medicine. Soc Exp Biol Med 77:323–324

Anand BK, Brobeck JR (1951b) Hypothalamic control of food intake in rats and cats. Yale J Biol Med 24:123–140

Arrieta-Cruz I, Gutiérrez-Juárez R (2016) The role of circulating amino acids in the hypothalamic regulation of liver glucose metabolism. Adv Nutr 7:790S–797S

Arruda AP, Milanski M, Romanatto T, Solon C, Coope A, Alberici LC et al (2010) Hypothalamic actions of tumor necrosis factor alpha provide the thermogenic core for the wastage syndrome in cachexia. Endocrinology 151:683–694

Arruda AP, Milanski M, Coope A, Torsoni AS, Ropelle E, Carvalho DP, Carvalheira JB, Velloso LA (2011) Low-grade hypothalamic inflammation leads to defective thermogenesis, insulin resistance, and impaired insulin secretion. Endocrinology 152:1314–1326

Aveleira CA, Botelho M, Carmo-Silva S, Pascoal JF, Ferreira-Marques M, Nóbrega C et al (2015) Neuropeptide Y stimulates autophagy in hypothalamic neurons. Proc Natl Acad Sci 112:E1642–E1651

Backholer K, Smith JT, Rao A, Pereira A, Iqbal J, Ogawa S, Li Q, Clarke IJ (2010) Kisspeptin cells in the ewe brain respond to leptin and communicate with neuropeptide Y and proopiomelanocortin cells. Endocrinology 151:2233–2243

Ballinger AB, Savage MO, Sanderson IR (2003) Delayed puberty associated with inflammatory bowel disease. Pediatr Res 53:205–210

Balsevich G, Uribe A, Wagner KV, Hartmann J, Santarelli S, Labermaier C, Schmidt MV (2014) Interplay between diet-induced obesity and chronic stress in mice: potential role of FKBP51. J Endocrinol 222:15–26

Bamshad M, Aoki VT, Adkison MG, Warren WS, Bartness TJ (1998) Central nervous system origins of the sympathetic nervous system outflow to white adipose tissue. Am J Phys 275:R291–R299

Banks WA (2001) Leptin transport across the blood-brain barrier: implications for the cause and treatment of obesity. Curr Pharm Des 7:125–133

Banks WA, Farrell CL (2003) Impaired transport of leptin across the blood-brain barrier in obesity is acquired and reversible. Am J Physiol Endocrinol Metab 285:E10–E15

Banks WA, Jaspan JB, Huang W, Kastin AJ (1997) Transport of insulin across the blood-brain barrier: saturability at euglycemic doses of insulin. Peptides 18:1423–1429

Banks AS, Davis SM, Bates SH, Myers MG (2000) Activation of downstream signals by the long form of the leptin receptor. J Biol Chem 275:14563–14572

Banks WA, Coon AB, Robinson SM, Moinuddin A, Shultz JM, Nakaoke R, Morley JE (2004) Triglycerides induce leptin resistance at the blood-brain barrier. Diabetes 53:1253–1260

Banks WA, Burney BO, Robinson SM (2008) Effects of triglycerides, obesity, and starvation on ghrelin transport across the blood-brain barrier. Peptides 29:2061–2065

Banting FG, Best CH, Macleod JJR (1922) The internal secretion of the pancreas. Am J Physiol 59:479

Baran K, Preston E, Wilks D, Cooney GJ, Kraegen EW, Sainsbury A (2002) Chronic central melanocortin-4 receptor antagonism and central neuropeptide-Y infusion in rats produce increased adiposity by divergent pathways. Diabetes 51:152–158

Barker-Gibb ML, Scott CJ, Boublik JH, Clarke IJ (1995) The role of neuropeptide Y (NPY) in the control of LH secretion in the ewe with respect to season, NPY receptor subtype and the site of action in the hypothalamus. J Endocrinol 147:565–579

Barsh GS, Schwartz MW (2002) Genetic approaches to studying energy balance: perception and integration. Nat Rev Genet 3:589–600

Bartness TJ, Kay Song C, Shi H, Bowers RR, Foster MT (2005) Brain-adipose tissue cross talk. Proc Nutr Soc 64:53–64

Baskin DG, Figlewicz Lattemann D, Seeley RJ, Woods SC, Porte D, Schwartz MW (1999) Insulin and leptin: dual adiposity signals to the brain for the regulation of food intake and body weight. Brain Res 848:114–123

Beard RS, Haines RJ, Wu KY, Reynolds JJ, Davis SM, Elliott JE et al (2014) Non-muscle Mlck is required for β-catenin- and FoxO1-dependent downregulation of Cldn5 in IL-1β-mediated barrier dysfunction in brain endothelial cells. J Cell Sci 127:1840–1853

Belda X, Fuentes S, Daviu N, Nadal R, Armario A (2015) Stress-induced sensitization: the hypothalamic–pituitary–adrenal axis and beyond. Stress 18:269–279

Belgardt BF, Bruning JC (2010) CNS leptin and insulin action in the control of energy homeostasis. Ann N Y Acad Sci 1212:97–113

Belgardt BF, Mauer J, Bruning JC (2010) Novel roles for JNK1 in metabolism. Aging (Albany NY) 2(9):621–626

Benani A, Troy S, Carmona MC, Fioramonti X, Lorsignol A, Leloup C, Casteilla L, Penicaud L (2007) Role for mitochondrial reactive oxygen species in brain lipid sensing: redox regulation of food intake. Diabetes 56:152–160

Benite-Ribeiro SA, Putt DA, Soares-Filho MC, Santos JM (2016) The link between hypothalamic epigenetic modifications and long-term feeding control. Appetite 107:445–453

Benoit SC, Air EL, Coolen LM, Strauss R, Jackman A, Clegg DJ, Seeley RJ, Woods SC (2002) The catabolic action of insulin in the brain is mediated by melanocortins. J Neurosci 22:9048–9052

Betley JN, Xu S, Cao ZFH, Gong R, Magnus CJ, Yu Y, Sternson SM (2015) Neurons for hunger and thirst transmit a negative-valence teaching signal. Nature 521:180–185

Biddinger SB, Kahn CR (2006) FROM MICE TO MEN: insights into the insulin resistance syndromes. Annu Rev Physiol 68:123–158

Bingham NC, Anderson KK, Reuter AL, Stallings NR, Parker KL (2008) Selective loss of leptin receptors in the ventromedial hypothalamic nucleus results in increased adiposity and a metabolic syndrome. Endocrinology 149(5):2138–2148

Bisschop PH, Fliers E, Kalsbeek A (2014) Autonomic regulation of hepatic glucose production. In: Comprehensive physiology. John Wiley & Sons, Inc., Hoboken, NJ, pp 147–165

Bjørbaek C, El-Haschimi K, Frantz JD, Flier JS (1999) The role of SOCS-3 in leptin signaling and leptin resistance. J Biol Chem 274:30059–30065

Blancas-Velazquez A, Mendoza J, Garcia AN, la Fleur SE (2017) Diet-induced obesity and circadian disruption of feeding behavior. Front Neurosci 11:23

Blouet C, Schwartz GJ (2010) Hypothalamic nutrient sensing in the control of energy homeostasis. Behav Brain Res 209:1–12

Borg WP, Sherwin RS, During MJ, Borg MA, Shulman GI (1995) Local ventromedial hypothalamus glucopenia triggers counterregulatory hormone release. Diabetes 44:180–184

Borg MA, Sherwin RS, Borg WP, Tamborlane WV, Shulman GI (1997) Local ventromedial hypothalamus glucose perfusion blocks counterregulation during systemic hypoglycemia in awake rats. J Clin Invest 99:361–365

Borg ML, Omran SF, Weir J, Meikle PJ, Watt MJ (2012) Consumption of a high-fat diet, but not regular endurance exercise training, regulates hypothalamic lipid accumulation in mice. J Physiol 590:4377–4389

Branecky KL, Niswender KD, Pendergast JS (2015) Disruption of daily rhythms by high-fat diet is reversible (N Cermakian, Ed). PLoS One 10:e0137970

Briggs DI, Enriori PJ, Lemus MB, Cowley MA, Andrews ZB (2010) Diet-induced obesity causes ghrelin resistance in arcuate NPY/AgRP neurons. Endocrinology 151:4745–4755

Briggs DI, Lemus MB, Kua E, Andrews ZB (2011) Diet-induced obesity attenuates fasting-induced hyperphagia. J Neuroendocrinol 23:620–626

Briggs DI, Lockie SH, Wu Q, Lemus MB, Stark R, Andrews ZB (2013) Calorie-restricted weight loss reverses high-fat diet-induced ghrelin resistance, which contributes to rebound weight gain in a ghrelin-dependent manner. Endocrinology 154:709–717

Brown SA, Kowalska E, Dallmann R (2012) (Re)inventing the circadian feedback loop. Dev Cell 22(3):477–487

Bruinstroop E, Fliers E, Kalsbeek A (2014) Hypothalamic control of hepatic lipid metabolism via the autonomic nervous system. Best Pract Res Clin Endocrinol Metab 28:673–684

Buckman LB, Thompson MM, Lippert RN, Blackwell TS, Yull FE, Ellacott KLJ (2015) Evidence for a novel functional role of astrocytes in the acute homeostatic response to high-fat diet intake in mice. Mol Metab 4:58–63

Buettner C, Camacho RC (2008) Hypothalamic control of hepatic glucose production and its potential role in insulin resistance. Endocrinol Metab Clin N Am 37:825–840

Buijs RM (2013) The autonomic nervous system. In: Handbook of clinical neurology, pp 1–11

Butler D, Bahr BA (2006) Oxidative stress and lysosomes: CNS-related consequences and implications for lysosomal enhancement strategies and induction of autophagy. Antioxid Redox Signal 8:185–196

Butler AA, Cone RD (2002) The melanocortin receptors: lessons from knockout models. Neuropeptides 36:77–84

Calegari VC, Torsoni AS, Vanzela EC, Araújo EP, Morari J, Zoppi CC, Sbragia L, Boschero AC, Velloso LA (2016) Inflammation of the hypothalamus leads to defective pancreatic islet function. J Biol Chem 291:26935–26935

Cancello R, Tounian A, Poitou C, Clément K (2004) Adiposity signals, genetic and body weight regulation in humans. Diabete Metab 30:215–227

Cano P, Jiménez-Ortega V, Larrad A, Reyes Toso CF, Cardinali DP, Esquifino AI (2008) Effect of a high-fat diet on 24-h pattern of circulating levels of prolactin, luteinizing hormone, testosterone, corticosterone, thyroid-stimulating hormone and glucose, and pineal melatonin content, in rats. Endocrine 33:118–125

Cao WH, Fan W, Morrison SF (2004) Medullary pathways mediating specific sympathetic responses to activation of dorsomedial hypothalamus. Neuroscience 126(1):229–240

Carvalheira JBC, Ribeiro EB, Araújo EP, Guimarães RB, Telles MM, Torsoni M, Gontijo JAR, Velloso LA, Saad MJA (2003) Selective impairment of insulin signalling in the hypothalamus of obese Zucker rats. Diabetologia 46:1629–1640

Catzeflis C, Pierroz DD, Rohner-Jeanrenaud F, Rivier JE, Sizonenko PC, Aubert ML (1993) Neuropeptide Y administered chronically into the lateral ventricle profoundly inhibits both the gonadotropic and the somatotropic axis in intact adult female rats. Endocrinology 132:224–234

Cavadas C, Aveleira CA, Souza GFP, Velloso LA (2016) The pathophysiology of defective proteostasis in the hypothalamus—from obesity to ageing. Nat Rev Endocrinol 12:723–733

Cazettes F, Cohen JI, Yau PL, Talbot H, Convit A (2011) Obesity-mediated inflammation may damage the brain circuit that regulates food intake. Brain Res 1373:101–109

Challet E (2015) Keeping circadian time with hormones. Diabetes Obes Metab 17:76–83

Chang A, Smith MC, Yin X, Fox RJ, Staugaitis SM, Trapp BD (2008) Neurogenesis in the chronic lesions of multiple sclerosis. Brain 131:2366–2375

Chao PT, Yang L, Aja S, Moran TH, Bi S (2011) Knockdown of NPY expression in the dorsomedial hypothalamus promotes development of brown adipocytes and prevents diet-induced obesity. Cell Metab 13:573–583

Chrousos GP, Torpy DJ, Gold PW (1998) Interactions between the hypothalamic-pituitary-adrenal axis and the female reproductive system: clinical implications. Ann Intern Med 129:229–240

Chun SK, Jo Y-H (2010) Loss of leptin receptors on hypothalamic POMC neurons alters synaptic inhibition. J Neurophysiol 104:2321–2328

Clarke IJ, Arbabi L (2016) New concepts of the central control of reproduction, integrating influence of stress, metabolic state, and season. Domest Anim Endocrinol 56:S165–S179

Clarke IJ, Backholer K, Tilbrook AJ (2005) Y2 receptor-selective agonist delays the estrogen-induced luteinizing hormone surge in ovariectomized ewes, but Y1-receptor-selective agonist stimulates voluntary food intake. Endocrinology 146:769–775

Cleary MP, Bergstrom HM, Dodge TL, Getzin SC, Jacobson MK, Phillips FC (2001) Restoration of fertility in young obese (Lep(ob) Lep(ob)) male mice with low dose recombinant mouse leptin treatment. Int J Obes Relat Metab Disord 25:95–97

Clegg DJ, Gotoh K, Kemp C, Wortman MD, Benoit SC, Brown LM et al (2011) Consumption of a high-fat diet induces central insulin resistance independent of adiposity. Physiol Behav 103:10–16

Clifford L, Dampney BW, Carrive P (2015) Spontaneously hypertensive rats have more orexin neurons in their medial hypothalamus than normotensive rats. Exp Physiol 100(4):388–398

Colombani A-L, Carneiro L, Benani A, Galinier A, Jaillard T, Duparc T et al (2009) Enhanced hypothalamic glucose sensing in obesity: alteration of redox signaling. Diabetes 58:2189–2197

Contreras C, Gonzalez F, Fernø J, Diéguez C, Rahmouni K, Nogueiras R, López M (2015) The brain and brown fat. Ann Med 47:150–168

Coomans CP, Lucassen EA, Kooijman S, Fifel K, Deboer T, Rensen PCN, Michel S, Meijer JH (2015) Plasticity of circadian clocks and consequences for metabolism. Diabetes Obes Metab 17:65–75

Coppari R, Ichinose M, Lee CE, Pullen AE, Kenny CD, McGovern RA et al (2005) The hypothalamic arcuate nucleus: a key site for mediating leptin's effects on glucose homeostasis and locomotor activity. Cell Metab 1:63–72

Cordero P, Campion J, Milagro FI, Goyenechea E, Steemburgo T, Javierre BM, Martinez JA (2011) Leptin and TNF-alpha promoter methylation levels measured by MSP could predict the response to a low-calorie diet. J Physiol Biochem 67:463–470

Correia MLG, Haynes WG, Rahmouni K, Morgan DA, Sivitz WI, Mark AL (2002) The concept of selective leptin resistance: evidence from agouti yellow obese mice. Diabetes 51:439–442

Coupé B, Ishii Y, Dietrich MO, Komatsu M, Horvath TL, Bouret SG (2012) Loss of autophagy in pro-opiomelanocortin neurons perturbs axon growth and causes metabolic dysregulation. Cell Metab 15:247–255

Cowley MA, Smith RG, Diano S, Tschöp M, Pronchuk N, Grove KL et al (2003) The distribution and mechanism of action of ghrelin in the CNS demonstrates a novel hypothalamic circuit regulating energy homeostasis. Neuron 37:649–661

Crujeiras AB, Goyenechea E, Abete I, Lage M, Carreira MC, Martínez JA, Casanueva FF (2010) Weight regain after a diet-induced loss is predicted by higher baseline leptin and lower ghrelin plasma levels. J Clin Endocrinol Metab 95:5037–5044

Crujeiras AB, Díaz-Lagares A, Abete I, Goyenechea E, Amil M, Martínez JA, Casanueva FF (2014) Pre-treatment circulating leptin/ghrelin ratio as a non-invasive marker to identify patients likely to regain the lost weight after an energy restriction treatment. J Endocrinol Investig 37:119–126

Crujeiras AB, Carreira MC, Cabia B, Andrade S, Amil M, Casanueva FF (2015) Leptin resistance in obesity: an epigenetic landscape. Life Sci 140:57–63

Cui H, López M, Rahmouni K (2017) The cellular and molecular bases of leptin and ghrelin resistance in obesity. Nat Rev Endocrinol 13:338–351

Cummings DE (2006) Ghrelin and the short- and long-term regulation of appetite and body weight. Physiol Behav 89:71–84

Cunningham PS, Ahern SA, Smith LC, da Silva Santos CS, Wager TT, Bechtold DA (2016) Targeting of the circadian clock via CK1δ/ε to improve glucose homeostasis in obesity. Sci Rep 6:29983

Czech MP, Tencerova M, Pedersen DJ, Aouadi M (2013) Insulin signalling mechanisms for triacylglycerol storage. Diabetologia 56:949–964

Dallman MF, Akana SF, Laugero KD, Gomez F, Manalo S, Bell ME, Bhatnagar S (2003) A spoonful of sugar: feedback signals of energy stores and corticosterone regulate responses to chronic stress. Physiol Behav 79:3–12

Dardente H, Cermakian N (2007) Molecular circadian rhythms in central and peripheral clocks in mammals. Chronobiol Int 24(2):195–213

De Bond J-AP, Smith JT (2014) Kisspeptin and energy balance in reproduction. Reproduction (Cambridge, England) 147:R53–R63

DeBoer MD, Li Y, Cohn S (2010) Colitis causes delay in puberty in female mice out of proportion to changes in leptin and corticosterone. J Gastroenterol 45:277–284

Della Corte C, Mosca A, Majo F, Lucidi V, Panera N, Giglioni E et al (2015) Nonalcoholic fatty pancreas disease and Nonalcoholic fatty liver disease: more than ectopic fat. Clin Endocrinol 83:656–662

Desai M, Li T, Ross MG (2011) Hypothalamic neurosphere progenitor cells in low birth-weight rat newborns: neurotrophic effects of leptin and insulin. Brain Res 1378:29–42

Diano S, Liu Z-W, Jeong JK, Dietrich MO, Ruan H-B, Kim E et al (2011) Peroxisome proliferation-associated control of reactive oxygen species sets melanocortin tone and feeding in diet-induced obesity. Nat Med 17:1121–1127

Dibner C, Schibler U, Albrecht U (2010) The mammalian circadian timing system: organization and coordination of central and peripheral clocks. Annu Rev Physiol 72:517–549

Dietrich MO, Horvath TL (2012) AgRP neurons: the foes of reproduction in leptin-deficient obese subjects. Proc Natl Acad Sci U S A 109:2699–2700

Dietrich MO, Liu Z-W, Horvath TL (2013) Mitochondrial dynamics controlled by mitofusins regulate Agrp neuronal activity and diet-induced obesity. Cell 155:188–199

Dimicco JA, Zaretsky DV (2007) The dorsomedial hypothalamus: a new player in thermoregulation. Am J Physiol Regul Integr Comp Physiol 292(1):R47–R63

Dufau ML (1998) The luteinizing hormone receptor. Annu Rev Physiol 60:461–496

Duong M, Cohen JI, Convit A (2012) High cortisol levels are associated with low quality food choice in type 2 diabetes. Endocrine 41:76–81

Duque-Guimarães DE, Ozanne SE (2013) Nutritional programming of insulin resistance: causes and consequences. Trends Endocrinol Metab 24:525–535

Eckel-Mahan K, Sassone-Corsi P (2013) Metabolism and the circadian clock converge. Physiol Rev 93:107–135

Egecioglu E, Stenström B, Pinnock SB, Tung LYC, Dornonville de la Cour C, Lindqvist A et al (2008) Hypothalamic gene expression following ghrelin therapy to gastrectomized rodents. Regul Pept 146:176–182

Elizondo-Vega R, Cortes-Campos C, Barahona MJ, Oyarce KA, Carril CA, García-Robles MA (2015) The role of tanycytes in hypothalamic glucosensing. J Cell Mol Med 19:1471–1482

English PJ, Ghatei MA, Malik IA, Bloom SR, Wilding JPH (2002) Food fails to suppress ghrelin levels in obese humans. J Clin Endocrinol Metab 87:2984

Esel E, Ozsoy S, Tutus A, Sofuoglu S, Kartalci S, Bayram F, Kokbudak Z, Kula M (2005) Effects of antidepressant treatment and of gender on serum leptin levels in patients with major depression. Prog Neuro-Psychopharmacol Biol Psychiatry 29:565–570

Farooqi IS, Jebb SA, Langmack G, Lawrence E, Cheetham CH, Prentice AM, Hughes IA, McCamish MA, O'Rahilly S (1999) Effects of recombinant leptin therapy in a child with congenital leptin deficiency. N Engl J Med 341(12):879–884

Farris W, Mansourian S, Chang Y, Lindsley L, Eckman EA, Frosch MP et al (2003) Insulin-degrading enzyme regulates the levels of insulin, amyloid beta-protein, and the beta-amyloid precursor protein intracellular domain in vivo. Proc Natl Acad Sci U S A 100:4162–4167

Farris W, Mansourian S, Leissring MA, Eckman EA, Bertram L, Eckman CB, Tanzi RE, Selkoe DJ (2004) Partial loss-of-function mutations in insulin-degrading enzyme that induce diabetes also impair degradation of amyloid beta-protein. Am J Pathol 164:1425–1434

Fishel MA, Watson GS, Montine TJ, Wang Q, Green PS, Kulstad JJ et al (2005) Hyperinsulinemia provokes synchronous increases in central inflammation and beta-amyloid in normal adults. Arch Neurol 62:1539–1544

Frank MG, Watkins LR, Maier SF (2013) Stress-induced glucocorticoids as a neuroendocrine alarm signal of danger. Brain Behav Immun 33:1–6

Friedman JM (2002) The function of leptin in nutrition, weight, and physiology. Nutr Rev 60(S1–14–84):85–87

Friedman JM, Halaas JL (1998) Leptin and the regulation of body weight in mammals. Nature 395:763–770

Froy O (2010) Metabolism and circadian rhythms—implications for obesity. Endocr Rev 31:1–24

Fuller CA, Horwitz BA, Horowitz JM (1975) Shivering and nonshivering thermogenic responses of cold-exposed rats to hypothalamic warming. Am J Phys 228:1519–1524

Gallant AR, Lundgren J, Drapeau V (2012) The night-eating syndrome and obesity. Obes Rev 13:528–536

Gamber KM, Huo L, Ha S, Hairston JE, Greeley S, Bjørbæk C (2012) Over-expression of leptin receptors in hypothalamic POMC neurons increases susceptibility to diet-induced obesity (N Irwin, Ed). PLoS One 7:e30485

Ganong WF (2000) Circumventricular organs: definition and role in the regulation of endocrine and autonomic function. Clin Exp Pharmacol Physiol 27:422–427

Gao Y, Sun T (2016) Molecular regulation of hypothalamic development and physiological functions. Mol Neurobiol 53:4275–4285

Gao J, Ghibaudi L, van Heek M, Hwa JJ (2002) Characterization of diet-induced obese rats that develop persistent obesity after 6 months of high-fat followed by 1 month of low-fat diet. Brain Res 936:87–90

Gao Q, Wolfgang MJ, Neschen S, Morino K, Horvath TL, Shulman GI, Fu X-Y (2004) Disruption of neural signal transducer and activator of transcription 3 causes obesity, diabetes, infertility, and thermal dysregulation. Proc Natl Acad Sci 101:4661–4666

García-Cáceres C, Quarta C, Varela L, Gao Y, Gruber T, Legutko B et al (2016) Astrocytic insulin signaling couples brain glucose uptake with nutrient availability. Cell 166:867–880

Gavini CK, Jones WC, Novak CM (2016) Ventromedial hypothalamic melanocortin receptor activation: regulation of activity energy expenditure and skeletal muscle thermogenesis. J Physiol 594:5285–5301

Gerber A, Saini C, Curie T, Emmenegger Y, Rando G, Gosselin P et al (2015) The systemic control of circadian gene expression. Diabetes Obes Metab 17(Suppl 1):23–32

Gerozissis K (2008) Brain insulin, energy and glucose homeostasis; genes, environment and metabolic pathologies. Eur J Pharmacol 585:38–49

Gibbons GF (1990) Assembly and secretion of hepatic very-low-density lipoprotein. Biochem J 268:1–13

Golden PL, Maccagnan TJ, Pardridge WM (1997) Human blood-brain barrier leptin receptor. Binding and endocytosis in isolated human brain microvessels. J Clin Invest 99:14–18

Gonnissen HKJ, Hulshof T, Westerterp-Plantenga MS (2013) Chronobiology, endocrinology, and energy- and food-reward homeostasis. Obes Rev 14:405–416

Gorina R, Font-Nieves M, Márquez-Kisinousky L, Santalucia T, Planas AM (2011) Astrocyte TLR4 activation induces a proinflammatory environment through the interplay between MyD88-dependent NFκB signaling, MAPK, and Jak1/Stat1 pathways. Glia 59:242–255

Götz M, Huttner WB (2005) The cell biology of neurogenesis. Nat Rev Mol Cell Biol 6:777–788

Gouaze A, Brenachot X, Rigault C, Krezymon A, Rauch C, Nedelec E, Lemoine A, Gascuel J, Bauer S, Penicaud L, Benani A (2013) Cerebral cell renewal in adult mice controls the onset of obesity. PLoS One 8(8):e72029

Guilding C, Hughes ATL, Brown TM, Namvar S, Piggins HD (2009) A riot of rhythms: neuronal and glial circadian oscillators in the mediobasal hypothalamus. Mol Brain 2:28

Guilherme A, Virbasius J V, Puri V, Czech MP (2008) Adipocyte dysfunctions linking obesity to insulin resistance and type 2 diabetes. Nat Rev Mol Cell Biol 9:367–377

Gumbs MCR, van den Heuvel JK, la Fleur SE (2016) The effect of obesogenic diets on brain neuropeptide Y. Physiol Behav 162:161–173

Gupta S, Knight AG, Gupta S, Keller JN, Bruce-Keller AJ (2012) Saturated long-chain fatty acids activate inflammatory signaling in astrocytes. J Neurochem 120:1060–1071

Hahn TM, Breininger JF, Baskin DG, Schwartz MW (1998) Coexpression of Agrp and NPY in fasting-activated hypothalamic neurons. Nat Neurosci 1:271–272

Halaas JL, Boozer C, Blair-West J, Fidahusein N, Denton DA, Friedman JM (1997) Physiological response to long-term peripheral and central leptin infusion in lean and obese mice. Proc Natl Acad Sci U S A 94:8878–8883

Han S-K, Gottsch ML, Lee KJ, Popa SM, Smith JT, Jakawich SK, Clifton DK, Steiner RA, Herbison AE (2005) Activation of gonadotropin-releasing hormone neurons by kisspeptin as a neuroendocrine switch for the onset of puberty. J Neurosci 25:11349–11356

Harlan SM, Morgan DA, Agassandian K, Guo DF, Cassell MD, Sigmund CD, Mark AL, Rahmouni K (2011) Ablation of the leptin receptor in the hypothalamic arcuate nucleus abrogates leptin-induced sympathetic activation. Circ Res 108(7):808–812

Harlan SM, Guo D-F, Morgan DA, Fernandes-Santos C, Rahmouni K (2013) Hypothalamic mTORC1 signaling controls sympathetic nerve activity and arterial pressure and mediates leptin effects. Cell Metab 17:599–606

Harrell CS, Gillespie CF, Neigh GN (2016) Energetic stress: the reciprocal relationship between energy availability and the stress response. Physiol Behav 166:43–55

Hastings MH, Maywood ES, O'Neill JS (2008) Cellular circadian pacemaking and the role of cytosolic rhythms. Curr Biol 18:R805–R815

Hatori M, Vollmers C, Zarrinpar A, DiTacchio L, Bushong EA, Gill S et al (2012) Time-restricted feeding without reducing caloric intake prevents metabolic diseases in mice fed a high-fat diet. Cell Metab 15:848–860

Hayden MS, Ghosh S (2008) Shared principles in NF-kappaB signaling. Cell 132:344–362

Hetherington AW, Ranson SW (1940) Hypothalamic lesions and adiposity in the rat. Anat Rec 78:24

Hirosumi J, Tuncman G, Chang L, Görgün CZ, Uysal KT, Maeda K, Karin M, Hotamisligil GS (2002) A central role for JNK in obesity and insulin resistance. Nature 420:333–336

van den Hoek AM, van Heijningen C, Schröder-van der Elst JP, Ouwens DM, Havekes LM, Romijn JA, Kalsbeek A, Pijl H (2008) Intracerebroventricular administration of neuropeptide Y induces hepatic insulin resistance via sympathetic innervation. Diabetes 57:2304–2310

Hogan S, Coscina DV, Himms-Hagen J (1982) Brown adipose tissue of rats with obesity-inducing ventromedial hypothalamic lesions. Am J Phys 243:E338–E344

Holland WL, Brozinick JT, Wang LP, Hawkins ED, Sargent KM, Liu Y, Narra K, Hoehn KL, Knotts TA, Siesky A, Nelson DH, Karathanasis SK, Fontenot GK, Birnbaum MJ, Summers SA (2007) Inhibition of ceramide synthesis ameliorates glucocorticoid-, saturated-fat-, and obesity-induced insulin resistance. Cell Metab 5(3):167–179

Holland WL, Bikman BT, Wang L-P, Yuguang G, Sargent KM, Bulchand S et al (2011) Lipid-induced insulin resistance mediated by the proinflammatory receptor TLR4 requires saturated fatty acid-induced ceramide biosynthesis in mice. J Clin Invest 121:1858–1870

Holt SJ, Wheal H V, York DA (1987) Hypothalamic control of brown adipose tissue in Zucker lean and obese rats. Effect of electrical stimulation of the ventromedial nucleus and other hypothalamic centres. Brain Res 405:227–233

Horvath TL (2006) Synaptic plasticity in energy balance regulation. Obesity 14:228S–233S

Horvath TL, Sarman B, García-Cáceres C, Enriori PJ, Sotonyi P, Shanabrough M et al (2010) Synaptic input organization of the melanocortin system predicts diet-induced hypothalamic reactive gliosis and obesity. Proc Natl Acad Sci U S A 107:14875–14880

Hosoi T, Sasaki M, Miyahara T, Hashimoto C, Matsuo S, Yoshii M, Ozawa K (2008) Endoplasmic reticulum stress induces leptin resistance. Mol Pharmacol 74:1610–1619

Hotamisligil GS (2006) Inflammation and metabolic disorders. Nature 444:860–867

Hotamisligil GS, Shargill NS, Spiegelman BM (1993) Adipose expression of tumor necrosis factoralpha: direct role in obesity-linked insulin resistance. Science 259(5091):87–91

Hur SSJ, Cropley JE, Suter CM (2017) Paternal epigenetic programming: evolving metabolic disease risk. J Mol Endocrinol 58:R159–R168

I'Anson H, Manning JM, Herbosa CG, Pelt J, Friedman CR, Wood RI, Bucholtz DC, Foster DL (2000) Central inhibition of gonadotropin-releasing hormone secretion in the growth-restricted hypogonadotropic female sheep. Endocrinology 141:520–527

Imai-Matsumura K, Matsumura K, Nakayama T (1984) Involvement of ventromedial hypothalamus in brown adipose tissue thermogenesis induced by preoptic cooling in rats. Jpn J Physiol 34:939–943

Ito Y, Banno R, Hagimoto S, Ozawa Y, Arima H, Oiso Y (2012) TNFα increases hypothalamic PTP1B activity via the NFκB pathway in rat hypothalamic organotypic cultures. Regul Pept 174:58–64

Jais A, Brüning JC (2017) Hypothalamic inflammation in obesity and metabolic disease. J Clin Invest 127:24–32

Joly-Amado A, Denis RGP, Castel J, Lacombe A, Cansell C, Rouch C et al (2012) Hypothalamic AgRP-neurons control peripheral substrate utilization and nutrient partitioning. EMBO J 31:4276–4288

Jones BA, Beamer M, Ahmed S (2010) Fractalkine/CX3CL1: a potential new target for inflammatory diseases. Mol Interv 10:263–270

Juruena MF (2014) Early-life stress and HPA axis trigger recurrent adulthood depression. Epilepsy Behav 38:148–159

Kahn BB, Flier JS (2000) Obesity and insulin resistance. J Clin Invest 106(4):473–481

Kaiyala KJ, Prigeon RL, Kahn SE, Woods SC, Schwartz MW (2000) Obesity induced by a high-fat diet is associated with reduced brain insulin transport in dogs. Diabetes 49:1525–1533

Kalra SP, Fuentes M, Fournier A, Parker SL, Crowley WR (1992) Involvement of the Y-1 receptor subtype in the regulation of luteinizing hormone secretion by neuropeptide Y in rats. Endocrinology 130:3323–3330

Kamegai J, Tamura H, Shimizu T, Ishii S, Sugihara H, Wakabayashi I (2000) Central effect of ghrelin, an endogenous growth hormone secretagogue, on hypothalamic peptide gene expression. Endocrinology 141:4797–4800

Kaushik S, Rodriguez-Navarro JA, Arias E, Kiffin R, Sahu S, Schwartz GJ, Cuervo AM, Singh R (2011) Autophagy in hypothalamic AgRP neurons regulates food intake and energy balance. Cell Metab 14:173–183

Kaushik S, Arias E, Kwon H, Lopez NM, Athonvarangkul D, Sahu S, Schwartz GJ, Pessin JE, Singh R (2012) Loss of autophagy in hypothalamic POMC neurons impairs lipolysis. EMBO Rep 13:258–265

Kleinridders A, Schenten D, Konner AC, Belgardt BF, Mauer J, Okamura T, Wunderlich FT, Medzhitov R, Bruning JC (2009) MyD88 signaling in the CNS is required for development of fatty acid-induced leptin resistance and diet-induced obesity. Cell Metab 10(4):249–259

Klöckener T, Hess S, Belgardt BF, Paeger L, Verhagen LAW, Husch A et al (2011) High-fat feeding promotes obesity via insulin receptor/PI3K-dependent inhibition of SF-1 VMH neurons. Nat Neurosci 14:911–918

Koch L, Wunderlich FT, Scibler J, Könner AC, Hampel B, Irlenbusch S et al (2008) Central insulin action regulates peripheral glucose and fat metabolism in mice. J Clin Invest 118:2132–2147

Kohsaka A, Laposky AD, Ramsey KM, Estrada C, Joshu C, Kobayashi Y, Turek FW, Bass J (2007) High-fat diet disrupts behavioral and molecular circadian rhythms in mice. Cell Metab 6:414–421

Kojima M, Hosoda H, Date Y, Nakazato M, Matsuo H, Kangawa K (1999) Ghrelin is a growth-hormone-releasing acylated peptide from stomach. Nature 402:656–660

Kokoeva MV, Yin H, Flier JS (2005) Neurogenesis in the hypothalamus of adult mice: potential role in energy balance. Science (New York, NY) 310:679–683

Kolb H, Mandrup-Poulsen T (2010) The global diabetes epidemic as a consequence of lifestyle-induced low-grade inflammation. Diabetologia 53:10–20

Könner AC, Brüning JC (2012) Selective insulin and leptin resistance in metabolic disorders. Cell Metab 16:144–152

Konner AC, Janoschek R, Plum L, Jordan SD, Rother E, Ma X, Xu C, Enriori P, Hampel B, Barsh GS, Kahn CR, Cowley MA, Ashcroft FM, Bruning JC (2007) Insulin action in AgRP-expressing neurons is required for suppression of hepatic glucose production. Cell Metab 5(6):438–449

Korner J, Chua SC, Williams JA, Leibel RL, Wardlaw SL (1999) Regulation of hypothalamic proopiomelanocortin by leptin in lean and obese rats. Neuroendocrinology 70:377–383

Kuroda H, Tahara Y, Saito K, Ohnishi N, Kubo Y, Seo Y et al (2012) Meal frequency patterns determine the phase of mouse peripheral circadian clocks. Sci Rep 2:711

Kyrou I, Tsigos C (2009) Stress hormones: physiological stress and regulation of metabolism. Curr Opin Pharmacol 9:787–793

Lambertsen KL, Clausen BH, Babcock AA, Gregersen R, Fenger C, Nielsen HH et al (2009) Microglia protect neurons against ischemia by synthesis of tumor necrosis factor. J Neurosci 29:1319–1330

Lamers F, Vogelzangs N, Merikangas KR, de Jonge P, Beekman ATF, Penninx BWJH (2013) Evidence for a differential role of HPA-axis function, inflammation and metabolic syndrome in melancholic versus atypical depression. Mol Psychiatry 18:692–699

Langlet F (2014) Tanycytes: a gateway to the metabolic hypothalamus. J Neuroendocrinol 26:753–760

Langlet F, Levin BE, Luquet S, Mazzone M, Messina A, Dunn-Meynell AA et al (2013) Tanycytic VEGF-A boosts blood-hypothalamus barrier plasticity and access of metabolic signals to the arcuate nucleus in response to fasting. Cell Metab 17:607–617

Lee JY, Sohn KH, Rhee SH, Hwang D (2001) Saturated fatty acids, but not unsaturated fatty acids, induce the expression of cyclooxygenase-2 mediated through toll-like receptor 4. J Biol Chem 276:16683–16689

Lee J-W, Swick AG, Romsos DR (2003) Leptin constrains phospholipase C-protein kinase C-induced insulin secretion via a phosphatidylinositol 3-kinase-dependent pathway. Exp Biol Med (Maywood) 228:175–182

Lee S, Chen TT, Barber CL, Jordan MC, Murdock J, Desai S et al (2007) Autocrine VEGF signaling is required for vascular homeostasis. Cell 130:691–703

Lee DA, Bedont JL, Pak T, Wang H, Song J, Miranda-Angulo A, Takiar V, Charubhumi V, Balordi F, Takebayashi H, Aja S, Ford E, Fishell G, Blackshaw S (2012) Tanycytes of the hypothalamic median eminence form a diet-responsive neurogenic niche. Nat Neurosci 15(5):700–702

Lee KY, Russell SJ, Ussar S, Boucher J, Vernochet C, Mori MA, Smyth G, Rourk M, Cederquist C, Rosen ED, Kahn BB, Kahn CR (2013) Lessons on conditional gene targeting in mouse adipose tissue. Diabetes 62(3):864–874

Lehr S, Hartwig S, Sell H (2012) Adipokines: a treasure trove for the discovery of biomarkers for metabolic disorders. Proteomics Clin Appl 6:91–101

Leloup C, Magnan C, Benani A, Bonnet E, Alquier T, Offer G et al (2006) Mitochondrial reactive oxygen species are required for hypothalamic glucose sensing. Diabetes 55:2084–2090

Levin BE, Dunn-Meynell AA (1997) Dysregulation of arcuate nucleus preproneuropeptide Y mRNA in diet-induced obese rats. Am J Phys 272:R1365–R1370

Liang L, Chen J, Zhan L, Lu X, Sun X, Sui H, Zheng L, Xiang H, Zhang F (2015) Endoplasmic reticulum stress impairs insulin receptor signaling in the brains of obese rats (C Scavone, Ed). PLoS One 10:e0126384

Lin S, Storlien LH, Huang XF (2000) Leptin receptor, NPY, POMC mRNA expression in the diet-induced obese mouse brain. Brain Res 875:89–95

Liu Y, Munro D, Layfield D, Dellinger A, Walter J, Peterson K, Rickman CB, Allingham RR, Hauser MA (2011) Serial analysis of gene expression (SAGE) in normal human trabecular meshwork. Mol Vis 17:885–893

Lledo P-M, Alonso M, Grubb MS (2006) Adult neurogenesis and functional plasticity in neuronal circuits. Nat Rev Neurosci 7:179–193

López M, Varela L, Vázquez MJ, Rodríguez-Cuenca S, González CR, Velagapudi VR et al (2010) Hypothalamic AMPK and fatty acid metabolism mediate thyroid regulation of energy balance. Nat Med 16:1001–1008

López M, Nogueiras R, Tena-Sempere M, Diéguez C (2016) Hypothalamic AMPK: a canonical regulator of whole-body energy balance. Nat Rev Endocrinol 12:421–432

Machluf Y, Gutnick A, Levkowitz G (2011) Development of the zebrafish hypothalamus. Ann N Y Acad Sci 1220:93–105

Maes HH, Neale MC, Eaves LJ (1997) Genetic and environmental factors in relative body weight and human adiposity. Behav Genet 27(4):325–351

Marazziti D, Rutigliano G, Baroni S, Landi P, Dell'Osso L (2014) Metabolic syndrome and major depression. CNS Spectr 19:293–304

Marchi M, Lisi S, Curcio M, Barbuti S, Piaggi P, Ceccarini G et al (2011) Human leptin tissue distribution, but not weight loss-dependent change in expression, is associated with methylation of its promoter. Epigenetics 6:1198–1206

Marco A, Kisliouk T, Weller A, Meiri N (2013) High fat diet induces hypermethylation of the hypothalamic Pomc promoter and obesity in post-weaning rats. Psychoneuroendocrinology 38:2844–2853

Marino JS, Xu Y, Hill JW (2011) Central insulin and leptin-mediated autonomic control of glucose homeostasis. Trends Endocrinol Metab 22:275–285

Mark AL (2013) Selective leptin resistance revisited. Am J Phys Regul Integr Comp Phys 305:R566–R581

van Marken Lichtenbelt WD, Vanhommerig JW, Smulders NM, Drossaerts JM, Kemerink GJ, Bouvy ND, Schrauwen P, Teule GJ (2009) Cold-activated brown adipose tissue in healthy men. N Engl J Med 360(15):1500–1508

Matsuzawa-Nagata N, Takamura T, Ando H, Nakamura S, Kurita S, Misu H et al (2008) Increased oxidative stress precedes the onset of high-fat diet-induced insulin resistance and obesity. Metabolism 57:1071–1077

McNay DEG, Briançon N, Kokoeva MV, Maratos-Flier E, Flier JS (2012) Remodeling of the arcuate nucleus energy-balance circuit is inhibited in obese mice. J Clin Invest 122:142–152

Meier-Ewert HK, Ridker PM, Rifai N, Regan MM, Price NJ, Dinges DF, Mullington JM (2004) Effect of sleep loss on C-reactive protein, an inflammatory marker of cardiovascular risk. J Am Coll Cardiol 43:678–683

Meng Q, Cai D (2011) Defective hypothalamic autophagy directs the central pathogenesis of obesity via the IkappaB kinase beta (IKKbeta)/NF-kappaB pathway. J Biol Chem 286:32324–32332

Mercer SW, Trayhurn P (1987) Effect of high fat diets on energy balance and thermogenesis in brown adipose tissue of lean and genetically obese ob/ob mice. J Nutr 117:2147–2153

Mercer AJ, Stuart RC, Attard CA, Otero-Corchon V, Nillni EA, Low MJ (2014) Temporal changes in nutritional state affect hypothalamic POMC peptide levels independently of leptin in adult male mice. Am J Phys 306:E904–E915

Michael MD, Kulkarni RN, Postic C, Previs SF, Shulman GI, Magnuson MA, Kahn CR (2000) Loss of insulin signaling in hepatocytes leads to severe insulin resistance and progressive hepatic dysfunction. Mol Cell 6:87–97

Michopoulos V, Toufexis D, Wilson ME (2012) Social stress interacts with diet history to promote emotional feeding in females. Psychoneuroendocrinology 37:1479–1490

Mifune H, Tajiri Y, Nishi Y, Hara K, Iwata S, Tokubuchi I, Mitsuzono R, Yamada K, Kojima M (2015) Voluntary exercise contributed to an amelioration of abnormal feeding behavior, locomotor activity and ghrelin production concomitantly with a weight reduction in high fat diet-induced obese rats. Peptides 71:49–55

Milanski M, Degasperi G, Coope A, Morari J, Denis R, Cintra DE, Tsukumo DM, Anhe G, Amaral ME, Takahashi HK, Curi R, Oliveira HC, Carvalheira JB, Bordin S, Saad MJ, Velloso LA (2009) Saturated fatty acids produce an inflammatory response predominantly through the activation of TLR4 signaling in hypothalamus: implications for the pathogenesis of obesity. J Neurosci 29(2):359–370

Milanski M, Arruda AP, Coope A, Ignacio-Souza LM, Nunez CE, Roman EA et al (2012) Inhibition of hypothalamic inflammation reverses diet-induced insulin resistance in the liver. Diabetes 61:1455–1462

Minokoshi Y (2017) Hypothalamic control of glucose and lipid metabolism in skeletal muscle. J Phys Fitness Sports Med 6(2):75–87

Minokoshi Y, Saito M, Shimazu T (1988) Sympathetic activation of lipid synthesis in brown adipose tissue in the rat. J Physiol 398:361–370

Minokoshi Y, Kim Y-B, Peroni OD, Fryer LGD, Müller C, Carling D, Kahn BB (2002) Leptin stimulates fatty-acid oxidation by activating AMP-activated protein kinase. Nature 415:339–343

Moore RY, Eichler VB (1972) Loss of a circadian adrenal corticosterone rhythm following suprachiasmatic lesions in the rat. Brain Res 42:201–206

Moraes JC, Coope A, Morari J, Cintra DE, Roman EA, Pauli JR et al (2009) High-fat diet induces apoptosis of hypothalamic neurons (X-Y Lu, Ed). PLoS One 4:e5045

Morris AA, Ahmed Y, Stoyanova N, Hooper WC, De Staerke C, Gibbons G, Quyyumi A, Vaccarino V (2012) The association between depression and leptin is mediated by adiposity. Psychosom Med 74:483–488

Morrison SF, Madden CJ, Tupone D (2014) Central neural regulation of brown adipose tissue thermogenesis and energy expenditure. Cell Metab 19:741–756

Münzberg H, Flier JS, Bjørbaek C (2004) Region-specific leptin resistance within the hypothalamus of diet-induced obese mice. Endocrinology 145:4880–4889

Muoio DM, Newgard CB (2008) Mechanisms of disease: molecular and metabolic mechanisms of insulin resistance and beta-cell failure in type 2 diabetes. Nat Rev Mol Cell Biol 9:193–205

Muroi Y, Ishii T (2016) A novel neuropeptide Y neuronal pathway linking energy state and reproductive behavior. Neuropeptides 59:1–8

Myers MG, Münzberg H, Leinninger GM, Leshan RL (2009) The geometry of leptin action in the brain: more complicated than a simple ARC. Cell Metab 9:117–123

Myers MG, Leibel RL, Seeley RJ, Schwartz MW (2010) Obesity and leptin resistance: distinguishing cause from effect. Trends Endocrinol Metab 21:643–651

Nagai K, Nagai N, Sugahara K, Niijima A, Nakagawa H (1994) Circadian rhythms and energy metabolism with special reference to the suprachiasmatic nucleus. Neurosci Biobehav Rev 18:579–584

Nakanishi M, Niidome T, Matsuda S, Akaike A, Kihara T, Sugimoto H (2007) Microglia-derived interleukin-6 and leukaemia inhibitory factor promote astrocytic differentiation of neural stem/progenitor cells. Eur J Neurosci 25:649–658

Nakazato M, Murakami N, Date Y, Kojima M, Matsuo H, Kangawa K, Matsukura S (2001) A role for ghrelin in the central regulation of feeding. Nature 409:194–198

Nguyen P, Leray V, Diez M, Serisier S, J Le B'h, Siliart B, Dumon H (2008) Liver lipid metabolism. J Anim Physiol Anim Nutr 92:272–283

Nicholas LM, Morrison JL, Rattanatray L, Zhang S, Ozanne SE, McMillen IC (2016) The early origins of obesity and insulin resistance: timing, programming and mechanisms. Int J Obes 40:229–238

Nitta T, Hata M, Gotoh S, Seo Y, Sasaki H, Hashimoto N, Furuse M, Tsukita S (2003) Size-selective loosening of the blood-brain barrier in claudin-5-deficient mice. J Cell Biol 161:653–660

Nogueiras R, Wiedmer P, Perez-Tilve D, Veyrat-Durebex C, Keogh JM, Sutton GM et al (2007) The central melanocortin system directly controls peripheral lipid metabolism. J Clin Invest 117:3475–3488

Nogueiras R, López M, Diéguez C (2010) Regulation of lipid metabolism by energy availability: a role for the central nervous system. Obes Rev 11:185–201 de Noronha SR, Campos GV, Abreu AR, de Souza AA, Chianca DA, de Menezes RC (2017) High fat diet induced-obesity facilitates anxiety-like behaviors due to GABAergic impairment within the dorsomedial hypothalamus in rats. Behav Brain Res 316:38–46

O'Neill JS, Reddy AB (2012) The essential role of cAMP/Ca2+ signalling in mammalian circadian timekeeping. Biochem Soc Trans 40:44–50

Obici S, Feng Z, Karkanias G, Baskin DG, Rossetti L (2002) Decreasing hypothalamic insulin receptors causes hyperphagia and insulin resistance in rats. Nat Neurosci 5:566–572

Odegaard JI, Chawla A (2013) Pleiotropic actions of insulin resistance and inflammation in metabolic homeostasis. Science (New York, NY) 339:172–177

Ogunnowo-Bada EO, Heeley N, Brochard L, Evans ML (2014) Brain glucose sensing, glucokinase and neural control of metabolism and islet function. Diabetes Obes Metab 16(Suppl 1):26–32

Oliveira PF, Sousa M, Silva BM, Monteiro MP, Alves MG (2017) Obesity, energy balance and spermatogenesis. Reproduction 153:R173–R185

Olofsson LE, Unger EK, Cheung CC, AW X (2013) Modulation of AgRP-neuronal function by SOCS3 as an initiating event in diet-induced hypothalamic leptin resistance. Proc Natl Acad Sci U S A 110:E697–E706

Opperhuizen A-L, Wang D, Foppen E, Jansen R, Boudzovitch-Surovtseva O, de Vries J, Fliers E, Kalsbeek A (2016) Feeding during the resting phase causes profound changes in physiology and desynchronization between liver and muscle rhythms of rats (R Silver, Ed). Eur J Neurosci 44:2795–2806

Ottaway N, Mahbod P, Rivero B, Norman LA, Gertler A, D'Alessio DA, Perez-Tilve D (2015) Diet-induced obese mice retain endogenous leptin action. Cell Metab 21:877–882

Ouchi N, Kihara S, Arita Y, Nishida M, Matsuyama A, Okamoto Y, Ishigami M, Kuriyama H, Kishida K, Nishizawa H, Hotta K, Muraguchi M, Ohmoto Y, Yamashita S, Funahashi T, Matsuzawa Y (2001) Adipocytederived plasma protein, adiponectin, suppresses lipid accumulation and class A scavenger receptor expression in human monocyte-derived macrophages. Circulation 103(8):1057–1063

Ozcan U, Cao Q, Yilmaz E, Lee A-H, Iwakoshi NN, Ozdelen E et al (2004) Endoplasmic reticulum stress links obesity, insulin action, and type 2 diabetes. Science (New York, NY) 306:457–461

Ozcan L, Ergin AS, Lu A, Chung J, Sarkar S, Nie D, Myers MG Jr, Ozcan U (2009) Endoplasmic reticulum stress plays a central role in development of leptin resistance. Cell Metab 9(1):35–51

Paranjape SA, Chan O, Zhu W, Horblitt AM, McNay EC, Cresswell JA, Bogan JS, McCrimmon RJ, Sherwin RS (2010) Influence of insulin in the ventromedial hypothalamus on pancreatic glucagon secretion in vivo. Diabetes 59:1521–1527

Paranjape SA, Chan O, Zhu W, Horblitt AM, Grillo CA, Wilson S, Reagan L, Sherwin RS (2011) Chronic reduction of insulin receptors in the ventromedial hypothalamus produces glucose intolerance and islet dysfunction in the absence of weight gain. Am J Phys Endocrinol Metab 301:E978–E983

Parton LE, Ye CP, Coppari R, Enriori PJ, Choi B, Zhang C-Y et al (2007) Glucose sensing by POMC neurons regulates glucose homeostasis and is impaired in obesity. Nature 449:228–232

Pearson CA, Placzek M (2013) Development of the medial hypothalamus: forming a functional hypothalamic-neurohypophyseal interface. Curr Top Dev Biol 106:49–88

Pecoraro N, Reyes F, Gomez F, Bhargava A, Dallman MF (2004) Chronic stress promotes palatable feeding, which reduces signs of stress: feedforward and feedback effects of chronic stress. Endocrinology 145:3754–3762

Pendergast JS, Branecky KL, Yang W, Ellacott KLJ, Niswender KD, Yamazaki S (2013) High-fat diet acutely affects circadian organisation and eating behavior. Eur J Neurosci 37:1350–1356

Peplonska B, Bukowska A, Sobala W (2015) Association of rotating night shift work with BMI and abdominal obesity among nurses and midwives (CR Sirtori, Ed). PLoS One 10:e0133761

Perez-Leighton CE, Billington CJ, Kotz CM (2014) Orexin modulation of adipose tissue. Biochim Biophys Acta 1842:440–445

Perreault M, Istrate N, Wang L, Nichols AJ, Tozzo E, Stricker-Krongrad A (2004) Resistance to the orexigenic effect of ghrelin in dietary-induced obesity in mice: reversal upon weight loss. Int J Obes Relat Metab Disord 28:879–885

Pierce AA, Xu AW (2010) De novo neurogenesis in adult hypothalamus as a compensatory mechanism to regulate energy balance. J Neurosci 30(2):723–730

Posey KA, Clegg DJ, Printz RL, Byun J, Morton GJ, Vivekanandan-Giri A et al (2009) Hypothalamic proinflammatory lipid accumulation, inflammation, and insulin resistance in rats fed a high-fat diet. Am J Phys Endocrinol Metab 296:E1003–E1012

Prevot V, Bellefontaine N, Baroncini M, Sharif A, Hanchate NK, Parkash J, Campagne C, de Seranno S (2010) Gonadotrophin-releasing hormone nerve terminals, tanycytes and neurohaemal junction remodelling in the adult median eminence: functional consequences for reproduction and dynamic role of vascular endothelial cells. J Neuroendocrinol 22:639–649

Puig J, Blasco G, Daunis-I-Estadella J, Molina X, Xifra G, Ricart W, Pedraza S, Fernández-Aranda F, Fernández-Real JM (2015) Hypothalamic damage is associated with inflammatory markers and worse cognitive performance in obese subjects. J Clin Endocrinol Metab 100:E276–E281

Purkayastha S, Zhang H, Zhang G, Ahmed Z, Wang Y, Cai D (2011) Neural dysregulation of peripheral insulin action and blood pressure by brain endoplasmic reticulum stress. Proc Natl Acad Sci U S A 108:2939–2944

Rahmouni K, Morgan DA, Morgan GM, Mark AL, Haynes WG (2005) Role of selective leptin resistance in diet-induced obesity hypertension. Diabetes 54:2012–2018

Ramnanan CJ, Edgerton DS, Kraft G, Cherrington AD (2011) Physiologic action of glucagon on liver glucose metabolism. Diabetes Obes Metab 13(Suppl 1):118–125

Raposinho PD, White RB, Aubert ML (2003) The melanocortin agonist Melanotan-II reduces the orexigenic and adipogenic effects of neuropeptide Y (NPY) but does not affect the NPY-driven suppressive effects on the gonadotropic and somatotropic axes in the male rat. J Neuroendocrinol 15:173–181

Reppert SM, Weaver DR (2002) Coordination of circadian timing in mammals. Nature 418(6901):935–941

Richter C, Woods IG, Schier AF (2014) Neuropeptidergic control of sleep and wakefulness. Annu Rev Neurosci 37:503–531

Rodriguez-Diaz R, Caicedo A (2014) Neural control of the endocrine pancreas. Best Pract Res Clin Endocrinol Metab 28:745–756

Roman EAFR, Reis D, Romanatto T, Maimoni D, Ferreira EA, Santos GA, Torsoni AS, Velloso LA, Torsoni MA (2010) Central leptin action improves skeletal muscle AKT, AMPK, and PGC1 alpha activation by hypothalamic PI3K-dependent mechanism. Mol Cell Endocrinol 314:62–69

Romanatto T, Cesquini M, Amaral ME, Roman EA, Moraes JC, Torsoni MA, Cruz-Neto AP, Velloso LA (2007) TNF-alpha acts in the hypothalamus inhibiting food intake and increasing the respiratory quotient-effects on leptin and insulin signaling pathways. Peptides 28(5):1050–1058

Romanatto T, Roman EA, Arruda AP, Denis RG, Solon C, Milanski M, Moraes JC, Bonfleur ML, Degasperi GR, Picardi PK, Hirabara S, Boschero AC, Curi R, Velloso LA (2009) Deletion of tumor necrosis factor-alpha receptor 1 (TNFR1) protects against diet-induced obesity by means of increased thermogenesis. J Biol Chem 284(52):36213–36222

Ropelle ER, Flores MB, Cintra DE, Rocha GZ, Pauli JR, Morari J et al (2010) IL-6 and IL-10 anti-inflammatory activity links exercise to hypothalamic insulin and leptin sensitivity through IKKbeta and ER stress inhibition (AJ Vidal-Puig, Ed). PLoS Biol 8:e1000465

Rosenwasser AM, Turek FW (2015) Neurobiology of circadian rhythm regulation. Sleep Med Clin 10:403–412

Rothwell NJ, Stock MJ (1981) A role for insulin in the diet-induced thermogenesis of cafeteria-fed rats. Metab Clin Exp 30:673–678

Rothwell NJ, Stock MJ (1997) A role for brown adipose tissue in diet-induced thermogenesis. Obes Res 5(6):650

Rudic RD, McNamara P, Curtis A-M, Boston RC, Panda S, Hogenesch JB, Fitzgerald GA (2004) BMAL1 and CLOCK, two essential components of the circadian clock, are involved in glucose homeostasis (Steve O'Rahilly, Ed). PLoS Biol 2:e377

Rudolph LM, Bentley GE, Calandra RS, Paredes AH, Tesone M, TJ W, Micevych PE (2016) Peripheral and central mechanisms involved in the hormonal control of male and female reproduction. J Neuroendocrinol 28. doi:10.1111/jne.12405

Ruffin M, Nicolaidis S (1999) Electrical stimulation of the ventromedial hypothalamus enhances both fat utilization and metabolic rate that precede and parallel the inhibition of feeding behavior. Brain Res 846:23–29

Ryan KK, Mul JD, Clemmensen C, Egan AE, Begg DP, Halcomb K, Seeley RJ, Herman JP, Ulrich-Lai YM (2014) Loss of melanocortin-4 receptor function attenuates HPA responses to psychological stress. Psychoneuroendocrinology 42:98–105

Sahar S, Sassone-Corsi P (2012) Regulation of metabolism: the circadian clock dictates the time. Trends Endocrinol Metab 23:1–8

Saltiel AR, Olefsky JM (2017) Inflammatory mechanisms linking obesity and metabolic disease. J Clin Invest 127:1–4

van de Sande-Lee S, FRS P, Cintra DE, Fernandes PT, Cardoso AR, Garlipp CR et al (2011) Partial reversibility of hypothalamic dysfunction and changes in brain activity after body mass reduction in obese subjects. Diabetes 60:1699–1704

Sapolsky RM, Romero LM, Munck AU (2000) How do glucocorticoids influence stress responses? Integrating permissive, suppressive, stimulatory, and preparative actions. Endocr Rev 21:55–89

Scherer T, O'Hare J, Diggs-Andrews K, Schweiger M, Cheng B, Lindtner C et al (2011) Brain insulin controls adipose tissue lipolysis and lipogenesis. Cell Metab 13:183–194

Schioth HB, Kakizaki Y, Kohsaka A, Suda T, Watanobe H (2001) Agouti-related peptide prevents steroid-induced luteinizing hormone and prolactin surges in female rats. Neuroreport 12:687–690

Schneeberger M, Dietrich MO, Sebastián D, Imbernón M, Castaño C, Garcia A et al (2013) Mitofusin 2 in POMC neurons connects ER stress with leptin resistance and energy imbalance. Cell 155:172–187

Schrader M, Fahimi HD (2006) Peroxisomes and oxidative stress. Biochim Biophys Acta 1763:1755–1766

Schwanzel-Fukuda M, Abraham S, Crossin KL, Edelman GM, Pfaff DW (1992) Immunocytochemical demonstration of neural cell adhesion molecule (NCAM) along the migration route of luteinizing hormone-releasing hormone (LHRH) neurons in mice. J Comp Neurol 321:1–18

Schwartz MW, Sipols AJ, Marks JL, Sanacora G, White JD, Scheurink A et al (1992) Inhibition of hypothalamic neuropeptide Y gene expression by insulin. Endocrinology 130:3608–3616

Schwartz MW, Woods SC, Porte D, Seeley RJ, Baskin DG (2000) Central nervous system control of food intake. Nature 404:661–671

Seimon R V, Hostland N, Silveira SL, Gibson AA, Sainsbury A (2013) Effects of energy restriction on activity of the hypothalamo-pituitary-adrenal axis in obese humans and rodents: implications for diet-induced changes in body composition. Horm Mol Biol Clin Invest 15:71–80

Seoane-Collazo P, Fernø J, Gonzalez F, Diéguez C, Leis R, Nogueiras R, López M (2015) Hypothalamic-autonomic control of energy homeostasis. Endocrine 50:276–291

Seydoux J, Rohner-Jeanrenaud F, Assimacopoulos-Jeannet F, Jeanrenaud B, Girardier L (1981) Functional disconnection of brown adipose tissue in hypothalamic obesity in rats. Pflugers Arch - Eur J Physiol 390:1–4

Sheffer-Babila S, Sun Y, Israel DD, Liu S-M, Neal-Perry G, Chua SC (2013) Agouti-related peptide plays a critical role in leptin's effects on female puberty and reproduction. Am J Physiol Endocrinol Metab 305:E1512–E1520

Shen J, Tanida M, Yao J-F, Niijima A, Nagai K (2008) Biphasic effects of orexin-A on autonomic nerve activity and lipolysis. Neurosci Lett 444:166–171

Shi H, Kokoeva MV, Inouyc K, Tzameli I, Yin H, Flier JS (2006) TLR4 links innate immunity and fatty acid-induced insulin resistance. J Clin Invest 116:3015–3025

Shi X, Wang X, Li Q, Su M, Chew E, Wong ET et al (2013) Nuclear factor κB (NF-κB) suppresses food intake and energy expenditure in mice by directly activating the Pomc promoter. Diabetologia 56:925–936

Shimazu T (1979) Nervous control of peripheral metabolism. Acta Phys Polon 30:1–18

Shimazu T, Ogasawara S (1975) Effects of hypothalamic stimulation on gluconeogenesis and glycolysis in rat liver. Am J Phys 228:1787–1793

Shimizu I, Walsh K (2015) The whitening of Brown fat and its implications for weight Management in Obesity. Curr Obes Rep 4:224–229

Shiuchi T, Haque MS, Okamoto S, Inoue T, Kageyama H, Lee S et al (2009) Hypothalamic orexin stimulates feeding-associated glucose utilization in skeletal muscle via sympathetic nervous system. Cell Metab 10:466–480

Silver IA, Erecińska M (1998) Glucose-induced intracellular ion changes in sugar-sensitive hypothalamic neurons. J Neurophysiol 79:1733–1745

Simonds SE, Pryor JT, Ravussin E, Greenway FL, Dileone R, Allen AM et al (2014) Leptin mediates the increase in blood pressure associated with obesity. Cell 159:1404–1416

Simoni M, Gromoll J, Höppner W, Nieschlag E (1997) Molecular pathophysiology of the pituitary-gonadal axis. Adv Exp Med Biol 424:89–97

Singh R, Cuervo AM (2011) Autophagy in the cellular energetic balance. Cell Metab 13:495–504

Sinha MK, Opentanova I, Ohannesian JP, Kolaczynski JW, Heiman ML, Hale J et al (1996) Evidence of free and bound leptin in human circulation. Studies in lean and obese subjects and during short-term fasting. J Clin Invest 98:1277–1282

Sipols AJ, Baskin DG, Schwartz MW (1995) Effect of intracerebroventricular insulin infusion on diabetic hyperphagia and hypothalamic neuropeptide gene expression. Diabetes 44:147–151

Solinas G, Karin M (2010) JNK1 and IKKbeta: molecular links between obesity and metabolic dysfunction. FASEB J 24(8):2596–2611

Sousa-Ferreira L, Garrido M, Nascimento-Ferreira I, Nobrega C, Santos-Carvalho A, Alvaro AR et al (2011) Moderate long-term modulation of neuropeptide Y in hypothalamic arcuate nucleus induces energy balance alterations in adult rats (D Tomé, Ed). PLoS One 6:e22333

Sousa-Ferreira L, de Almeida LP, Cavadas C (2014) Role of hypothalamic neurogenesis in feeding regulation. Trends Endocrinol Metab 25:80–88

Steculorum SM, Ruud J, Karakasilioti I, Backes H, Engström Ruud L, Timper K et al (2016) AgRP neurons control systemic insulin sensitivity via myostatin expression in brown adipose tissue. Cell 165:125–138

Stephan FK, Zucker I (1972) Circadian rhythms in drinking behavior and locomotor activity of rats are eliminated by hypothalamic lesions. Proc Natl Acad Sci U S A 69(6):1583

Stütz AM, Staszkiewicz J, Ptitsyn A, Argyropoulos G (2007) Circadian expression of genes regulating food intake*. Obesity 15:607–615

Sudo M, Minokoshi Y, Shimazu T (1991) Ventromedial hypothalamic stimulation enhances peripheral glucose uptake in anesthetized rats. Am J Phys 261:E298–E303

Summers SA, Nelson DH (2005) A role for sphingolipids in producing the common features of type 2 diabetes, metabolic syndrome X, and Cushing's syndrome. Diabetes 54:591–602

Swerdloff RS, Batt RA, Bray GA (1976) Reproductive hormonal function in the genetically obese (ob/ob) mouse. Endocrinology 98:1359–1364

Tajima D, Masaki T, Hidaka S, Kakuma T, Sakata T, Yoshimatsu H (2005) Acute central infusion of leptin modulates fatty acid mobilization by affecting lipolysis and mRNA expression for uncoupling proteins. Exp Biol Med 230:200–206

Tanida M, Yamamoto N, Shibamoto T, Rahmouni K (2013) Involvement of hypothalamic AMP-activated protein kinase in leptin-induced sympathetic nerve activation. PLoS One 8:e56660

Tanti JF, Jager J (2009) Cellular mechanisms of insulin resistance: role of stress-regulated serine kinases and insulin receptor substrates (IRS) serine phosphorylation. Curr Opin Pharmacol 9(6):753–762

Teff KL, Townsend RR (2004) Prolonged mild hyperglycemia induces vagally mediated compensatory increase in C-Peptide secretion in humans. J Clin Endocrinol Metab 89(11):5606–5613

Tengstrand B, Carlstrom K, Hafstrom I (2009) Gonadal hormones in men with rheumatoid arthritis—from onset through 2 years. J Rheumatol 36:887–892

Thaler JP, Yi C-X, Schur EA, Guyenet SJ, Hwang BH, Dietrich MO et al (2012) Obesity is associated with hypothalamic injury in rodents and humans. J Clin Investig 122:153–162

Thon M, Hosoi T, Ozawa K (2016) Possible integrative actions of Leptin and insulin signaling in the hypothalamus targeting energy homeostasis. Front Endocrinol 7

Thorens B (2011) Brain glucose sensing and neural regulation of insulin and glucagon secretion. Diabetes Obes Metab 13(Suppl 1):82–88

Thurlby PL, Trayhurn P (1979) The role of thermoregulatory thermogenesis in the development of obesity in genetically-obese (ob/ob) mice pair-fed with lean siblings. Br J Nutr 42(3):377–385

Tiesjema B, Adan RAH, Luijendijk MCM, Kalsbeek A, la Fleur SE (2007) Differential effects of recombinant adeno-associated virus-mediated neuropeptide Y overexpression in the hypothalamic paraventricular nucleus and lateral hypothalamus on feeding behavior. J Neurosci 27:14139–14146

Tilbrook AJ, Turner AI, Clarke IJ (2002) Stress and reproduction: central mechanisms and sex differences in non-rodent species. Stress 5(2):83–100

Torri C, Pedrazzi P, Leo G, Müller EE, Cocchi D, Agnati LF, Zoli M (2002) Diet-induced changes in hypothalamic pro-opio-melanocortin mRNA in the rat hypothalamus. Peptides 23:1063–1068

Tortoriello DV, McMinn J, Chua SC (2004) Dietary-induced obesity and hypothalamic infertility in female DBA/2J mice. Endocrinology 145(3):1238–1247

Tsaousidou E, Paeger L, Belgardt BF, Pal M, Wunderlich CM, Brönneke H et al (2014) Distinct roles for jnk and ikk activation in agouti-related peptide neurons in the development of obesity and insulin resistance. Cell Rep 9:1495–1506

Tsatsanis C, Dermitzaki E, Avgoustinaki P, Malliaraki N, Mytaras V, Margioris AN (2015) The impact of adipose tissue-derived factors on the hypothalamic-pituitary-gonadal (HPG) axis. Hormones 14:549–562

Tschöp M, Weyer C, Tataranni PA, Devanarayan V, Ravussin E, Heiman ML (2001) Circulating ghrelin levels are decreased in human obesity. Diabetes 50:707–709

Tsukumo DML, Carvalho-Filho MA, Carvalheira JBC, Prada PO, Hirabara SM, Schenka AA et al (2007) Loss-of-function mutation in toll-like receptor 4 prevents diet-induced obesity and insulin resistance. Diabetes 56:1986–1998

Turek FW, Joshu C, Kohsaka A, Lin E, Ivanova G, McDearmon E et al (2005) Obesity and metabolic syndrome in circadian clock mutant mice. Science 308:1043–1045

Ueno H, Nakazato M (2016) Mechanistic relationship between the vagal afferent pathway, central nervous system and peripheral organs in appetite regulation. J Diabetes Investig 7:812–818

Ulrich-Lai YM, Herman JP (2009) Neural regulation of endocrine and autonomic stress responses. Nat Rev Neurosci 10:397–409

Urayama A, Banks WA (2008) Starvation and triglycerides reverse the obesity-induced impairment of insulin transport at the blood-brain barrier. Endocrinology 149(7):3592

Uysal KT, Wiesbrock SM, Marino MW, Hotamisligil GS (1997) Protection from obesity-induced insulin resistance in mice lacking TNF-alpha function. Nature 389:610–614

Valdearcos M, AW X, Koliwad SK (2015) Hypothalamic inflammation in the control of metabolic function. Annu Rev Physiol 77:131–160

Valenzano A, Moscatelli F, Triggiani AI, Capranica L, De Ioannon G, Piacentini MF, Mignardi S, Messina G, Villani S, Cibelli G (2016) Heart-rate changes after an ultraendurance swim from Italy to Albania: a case report. Int J Sports Physiol Perform 11(3):407–409

Wahab F, Atika B, Shahab M (2013) Kisspeptin as a link between metabolism and reproduction: evidences from rodent and primate studies. Metab Clin Exp 62:898–910

Wahlestedt C, Skagerberg G, Ekman R, Heilig M, Sundler F, Håkanson R (1987) Neuropeptide Y (NPY) in the area of the hypothalamic paraventricular nucleus activates the pituitary-adrenocortical axis in the rat. Brain Res 417:33–38

Waise TMZ, Toshinai K, Naznin F, NamKoong C, Md Moin AS, Sakoda H, Nakazato M (2015) One-day high-fat diet induces inflammation in the nodose ganglion and hypothalamus of mice. Biochem Biophys Res Commun 464:1157–1162

Wang X, Ge A, Cheng M, Guo F, Zhao M, Zhou X, Liu L, Yang N (2012) Increased hypothalamic inflammation associated with the susceptibility to obesity in rats exposed to high-fat diet. Exp Diabetes Res 2012:847246

Waterson MJ, Horvath TL (2015) Neuronal regulation of energy homeostasis: beyond the hypothalamus and feeding. Cell Metab 22:962–970

Whittle AJ, López M, Vidal-Puig A (2011) Using brown adipose tissue to treat obesity—the central issue. Trends Mol Med 17:405–411

Wiater MF, Mukherjee S, Li A-J, Dinh TT, Rooney EM, Simasko SM, Ritter S (2011) Circadian integration of sleep-wake and feeding requires NPY receptor-expressing neurons in the mediobasal hypothalamus. Am J Phys Regul Integr Comp Phys 301:R1569–R1583

Williams LM (2012) Hypothalamic dysfunction in obesity. Proc Nutr Soc 71:521–533

Williams KW, Elmquist JK (2012) From neuroanatomy to behavior: central integration of peripheral signals regulating feeding behavior. Nat Neurosci 15:1350–1355

Williams KW, Liu T, Kong X, Fukuda M, Deng Y, Berglund ED et al (2014) Xbp1s in Pomc neurons connects ER stress with energy balance and glucose homeostasis. Cell Metab 20:471–482

Wu J, Cohen P, Spiegelman BM (2013) Adaptive thermogenesis in adipocytes: is beige the new brown? Genes Dev 27:234–250

Yan J, Zhang H, Yin Y, Li J, Tang Y, Purkayastha S, Li L, Cai D (2014) Obesity- and aging-induced excess of central transforming growth factor-β potentiates diabetic development via an RNA stress response. Nat Med 20:1001–1008

Yang JL, Liu DX, Jiang H, Pan F, Ho CS, Ho RC (2016) The effects of high-fat-diet combined with chronic unpredictable mild stress on depression-like behavior and leptin/leprb in male rats. Sci Rep 6:35239

Yorimitsu T, Nair U, Yang Z, Klionsky DJ (2006) Endoplasmic reticulum stress triggers autophagy. J Biol Chem 281:30299–30304

Zabolotny JM, Bence-Hanulec KK, Stricker-Krongrad A, Haj F, Wang Y, Minokoshi Y et al (2002) PTP1B regulates leptin signal transduction in vivo. Dev Cell 2:489–495

Zhang Y, Proenca R, Maffei M, Barone M, Leopold L, Friedman JM (1994) Positional cloning of the mouse obese gene and its human homologue. Nature 372:425–432

Zhang X, Zhang G, Zhang H, Karin M, Bai H, Cai D (2008) Hypothalamic IKKbeta/NF-kappaB and ER stress link overnutrition to energy imbalance and obesity. Cell 135:61–73

Zhou Q, Chen H, Yang S, Li Y, Wang B, Chen Y, Wu X (2014) High-fat diet decreases the expression of Kiss1 mRNA and kisspeptin in the ovary, and increases ovulatory dysfunction in postpubertal female rats. Reprod Biol Endocrinol 12:127

Zigman JM, Bouret SG, Andrews ZB (2016) Obesity impairs the action of the neuroendocrine ghrelin system. Trends Endocrinol Metab 27:54–63

Chapter 5
Diabesity and Brain Energy Metabolism: The Case of Alzheimer's Disease

Susana Cardoso, Raquel Seiça, and Paula I. Moreira

Abstract It is widely accepted that high calorie diets and a sedentary lifestyle sturdily influence the incidence and outcome of type 2 diabetes and obesity, which can occur simultaneously, a situation called diabesity. Tightly linked with metabolic and energy regulation, a close association between diabetes and Alzheimer's disease (AD) has been proposed. Among the common pathogenic mechanisms that underpin both conditions, insulin resistance, brain glucose hypometabolism, and metabolic dyshomeostasis appear to have a pivotal role. This century is an unprecedented diabetogenic period in human history, so therapeutic strategies and/or approaches to control and/or revert this evolving epidemic is of utmost importance. This chapter will make a brief contextualization about the impact that diabetes and obesity can exert in brain structure and function alongside with a brief survey about the role of insulin in normal brain function, exploring its roles in cognition and brain glucose metabolism. Later, attention will be given to the intricate relation of diabesity, insulin resistance, and AD. Finally, both pharmacological and lifestyle interventions will also be reviewed as strategies aimed at fighting diabesity and/or AD-related metabolic effects.

Keywords Alzheimer's disease • Brain • Diabesity • Glucose metabolism • Insulin resistance

S. Cardoso (✉)
Institute for Interdisciplinary Research, University of Coimbra, Coimbra, Portugal

CNC—Center for Neuroscience and Cell Biology, University of Coimbra, Coimbra, Portugal
e-mail: susana.t.cardoso@gmail.com; susana.cardoso@cnc.uc.pt

R. Seiça
Institute of Physiology, Institute for Biomedical Imaging and Life Sciences—IBILI,
Faculty of Medicine, University of Coimbra, Coimbra, Portugal

P.I. Moreira
CNC—Center for Neuroscience and Cell Biology, University of Coimbra, Coimbra, Portugal

Institute of Physiology, Institute for Biomedical Imaging and Life Sciences—IBILI,
Faculty of Medicine, University of Coimbra, Coimbra, Portugal

© Springer International Publishing AG 2017
L. Letra, R. Seiça (eds.), *Obesity and Brain Function*, Advances in
Neurobiology 19, DOI 10.1007/978-3-319-63260-5_5

5.1 Introduction

If one consider that obesity is an escalating worldwide problem, when it comes together with the occurrence of type 2 diabetes (T2D) we will be in front of a major health crisis. And this is the actual health scenario. As numbers reveal, 80–90% of overweight patients with abdominal fat deposition develop T2D whereas it is estimated that about 90% of T2D is attributed to excess weight leading to the establishment of a new concept, *diabesity* (Hossain et al. 2007). Firstly coined a few years ago by Sims et al. (1973), diabesity comprises the occurrence of obesity and T2D in the same individual, and reflects the intricate relationship between both metabolic conditions (Verma and Hussain 2017).

From a clinical point of view, obesity, aside from a genetic component, is mostly linked with sedentary lifestyle and energy-rich diets that contribute to a state of insulin resistance and an impairment in glycemic control, constituting a major risk factor for metabolic syndrome, T2D, and cardiovascular disease (Verma and Hussain 2017). In turn, affecting over 90% of diabetic patients, and with a projection of 366 million affected by 2030 (Wild et al. 2004), T2D is frequently associated with overweight, being characterized by relative insulin deficiency due to decreased insulin secretion by the pancreatic β-cells and/or the decreased effect of insulin in target tissues, a condition known as insulin resistance (Zimmet et al. 2001). Even though predictions reveal that only about 20–25% of the overweight and obese population is expected to develop diabetes (Lean 2010), it has been shown that the increase in the prevalence of T2D is closely linked to the upsurge in obesity (Webber et al. 2014; Hossain et al. 2007).

In this scenario, insulin resistance stands as a common root for both morbidities (Verma and Hussain 2017). Insulin resistance is a result of not only a decrease of insulin receptor signaling in classical target organs (e.g., liver, skeletal muscle, and adipose tissue) but also a result of an impaired insulin signaling in nonclassical organs like the brain where insulin is believed to act as an important neuromodulator, contributing to several neurobiological processes in particular energy homeostasis and cognition (Cardoso et al. 2009; Wilcox 2005). In this regard, compelling evidence suggests that insulin resistance, hallmark of T2D and obesity, may greatly contribute to age-related cognitive deficits. In fact, in the last decades, insulin resistance is in the spotlight being considered a major risk factor for dementia including Alzheimer's disease (AD) (Chen and Zhong 2013; Profenno et al. 2010). As demonstrated by epidemiologic studies in elderly, experimental investigations in humans and animal models consistently establish that dysfunctional brain insulin signaling promotes and accelerates cognitive dysfunction and AD progression by contributing to a reduced cerebral glucose metabolic rate, among other alterations (Sebastiao et al. 2014).

The increased awareness of diabetes as a risk factor for AD together with the metabolic alterations shared between both diseases culminated in the proposal of the term "type 3 diabetes" to designate sporadic AD, which represents more than 95% of all AD cases. The concept type 3 diabetes is based on the notion that sporadic

AD is a form of brain diabetes characterized by cerebral glucose hypometabolism and metabolic dyshomeostasis that manifest in the elderly as a result of a cumulative, lifelong impact in the brain, with molecular and biochemical features similar to those of diabetes and other peripheral insulin resistance disorders (de la Monte 2014; Lester-Coll et al. 2006).

This chapter will summarize evidence concerning the effects of diabesity in the brain alongside with a brief description of physiological effects of insulin in this organ. Later, attention will be given to the intricate relation of diabesity, insulin resistance, and AD. Finally, both pharmacological and lifestyle interventions will also be discussed as possible strategies to fight diabesity and/or AD-related metabolic effects.

5.2 Diabesity and Brain Health/Cognition: A Brief Overview

Due to the escalating changes in human behavior due to the adoption of a Western diet and a sedentary lifestyle, we are now observing an unprecedented burst in obesity and related comorbidities, in particular T2D (Hendrickx et al. 2005), which if left uncorrected will have a significant health impact leading to a reduction in individuals quality of life and overall life expectancy (Verma and Hussain 2017) (Fig. 5.1). In particular, it is now widely accepted that the occurrence of both metabolic conditions can evolve to several comorbidity concerns at late stages of life, including a high risk for cognitive decline and dementia such as AD (Irie et al. 2008; Wolf et al. 2007; Elias et al. 2005; Whitmer et al. 2005; Leibson et al. 1997).

By definition, AD is an age-related progressive neurodegenerative disorder mainly affecting elderly individuals and is characterized by impairment of memory and cognition. It is currently estimated that AD affects nearly 10% of individuals over the age of 65 and nearly 50% of those over the age of 85 (Cardoso et al. 2016; Morris et al. 2014). Often diagnosed in people aged 65 (about 95%) and older, the disease is typically referred to as sporadic, late-onset AD, occurring due to a sporadic component and often associated with the presence of the allele ε4 of the apolipoprotein E (ApoE4) gene on chromosome 19. On the other hand, a small proportion of subjects (accounting for 1–5% of all cases) present an early-onset (where initial symptoms can be observed between 30 and 65 years of age) of the disease due to a familiar genetic cause involving mutations in the amyloid-β protein precursor (APP) and presenilins 1 and 2 (PS1 and PS2) genes (Hampel et al. 2011).

As recently confirmed by a meta-analysis, the high prevalence of obesity and diabetes may result in an increased incidence of AD (Profenno et al. 2010), so it is of the utmost importance to explore the association between diabesity and brain disorders (Prickett et al. 2015). Indeed, several studies have emphasized the brain volume decline as a function of obesity (Pannacciulli et al. 2006; Ward et al. 2005; Gustafson et al. 2004). As shown, there is a strong relation between obesity and brain atrophy in cognitively normal elderly population (Raji et al. 2010), and a direct association of waist-hip ratio with decreased hippocampal volume (Jagust et al. 2005). Likewise, others have shown the existence of a link between obesity/

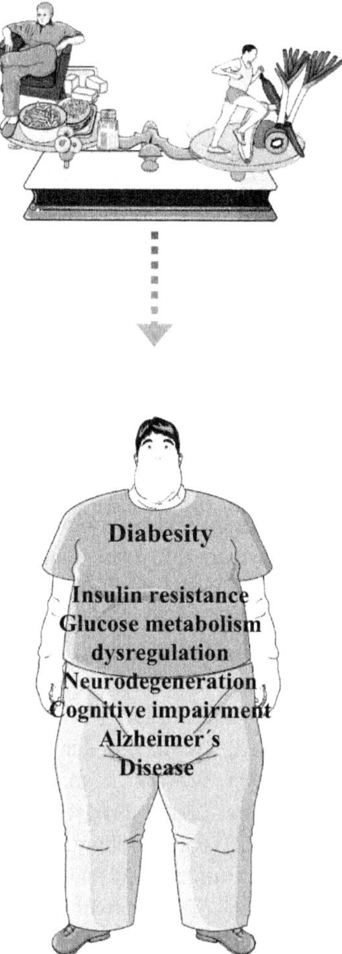

Fig. 5.1 Diabesity and Alzheimer's disease: how our lifestyle contributes to our brain health. With the adoption of a sedentary lifestyle and alteration of diet habits to energy-rich diets, worldwide population is now under a tick-tack clock to become obese subjects. In the shadow of obesity appears type 2 diabetes, a major metabolic disease of our times. The occurrence of both pathologies in the same individual is now designated *diabesity*, which overtime will significantly impact body health compromising quality of life. Particularly, it is now widely accepted that the occurrence of diabesity can evolve to several comorbidity concerns at late stages of life, including a high risk for cognitive decline and dementia, and may lead to a drastic increase in the prevalence of neurodegenerative diseases such as Alzheimer's disease. In this scenario, among the common pathogenic mechanisms that underpin both conditions, insulin resistance, brain glucose hypometabolism, and metabolic dyshomeostasis appear to have a pivotal role. Approaches based on health promotion could therefore help to maintain functional integrity in middle-aged and older adults and hence contribute to successful aging and brain health. Such challenges can be faced through lifestyle changes and pharmacological approaches

overweight in midlife and the risk for developing dementia (Anstey et al. 2011; Hassing et al. 2009; Whitmer et al. 2008), in particular AD (Gustafson et al. 2003). As suggested, elderly persons with higher adiposity show an increased risk for brain atrophy and consequently dementia, and even elderly subjects, who were healthy and confirmed to be cognitively stable for at least 5 years after baseline scanning, were afflicted with obesity-associated brain atrophy in areas targeted by neurodegeneration: hippocampus, frontal lobe, and thalamus (Raji et al. 2010). In close agreement, other study revealed that an increased body mass index (BMI) is associated with specific anterior hippocampal atrophy in the early course of AD pathophysiology (Ho et al. 2011). By using proton magnetic resonance spectroscopy (MRS), studies also showed that higher BMI lowers neuronal viability in several brain regions including frontal, parietal, and temporal lobes (Gazdzinski et al. 2010). Further, every unit increase in BMI was associated with a 0.5–1.5% average brain tissue reduction in mild cognitive impairment (MCI) and AD subjects (after controlling the variables age, sex, and education) (Verstynen et al. 2012; Ho et al. 2010). Overall, all anthropometric measures of obesity, such as body weight, BMI, or waist circumference show a negative association with the cognitive performance of patients (Elias et al. 2012), in particular worse executive function (Reinert et al. 2013; Gunstad et al. 2007). As observed in rodent models of obesity and obese humans, this metabolic disorder is strongly related with reduced memory performance (Jurdak et al. 2008; Popovic et al. 2001), such as delayed recall and recognition (Cournot et al. 2006; Gunstad et al. 2006) along with visual what-where-when episodic memory tasks (Cheke et al. 2016). However, as pinpointed elsewhere, the course of dementia in the elderly population can be often associated with a decrease in body weight due to disease itself, to healthcare and self-care as the disease progresses, etc., and so, there has to be caution in interpreting the studies data to what concerns the age and disease stage of the participating subjects (Pugazhenthi et al. 2016; Qizilbash et al. 2015; Ho et al. 2011; Gustafson et al. 2003).

As evidenced in literature, besides the commonly associated chronic complications such as nephropathy, angiopathy, retinopathy, and peripheral neuropathy, diabetes is also intimately related with alterations in brain function and structure (Chen et al. 2012a, b). In fact, back in 1922, Miles and Root (1922) showed that people with diabetes performed poorly on cognitive tasks examining memory and attention (Biessels et al. 1994; Miles and Root 1922), a condition known to be associated with a significant brain atrophy in temporal lobes and in cortical, subcortical, and hippocampal areas (Korf et al. 2007; Biessels et al. 2002). Using magnetic resonance imaging (MRI) techniques and cognitive tests, Gold et al. (2007) also observed that middle-aged individuals with well-controlled T2D have clear deficits in hippocampal-based memory (recent or declarative) and selective MRI-based atrophy of the hippocampus relative to matched control subjects. As authors suggest, hippocampal damage and memory impairments can be viewed as possible early brain complications of T2D and, as disease progresses, other more resilient brain areas are also affected leading to other cognitive deficits (Gold et al. 2007).

Nowadays, it is known that people with diabetes, especially T2D, are approximately 1.5-fold more likely to experience cognitive decline and 1.6-fold more likely to develop dementia than individuals without diabetes (Scheen 2010). The incidence of other neurological disorders, such as vascular pathology and stroke, also appears to be doubled in T2D individuals compared with nondiabetic subjects (Patrone et al. 2014; Peters et al. 2014). With the aging of the population, T2D and dementia become progressively more common and, since the original Rotterdam study (Ott et al. 1999), epidemiological/clinical observations have accumulated showing that diabetic patients are significantly more likely to develop cognitive deterioration and exhibit increased susceptibility to AD compared to age-matched subjects (Crane et al. 2013; Zhao and Townsend 2009). Moreover, about 80% of AD patients have diabetes or abnormal blood glucose levels (Janson et al. 2004) and animal models of diabetes present cognitive dysfunction and AD-like pathological features in the cortex and hippocampus, including tau protein phosphorylation (Carvalho et al. 2012; Cao et al. 2007; Li et al. 2007). In addition, Hokama et al. (2014) by using a microarray analysis of diabetes-related genes in the brains of postmortem AD patients and in a mouse model of AD observed significant alterations in the mRNA expression profiles of genes related to insulin signaling, obesity, and diabetes in the frontal cortex, temporal cortex, and, in particular, in the hippocampus (Hokama et al. 2014). The same authors noticed that those alterations were independent of peripheral diabetes-related abnormalities. Such observations prompted authors to suggest that altered expression of genes related to diabetes in AD brains is a result of AD pathology, which in turn may be aggravated by peripheral insulin resistance or diabetes (Hokama et al. 2014).

Hence, as a compelling body of research establishes, there is an undoubted certainty that the brain is strongly affected by the increasing prevalence of both obesity and diabetes, contributing to an increased risk for AD. Understanding the role of diabesity in AD is important as the insights gathered could allow to an earlier diagnosis of the disease and to lifestyle and clinical interventions to slow and/or prevent AD (Profenno et al. 2010).

5.3 Diabesity, Brain Energy Metabolism, and Alzheimer's Disease: Is Insulin Resistance the Common Link?

Despite its relative small size, the brain is the organ with the most abundant energy metabolism in the human body. Due to its high glucose demand (primary fuel) and its inability to store glucose, a continuous supply from cerebral blood flow and its transport into the brain through the blood-brain barrier (BBB) to reach neurons and glial cells must be assured. Given that cerebral blood flux (CBF) depends on glucose levels in the blood stream, alterations in circulating glucose levels can negatively affect brain functioning (Neumann et al. 2008). And so, to ensure the proper delivery of glucose to the brain, it is necessary that the processes of glucose transportation and intracellular glucose metabolism are well regulated. While the former is mainly dependent on insulin signaling pathway (Apelt et al. 1999) and levels of glucose transporters (Simpson et al. 1994a), the latter is intimately dependent on mitochondrial function (Chen and Zhong 2013).

In the simplest terms, diabetes is a syndrome of dysfunctional metabolism in which a person has hyperglycemia or abnormal high blood glucose levels, either because the body does not produce enough insulin or because cells do not respond to the insulin that is produced, hence higher than normal insulin concentrations are required to maintain glucose homeostasis reaching a state of insulin resistance (Williamson et al. 2012). In periphery as well as in the brain, insulin resistance is mainly mediated by abnormalities in several steps of the insulin signaling pathway, such as the number of insulin receptors (IR), the binding efficiency of their ligands, or the reduced activation of downstream signaling molecules. These defects in insulin signaling can lead to reduced glucose utilization and energy metabolism, increased oxidative stress, as well as impaired expression and function of insulin-responsive genes required for cognitive-motor functions and plasticity (Gerozissis 2008). That said, due to its intricate link to obesity, T2D, and AD, insulin resistance is now considered a major public health problem.

In the next subsections, it will be made a brief discussion of the role of insulin signaling pathway in the normal brain and later it will be discussed the several nodes of insulin pathway dysregulation as a functional link between diabesity and AD.

5.3.1 A Brief Insight into Insulin Signaling Pathway in the Normal Brain: From Glucose Metabolism to Cognitive Processes

For a long time, insulin-related studies remained limited to peripheral tissues and glucose metabolism (Blazquez et al. 2014). However, the table turned with the demonstration of neuronal insulin synthesis (Schechter et al. 1998; Devaskar et al. 1994), the existence of IR in the brain (Havrankova et al. 1979), and the responsiveness of hippocampal glucose metabolism to the application of exogenous insulin (Hoyer et al. 1996). Nowadays, it is known that IR have a broad distribution within the brain, being highly present in the olfactory bulb, hypothalamus, cerebral cortex, cerebellum, and hippocampus (Cardoso et al. 2009; Havrankova et al. 1978); brain areas are tightly involved in several neurobiological processes, in particular energy homeostasis and cognition. Hence, although the control of peripheral glucose homeostasis is one of the main functions of insulin signaling, its action on the brain cannot go unnoticed.

Like in the periphery, brain insulin and the closely related insulin-like growth factor 1 (IGF-1) initiate a cascade of events by binding to cell surface receptors promoting their activation, and subsequent tyrosine phosphorylation of insulin receptor substrates (IRS) 1 and 2, the signaling of which promote brain cells growth, survival, and energy metabolism by the activation of two canonical pathways, the phosphoinositide-3 kinase (PI3K-PKB)/Akt and the Ras/mitogen-activated kinase (MAPK) pathways (Calvo-Ochoa and Arias 2015; Kleinridders et al. 2014; de la Monte 2009).

Briefly, insulin activation of PI3K results in its relocation from the cytoplasm to the plasma membrane where it phosphorylates the membrane phospholipid phosphatidylinositol 3,4-biphosphate (PIP2) to phosphatidylinositol 3,4,5-triphosphate (PIP3). This signaling molecule then attracts more PH domain-containing proteins to the membrane, altering their activity or localization, and promoting downstream activation of serine/threonine kinases, such as protein kinase B (PKB, also known as Akt). Activated PKB/Akt moves from the plasma membrane to the cytosol and nucleus where it phosphorylates serine/threonine residues in target proteins eliciting a neuronal survival response. In particular, data show that PKB/Akt overexpression protects against apoptosis by inactivating proteins such as the Bcl-2 family member Bad and caspase 9, glycogen synthase kinase-3β (GSK-3β), transcriptions factors from the forkhead box O (FoxO) family, CREB and IκB kinase (IKK) (Lizcano and Alessi 2002).

The second major pathway downstream of the insulin/IR pathway is the Ras/MAPK/ERK pathway. In short, the cytoplasmic intermediate protein (shc) binds to IR promoting its phosphorylation and binding to Grb2, which is associated with son of sevenless (SOS), a guanylnucleotide exchange factor for GTP-binding protein Ras. Binding of Grb2/SOS complex to IR activates Ras that, in turn, recruits Raf leading to MAPK/ERK kinase (MEK) activation. Activated MEK phosphorylates ERK1/2 on its threonine/tyrosine residues that thereby becomes activated modulating memory and learning processes (Selcher et al. 1999; Atkins et al. 1998), as well as long-term potentiation (LTP) (Toyoda et al. 2007) and long-term depression (LTD) (Ito-Ishida et al. 2006).

Overall, the specific brain localization of IR together with the coordinated activation of both pathways favor the notion that insulin/IR influences memory and learning processes and modulates synaptic activities in both pre- and postsynaptic sites (Zhao et al. 2004a, b). Evidence from rodent studies showed that an acute intracerebroventricular injection or an intrahippocampal administration of insulin enhances memory in a passive-avoidance task (Babri et al. 2007; Park et al. 2000), and when intranasally administered to humans insulin is transported into hypothalamus and hippocampus exerting a cognition-enhancing effect independently of changes in peripheral glucose levels (Benedict et al. 2007, 2004). Interestingly, when rats are trained on a spatial memory task, an increase in IR mRNA in the dentate gyrus and hippocampal CA1 area is observed (Zhao et al. 1999). Thereby, IR expression and/or function are also influenced by learning supporting the notion that insulin contributes to normal memory function (Craft and Watson 2004). Compelling data also show that insulin/IR can modulate brain concentration of neurotransmitters associated with important roles in cognition such as acetylcholine, norepinephrine, and dopamine (Figlewicz et al. 1999; Zhao et al. 2004a, b). Insulin also plays a modulatory role on synaptic LTP and LTD by exerting direct electrophysiological effects on central neurons that are highly influenced by gamma-aminobutyric acid (GABA)-inputs (Ma et al. 2003) and by influencing glutamate-mediated

N-methyl D-aspartate (NMDA) receptors activity (Zheng and Quirion 2009; Joseph et al. 2008; Skeberdis et al. 2001).

Besides the abovementioned properties, another function of insulin/IR signaling pathway is related to glucose metabolism/uptake (Grillo et al. 2009). PI3K/Akt activation promotes the translocation of the insulin-dependent glucose transporter 4 (GLUT4) from the endosomal pool into the plasma membrane, leading to an enhancement in glucose transport to the cell and facilitating glucose utilization by neuronal cells (Bondy and Cheng 2004; McEwen and Reagan 2004). However, despite such evidence, there is still the assumption that the majority of glucose utilization in the brain is mediated through non-insulin-dependent transporters such as GLUT1 expressed in endothelial and glial cells and GLUT3 expressed in neurons (Duelli et al. 2001). Therefore, there is some controversy to whether insulin mediates brain glucose metabolism as occurs in periphery. Still, as discussed elsewhere, while these GLUT1 and GLUT3 are responsible for the majority of glucose uptake, they cannot account for total glucose metabolism in the hippocampus, for instance. Thus, additional transport systems localized in neuronal cells may contribute to neuronal glucose uptake and utilization (Grillo et al. 2009). In fact, even though insulin does not seem to affect basal cerebral glucose metabolism or the transport of glucose into the brain (Tomlinson and Gardiner 2008), evidence suggests that insulin has specific effects on glucose metabolism of certain brain areas. For instance, basal insulin levels have the ability to increase cerebral glucose metabolism/uptake, particularly in the cortex (Bingham et al. 2002). Also, animal studies show that hyperinsulinemia does not affect whole-brain glucose utilization but affects glucose metabolism in specific brain regions (Doyle et al. 1995; Lucignani et al. 1987). These effects are mainly due to the distribution and presence of the insulin-sensitive glucose transporters GLUT4 and GLUT8 (also known as GLUT-x1), which overlap with the distribution of insulin and IRs in the brain. As shown, GLUT4 and GLUT8 are localized in neuronal cell bodies in the cortex and cerebellum, but mainly in the hippocampus and amygdala, where they maintain hippocampus-dependent cognitive functions (Craft and Watson 2004). As previously reported, the intracerebroventricular administration of insulin stimulates the translocation of GLUT4 to the plasma membrane in the rat hippocampus in a time- and PI3-kinase-dependent manner (Grillo et al. 2009). That said, by also regulating brain glucose uptake, brain insulin signaling allows neurons to rapidly increase glucose utilization during increases in neuronal activity associated with learning and memory processes (Grillo et al. 2009; Fulop et al. 2003).

In summary, insulin-signaling pathways, through an intrinsic regulation, coordinate themselves to ensure cell survival, energy metabolism, synaptic plasticity, and memory and learning processes.

5.3.2 *Diabesity, (Brain) Insulin Resistance, and Alzheimer's Disease: A Dangerous Triad*

Understanding the role of insulin in the brain has gradually expanded from initial conceptions of the brain as insulin-insensitive to recent demonstration of insulin as a key component of hippocampal memory processes. In this regard, compelling evidence suggests that insulin resistance, hallmark of T2D and obesity, may greatly contribute to age-related cognitive deficits and AD (Craft et al. 2013, Liu et al. 2011). And this is of particular interest since the underlying neuropathology may persist for decades before the clinical manifestation of the disease (Pugazhenthi et al. 2016).

Clinically, insulin resistance may be manifested by glucose intolerance and a declining insulin sensitivity for years before the diagnosis of diabetes, due to the effort of the endocrine pancreas to increase insulin secretion to maintain normal glucose levels (Williamson et al. 2012; Shulman 2000). Furthermore, in addition to reduced transport of insulin to the brain, peripheral insulin resistance may induce a state of central insulin resistance (Hildreth et al. 2012). Such notion was confirmed by the observation that in obese subjects, reductions in stimulated and spontaneous cerebrocortical activity are directly correlated with the degree of peripheral insulin resistance during hyperinsulinemic–euglycemic clamp studies (Bingham et al. 2002). Besides, compared with insulin-sensitive subjects, obese insulin-resistant individuals revealed significantly smaller increases in whole brain and regional glucose metabolism in response to insulin infusion (Bingham et al. 2002), thus pointing to a real connection between peripheral and central insulin resistance and suggesting that peripheral insulin sensitivity may impact the brain (Hildreth et al. 2012). Actually, as recently underscored, healthy middle-aged and elderly subjects show a negative correlation between HOMA-IR score (a measure of peripheral insulin resistance) and verbal fluency performance, total brain size and regional gray matter volume in bilateral areas of the middle and superior temporal gyri, i.e., typical speech-processing areas (Benedict et al. 2012; Willette et al. 2013). Similarly, a study performed in middle-aged subjects shows an inverse correlation of peripheral insulin resistance with total cerebral volume, executive functions, and verbal and visuospatial memory (Tan et al. 2011).

Insulin transport to the brain is reduced in aging and in some animal models of T2D (Messier and Teutenberg 2005), whereas insulin resistance could promote reduction in both insulin uptake and capacity of the hormone to stimulate its receptors in the brain (Messier and Teutenberg 2005). In fact, experimental evidence from animal models of T2D show impaired hippocampal translocation of GLUT4 (Reagan 2005; Winocur et al. 2005), reduced hippocampal synaptic plasticity (Mielke et al. 2005), and reduced temporal lobe insulin signaling (Moroz et al. 2008). To further elucidate the role of brain insulin system dysfunction in the neurodegenerative events that occur in sporadic AD, the intracerebroventricular (icv) administration of streptozotocin (STZ) in rats emerged as a suitable experimental approach for studying sporadic AD (Correia et al. 2013). Mounting evidence validates that single or multiple injections of low doses of the diabetogenic drug STZ, either uni- or bilaterally into the lateral cerebral ventricles produce(s) neurochemical and brain glucose metabolism changes, alongside with long-term and progressive deficits in learning, memory, and cognitive behavior, which resemble features of AD patients (Grunblatt et al. 2007; Salkovic-Petrisic and Hoyer 2007). Besides, others have shown that this icvSTZ model is also characterized by a significant decrease in IRs expression in cortex and hippocampus, insulin-1 mRNA in hippocampus, insulin-2 mRNA in cortex and a significant increase of tau phosphorylation in hippocampus, those alterations being associated with impaired memory and learning (Grunblatt et al. 2007). In close agreement, time-dependent changes in the phosphorylated and total GSK-3α/β were found in rat brain after icvSTZ administration, which have been suggested to be related to the formation of Aβ peptide-like aggregates in brain capillaries (Salkovic-Petrisic et al. 2006). Likewise, de la Monte et al. (2006) also reported an increase of GSK-3β, phosphorylated tau, ubiquitin, APP, and Aβ and decreased levels of tau protein in icvSTZ-treated animals, all of which are features of AD brains.

Likewise, studies performed in an animal model of obesity revealed that the ingestion of high-fat diet (HFD) for several weeks induces a systemic insulin resistance and a consequent dysregulation of brain insulin signaling pathway, including a decrease in the expression and plasma membrane localization of GLUT3/GLUT4, culminating in a substantial decrease in long-term potentiation in the CA1 region of the hippocampus (indicative of impaired synaptic plasticity) (Liu et al. 2015). Other studies confirm that obesity-induced brain insulin resistance is deeply correlated with deleterious effects on synaptic integrity and cognitive behaviors (Arnold et al. 2014; McNay et al. 2010). More recently, HFD-induced alterations in peripheral insulin sensitivity was also associated with a central insulin resistance and biochemical changes related to increased Aβ deposition and neurofibrillary tangle formation, as well as a decreased synaptic plasticity contributing to a decline in cognitive function (Kothari et al. 2017). Similarly, the administration of a HFD to an animal model of AD, the Tg2576 mice, was shown to evoke a state of obesity and insulin resistance as well as a burst in Aβ formation in the brain (Kohjima et al. 2010). In close agreement, a previous study also found that diet-induced insulin resistance in Tg2576 mice is associated with reduced neuronal insulin receptor signaling, impaired performance in a spatial water maze, and increased Aβ levels in the brain resulting from enhanced γ-secretase activity and reduced insulin degrading enzyme (IDE) activity (Ho et al. 2004). Similarly, others showed that deficient insulin signaling is correlated with reduced IDE in AD brains and in Tg2576 Swedish APP transgenic mice (Zhao et al. 2004a, b).

As outlined earlier, insulin signaling-mediated neuroprotection is closely linked with the activation of PKB/Akt as its overexpression in PC12 cells confers protection against Aβ induced cell death (Martin et al. 2001) and, conversely, the intracellular Aβ expression inhibits both insulin-induced Akt-phosphorylation and activity (Lee et al. 2009). Moreover, PKB/Akt signaling induces the phosphorylation and inhibition of GSK-3β, a serine/threonine protein kinase ubiquitously expressed throughout the body and that has as substrate the protein tau (Cole et al. 2007). In

T2D and AD brains GSK-3β expression and activity is deregulated and, consequently, tau phosphorylation is increased (Freude et al. 2005; Leroy et al. 2002). Contrariwise, GSK-3β activity can be downregulated in response to insulin or IGF-1 through the activation of the PI3K/Akt pathway and consequent activation of neuronal survival pathways (Duarte et al. 2005, 2008). Impaired insulin signaling has so been proposed as one of the upstream impairments that may lead to tau hyperphosphorylation (Maccioni et al. 2010).

As postulated, there seems to be a strong correlation between diabetes/insulin resistance and the risk of AD in subjects carrying ApoE4, as patients with diabetes who carry ApoE4 are twofold more likely to develop AD than nondiabetic ApoE4 carriers (Peila et al. 2002), a situation that appears to involve IDE, a metalloprotease enzyme suggested to be a key player contributing to insulin signaling dysfunction and accumulation of Aβ (Mittal et al. 2016; Hoyer 2004). As mentioned, Aβ is in part cleared by IDE whereas AD patients carrying ApoE4 have been reported to express reduced IDE protein and mRNA levels in the hippocampus (Biessels et al. 2005; Schipper 2011) suggesting a causal link between impaired insulin metabolism or insulin resistance, and the pathogenesis of AD (Schipper 2011). Also, Segev et al. (2016) recently demonstrated that ApoE4 mouse model of AD fed with HFD for several weeks develop a diabetes-like metabolism, as well as changes in β-site amyloid precursor protein-cleaving enzyme 1 (BACE1) protein levels. The authors suggest that environmental factors (i.e., diet) may converge with genetic factors in the onset and progression of disease symptoms (Segev et al. 2016). However, this is a controversial theme as others have reported the association between insulin resistance or diabetes and the risk of AD, albeit independent of ApoE (Blazquez et al. 2014; de la Monte 2014; Kuusisto et al. 1997).

Accordingly to literature, insulin resistance entails disruption of metabolic homeostasis, largely due to mitochondrial dysfunction; this further impairs cellular function at multiple levels with a broad range of consequences, from increased oxidative stress, DNA damage, to several forms of cell death (Yin et al. 2014). As demonstrated, deficits in insulin/IGF signaling and energy production strongly correlate with mitochondrial dysfunction, oxidative injury, and compensatory cytoprotective responses in brains with different Braak stage severities of AD (de la Monte et al. 2006). More recently, with the objective of dissecting the molecular mechanisms shared by obesity and AD, Nuzzo et al. (2015) analyzed the effect of HFD intervention on brain metabolism, redox balance, and mitochondrial homeostasis. Results obtained revealed that brains of HFD-fed mice show not only markers of insulin resistance but also elevated levels of APP and Aβ40/Aβ42 together with BACE, GSK3β, and tau proteins involved in APP processing and Aβ accumulation. Of note, authors found that those alterations were accompanied by markers of increased oxidative stress and mitochondrial dysfunction and dynamics (Nuzzo et al. 2015). Moreover, concomitant studies have demonstrated that obesity and excess energy intake shift the balance of mitochondrial dynamics, contributing to mitochondrial dysfunction and metabolic deterioration, all these alterations leading to insulin resistance (Jheng et al. 2012). In fact, Santos et al. (2014) recently found that during the early stages of T2D, brain mitochondrial function is spared through an adaptive metabolic strategy between mitochondrial fusion–fission and biogenesis and autophagy. Particularly, the study revealed that in the early stages of T2D, mitochondrial fission prevails, mitochondrial biogenesis is maintained, and autophagy decreases. According to authors, mitochondrial fission could be involved in the recruitment and transport of mitochondria to critical subcellular compartments with high energy demand, such as synaptic terminals, where these organelles remain stationary and preserve synaptic and neuronal function and integrity, in part by supplying ATP (Santos et al. 2014). In this regard, a frequently observed characteristic in tissues of diabetic patients and animals is mitochondrial deformation, most notably mitochondrial swelling or accumulation of small mitochondria (Vincent et al. 2010; Kelley et al. 2002), probably due to an imbalanced mitochondrial fusion and fission processes in favor of excessive mitochondrial fission which, as described, can occur via a GSK3β/dynamin-related protein (DRP)1-dependent mechanism (Yan et al. 2015). Of note, mitochondrial deficits that occur in T2D animals are potentiated in a presence of an additional stress, i.e., the presence of Aβ (Moreira et al. 2003). In this line, Carvalho et al. (2012) nicely demonstrated that the metabolic alterations associated to diabetic or prediabetic conditions induce mitochondrial abnormalities, oxidative imbalance, and an increase in Aβ protein levels, all of which closely resemble what happens in AD brains. Such findings prompted authors to suggest that the metabolic alterations associated to diabetes contribute to the development of AD-like pathologic features and that mitochondria lingers in this scenario as a functional link between both pathologies (Carvalho et al. 2012). These observations were recently corroborated by the work of Petrov et al. (2015), in which authors demonstrate that the metabolic alterations induced by HFD have direct effects on brain insulin regulation and mitochondrial function contributing to AD pathology, with mitochondria as a key culprit leading to cognitive decline in both the HFD-treated and AD-like rodents at a relatively young age (Petrov et al. 2015).

To sum up, as highlighted, the connection between peripheral and central insulin resistance and associated factors such as mitochondrial (dys)function suggest that preventing the cellular and metabolic dysfunctions common to both diabesity and AD will not only attenuate pathological events in peripheral organs, but will also benefit the brain.

5.3.3 Alzheimer's Disease as a (Brain) Metabolic Disorder

Growing evidence supports the concept that AD fundamentally represents a metabolic disease in which brain glucose utilization and energy production are impaired, evoking a state of cerebral glucose hypometabolism (Chen and Zhong 2013). Actually, evidence establishes that in AD progressive decline in cerebral glucose utilization and deficits in insulin signaling and insulin-responsive gene expression underlie and worsen severity of disease (Chen and Zhong 2013; Correia et al. 2012).

In this context, AD patients present decreased GLUT1 and GLUT3 expression especially in the cerebral cortex and in the dentate gyrus of the hippocampus (Simpson et al. 1994b), whereas insulin signaling-mediated action seems to be compromised during AD development. For instance, AD patients show impaired brain insulin signaling transduction with reduced tyrosine kinase activity of the IR (Talbot et al. 2012; Frolich et al. 1998) and a decrease in mRNA and protein levels of insulin, IGF1, their receptors, and downstream signaling elements such as the IRS1 (Talbot et al. 2012; Moloney et al. 2010; Rivera et al. 2005; Frolich et al. 1999), these defects being correlated with the magnitude of cognitive impairments (Talbot et al. 2012) and appear to deteriorate with progression of the disease (Rivera et al. 2005). As confirmed, one of the pathophysiological features of AD is a consistent reduction in the regional cerebral glucose utilization, which may precede cognitive dysfunction and pathological alterations for decades (Andersen et al. 2016; Zhang et al. 2016; Cunnane et al. 2011; Fukuyama et al. 1994). The reduction in cerebral glucose metabolism is attributable not only to an insufficient supply of glucose to the brain, but also to diminished glucose breakdown in brain tissue (Hoyer 1998, 1991) caused by a disturbance in the control of glucose utilization at the level of the insulin signaling pathway (Hoyer 2004). Indeed, as shown in the early stages of AD, cerebral glucose utilization is reduced by 45%, and CBF by approximately 20% whereas in the later stages of the disease metabolic and physiological abnormalities aggravate, resulting in 55–65% reductions in CBF (Hoyer and Nitsch 1989). Accordingly, recent data by Baker et al. (2011) show that cognitively intact adults with pre-diabetes/T2D present a strong association between peripheral insulin resistance and disturbances in cerebral glucose metabolic rate in frontal, temporal-parietal, and cingulate regions, brain areas which are known to be affected in AD. Considering that reductions in regional cerebral glucose metabolic rate, as measured by fluorodeoxyglucose positron emission tomography (FDG-PET), are intimately associated with increased AD risk and can be observed years before dementia onset (Mistur et al. 2009; Minoshima et al. 1997), it has been suggested that peripheral insulin resistance may be a marker of AD risk that is associated with reduced cerebral glucose metabolism and subtle cognitive impairments at the earliest stage of disease, even before the onset of MCI (Baker et al. 2011). Accordingly, a recent study demonstrated similar associations in asymptomatic late middle-aged participants presenting amyloid deposition and in stable MCI participants enriched for amyloid status (Willette et al. 2015a, b). The authors observed that higher HOMA-IR predicted less prefrontal glucose metabolism only in the amyloid-positive patients (Willette et al. 2015b). Importantly, the HOMA-IR score of patients at risk for AD was also negatively associated with right and total hippocampal volume as well as overall cognitive performance, and verbal and nonverbal memory tests (Rasgon et al. 2011). Thus, insulin resistance not only disrupts brain glucose metabolism in cognitive healthy and impaired subjects, but also promotes significant alterations in total brain and hippocampal volume further compromising cognitive functions (Freiherr et al. 2013). So, currently, low glucose metabolism at baseline and longitudinal glucose metabolism decline are viewed as sensitive measures to monitor changes in cognition and functionality in AD and

MCI, and are being increasingly adopted to assist diagnosis and used to predict future cognitive decline (Atamna and Frey 2007).

The brain utilizes a vast amount of energy to sustain its basic functions (e.g., maintaining or re-establishing membrane potentials, signaling, and other essential cellular activities), relying mostly on ATP production by mitochondria (Chen and Zhong 2013). Current data indicate that mitochondria, the primary metabolic platform, are affected during insulin resistance (Kleinridders et al. 2014; Cheng et al. 2010). In fact, using the icvSTZ rat, as an animal model of sporadic AD, Correia et al. (2013) nicely demonstrated that the insulin-resistant brain state that characterizes icvSTZ rats is associated with mitochondrial abnormalities and oxidative status, considered early events in AD. IcvSTZ rats presented a decline in mitochondrial bioenergetics function, a decrease in the activity of the three mitochondrial enzymes, pyruvate dehydrogenase (PDH) and α-ketoglutarate dehydrogenase (α-KGDH)—two enzymes in the rate-limiting step of the tricarboxylic acid (TCA) cycle, and cytochrome oxidase (COX)—the terminal enzyme in the mitochondrial respiratory chain that is responsible for reducing molecular oxygen, and an increase in mitochondrial oxidative stress and damage (Correia et al. 2013). Those alterations were associated with a marked cognitive impairment and an increase in the two neuropathological markers of AD (Correia et al. 2013). Those results encouraged authors to speculate that the "mitoenergetic failure" induced by icvSTZ is intimately associated with central insulin resistance (Correia et al. 2013). In fact, it is of general opinion that the decrease in the cerebral glucose metabolism previously documented in AD brains is tightly correlated with the altered expression and decreased activity of several key mitochondrial energy-related proteins, including PDH, isocitrate dehydrogenase, and α-KGDH, which can occur prior to the onset of memory deficits and the appearance of the two histopathological culprits of the disease (Manczak and Reddy 2012, Manczak et al. 2004; Bubber et al. 2005; Aksenov et al. 1999). Further, as reported by Bubber and collaborators (2005), all the changes in TCA cycle enzymatic activities (specifically that of PDH complex) are positively correlated with the degree of clinical disability in AD, suggesting a coordinated mitochondrial alteration. As shown, the reduced activity of PDH promotes a decrease in acetyl coenzyme A levels, which in turn reduces the acetylcholine synthesis (Sims et al. 1983). Indeed, alterations in cholinergic neurons together with disturbances in the serotoninergic, noradrenergic, and dopaminergic systems have been reported to be correlated with the progression of mental impairment in AD patients (Wang et al. 2007; Baskin et al. 1999). Nonetheless, a major consequence of altered brain glucose metabolism is a decrease in glucose-derived ATP production by around 50% in the beginning of AD, further compromising the ATP-dependent processes crucial for the normal cell functioning (Moreira et al. 2007; Mattson et al. 2001).

In further support of such findings, it was recently proposed that decreased expression and function of PI3K/Akt-mediated GLUTs in AD brain could also contribute to brain glucose hypometabolism and the subsequent decline in mitochondrial ATP production (Bosco et al. 2011). Previous studies revealed that a decreased brain glucose metabolism is strongly linked with a decreased O-GlcNAcylation, a recently recognized posttranslational modification of numerous cytoplasmic and nuclear proteins including tau protein (Liu et al. 2009; Gong et al. 2006). That said, it has been demonstrated that tau phosphorylation is inversely regulated by O-GlcNAcylation and that decreased O-GlcNAcylation induces hyperphosphorylation of tau (Gatta et al. 2016; Liu et al. 2004) suggesting that impaired brain glucose metabolism leads to abnormal phosphorylation of tau and neurofibrillary degeneration via downregulation of tau O-GlcNAcylation (Liu et al. 2009). In effect, the levels and the activation of the insulin-PI3K-Akt signaling components showed to be negatively correlated with the level of tau phosphorylation and positively correlated with tau O-GlcNAcylation suggesting that impaired insulin-PI3KAkt signaling might contribute to neurodegeneration in AD through decreased O-GlcNAcylation and consequent tau hyperphosphorylation (Liu et al. 2011).

Overall, compelling evidence clearly states that impaired glucose metabolism and mitochondrial dysfunction are mechanistically involved with the brain insulin-resistant state that characterizes AD, hence suggesting that regulation of glucose/energy metabolism is a critical checkpoint for brain function.

5.4 Challenges and Strategies for Prevention and Control: Is There a Window of Opportunity to Revert Diabesity and/or Alzheimer's Disease-Associated Brain Metabolic Alterations?

Mounting evidence shows that the accumulation of diabesity-associated deleterious effects over time ultimately causes detrimental effects in the central nervous system, and even mild forms of cognitive dysfunction can affect daily living activities (Pugazhenthi et al. 2016). Together with the aging of population, diabesity increases the risk of cognitive impairment and dementia (Katsiardanis et al. 2013). So, the maintenance of brain function and the reduction of risk of neurological disorders have become key issues for the society (Smith 2016). Importantly, cognitive deficits associated with metabolic dysregulation are amenable to change or even reversible by specific interventions (Hendrickx et al. 2005).

Next, current therapeutic approaches directed against insulin resistance, a hallmark of diabesity and AD, and strategies focused on glycemic control and weight loss will be discussed.

5.4.1 Drugs Targeting Insulin Resistance: From PPARs to Incretins

Considering the common pathogenic mechanisms between diabesity and AD, attention has been drawn to the possibility that therapeutics for one disease can be effective for the other (Sebastiao et al. 2014; Zhong and Weisgraber 2009). Thus, correcting insulin resistance and repairing insulin signaling dysfunction would not only target diabesity-mediated effects but also be potentially beneficial for AD. One of the most popular targets is peroxisome proliferator-activated receptors (PPARs), which belong to the steroid hormone superfamily ligand-inducible transcription factors. Mostly known as antidiabetic drugs, thiazolidinediones (TZDs; one class of PPAR-γ agonists) are described to promote an enhancement in insulin sensitivity, improve mitochondrial function, modulate glucose metabolism, and reduce inflammatory responses (Feinstein 2003). One of the most frequently used TZD is rosiglitazone. In a small-scale clinical trial involving 30 patients with mild AD or MCI, rosiglitazone was showed to improve subjects' memory and selective attention while preserving performance on delayed recall and attention tasks compared with placebo group (Watson et al. 2005). In turn, even though a larger clinical trial involving 500 patients with mild to moderate AD revealed that rosiglitazone treatment resulted in a significant improvement in cognition, this only occurred in patients without ApoEε4 whereas patients with ApoEε4 showed no alterations in the cognitive tests (Risner et al. 2006). However, in a recent phase III clinical trial rosiglitazone did not promote any effect on objective cognitive performance in AD patients (Gold et al. 2010). In in vitro studies, rosiglitazone demonstrated to potentiate the ability of insulin to protect synapses against Aβ derived diffusible ligands (ADDLs)-induced IR loss (De Felice et al. 2009). Other studies revealed that PPAR-γ activation protects rat hippocampal neurons against Aβ toxicity (Inestrosa et al. 2005; Combs et al. 2000), induces upregulation of Bcl-2 pathway, protects mitochondrial function, and prevents neuronal degeneration induced by Aβ exposure and oxidative stress (Fuenzalida et al. 2007). Indeed, rosiglitazone beneficial effects in memory and cognition seem to be mediated by the improvement of mitochondrial function (Landreth et al. 2008), since it leads to an increase in mitochondria number and metabolic efficiency (Kummer and Heneka 2008). However, despite some promising results, TZDs treatment was soon associated with several secondary complications and its prescription was suspended (Sebastiao et al. 2014).

One of the difficulties often encountered by obese diabetic patients in having an appropriate medical management is the fact that most of the conventional antidiabetic medications used in the diabetes control are associated with weight gain (Deol et al. 2017). This leaves obese diabetic patients with the option of bariatric surgery, which in a significant percentage of cases results in marked weight loss and diabetes lessening (Schauer et al. 2014). However, this choice has some drawbacks, is an invasive and expensive procedure, often associated with short-term and long-term adverse consequences (Paulus et al. 2015). To overcome those difficulties, nowadays a new class of antidiabetic agents, the incretins, are usually prescribed by physicians in the management of diabesity.

Among those antidiabetic agents, the ones preferred are glucagon-like peptide-1 (GLP-1) agonists and analogues, sodium-glucose co-transporter type-2 (SGLT-2) inhibitors and dipeptidyl peptidase-4 (DPP-4) inhibitors, mainly due to their weight loss (GLP-1 agonists and analogues and SGLT-2) or weight neutral (DPP-4 inhibitors) effects. As observed, the combination therapy with GLP-1 analogues and SGLT-2 inhibitors revealed promising effects in promoting a modest level of weight loss and improvement in glycated hemoglobin (HbA1c) (Deol et al. 2017; Saroka et al. 2015). Likewise, exendin-4 (a GLP-1R agonist) is currently used in clinical context as an important complement to diet and exercise in the improvement of glycemic control in T2D adults (Campbell 2011), either as monotherapy or in combination with other oral anti-T2D agents (Chen et al. 2012a, b). In particular, Chen and colleagues (2012a, b) observed that exendin-4 has in vitro neuroprotective effects against diabetes-associated glucose metabolic dysregulation via the PI3-kinase pathway, as well as a downregulatory effect in icvSTZ-induced tau hyperphosphorylation through downregulation of GSK-3β activity, a key kinase in both diabetes and AD. Related studies showed that liraglutide (a GLP-1 analogue) ameliorates aberrant IR localization and signaling in parallel with decreasing both Aβ plaque and glial pathology in a mouse model of AD (Long-Smith et al. 2013).

In a similar way, DPP-4 inhibitors were shown to delay the development of AD neuropathological hallmarks in an adult, double transgenic mouse model of AD (D'Amico et al. 2010), and to improve learning behavior in adult, insulin-resistant rats (Pintana et al. 2013; D'Amico et al. 2010). More specifically, chronic administration of sitagliptin (a DPP-4 inhibitor) reduced both hippocampal APP and Aβ deposition in transgenic AD mice (D'Amico et al. 2010), whereas in HFD-induced insulin-resistant rats, it promoted a decrease in plasma insulin, cholesterol, and high-density lipoproteins (HDL) levels, and ameliorated HOMA values (Pintana et al. 2013). Alongside with those effects, DPP-4 inhibitors completely prevented brain and hippocampal mitochondrial dysfunction while improving the learning behaviors impaired by the HFD (Pintana et al. 2013).

Noteworthy, many studies are now starting to demonstrate the intranasal approach as a preferable route of administration as it can directly deliver the drugs to the brain (Dhuria et al. 2010; Hanson and Frey 2008) in a noninvasive and safe way. For instance, it has been demonstrated that intranasal insulin is able to improve memory in normal adults without side effects (Ott et al. 2012), to prevent cognitive decline, cerebral atrophy, Aβ accumulation, and white matter lesions in type 1 diabetic animals (Subramanian and John 2012; Francis et al. 2008), and to reduce tau hyperphosphorylation in a rat model of T2D (Yang et al. 2013). Similarly, intranasal GLP-1 administration improves the glycemic control in T2D patients without any adverse effects (Ueno et al. 2014), thus demonstrating the potential of intranasal approach in the treatment and prevention of metabolic diseases (Freiherr et al. 2013).

Hence, given their impressive body of positive outcomes in both T2D/obesity and AD and their minimal risk of damage associated with hypoglycemia and weight gain, incretins represent potential therapeutics for the treatment of diabesity-associated neurodegeneration and AD.

5.4.2 Lifestyle Alterations: "Move More and Eat Well"

Despite all the indications reporting the beneficial effects of exercise, numbers show that in 2012 one in every three adults worldwide was inactive and it was proposed that this endemic inactivity starts early in life (Hallal et al. 2012). And this gains an extra importance when considering that physical inactivity is a primary causal mechanism of every risk factor for T2D and/or obesity (Booth et al. 2012).

For many years, exercise has been an integral component of T2D and obesity management. As reported, it seems that nondrug interventions (diet or exercise) prevent more T2D than drug approaches (Hopper et al. 2011). Exercise training reduces insulin resistance and improves glucose intolerance in obese persons (Kelley and Goodpaster 1999), and also promotes a reduction in HbA1c levels, improves glycemia, and reduces visceral adipose tissue and plasma triglycerides in T2D subjects (Boule et al. 2001; Thomas et al. 2006).

It is currently recognized that the beneficial effects of exercise can go beyond its direct effects in physical indexes and exert important benefits for both affective experience and cognitive performance regardless of age (Hogan et al. 2013). Indeed, evidence suggests that in humans there is a link between physical fitness and cognitive performance (Chodzko-Zajko and Moore 1994). In fact, physical inactivity and a sedentary lifestyle are currently considered significant risk factors to develop dementia and AD (Radak et al. 2010; Laurin et al. 2001). In this picture, epidemiological studies suggest that simple lifestyle changes may be sufficient to experience better brain health during aging and to slow the onset and progression of AD (Pope et al. 2003). Additionally, as projection models show, dementia in old age could be lowered by 10% in 2050 if midlife obesity was decreased by 20% (Nepal et al. 2014). In accordance, over the past years, retrospective and prospective epidemiological studies documented brain health benefits of exercise in relation to the development of AD and dementia of any type (Palleschi et al. 1996). In close agreement with those studies, data obtained from several animal models of AD show that physical exercise can be considered as a simple behavioral intervention sufficient to inhibit the development of AD-like neuropathology and to improve cognitive behavior (Garcia-Mesa et al. 2011; Cho et al. 2010; Um et al. 2008). As recently verified, an acute bout of exercise in HFD-fed mice is enough to promote the normalization of the observed alterations in Akt/insulin-signaling in the brain alongside with a reduction in BACE1 content and activity independent of changes in adiposity (MacPherson et al. 2015). These results highlight the therapeutic potential of exercise, independent of alterations in body mass or adiposity, as a tool to ameliorate HFD-induced markers of energetic stress in brain and early AD-associated pathology (MacPherson et al. 2015).

Of note, at least in AD mouse models, it has to be taken into account the age of the animal and thus the level of pathology when selecting an appropriate exercise regime, as the maximal effects on AD pathology are observed when exercise is initiated prior to the appearance of Aβ plaques or at an early-mild stage of plaque deposition (Ryan and Kelly 2016). On the other hand, when it comes to human studies, data suggest that the window of opportunity for physical activity interventions to prevent dementia may extend from midlife to older ages as staying physically active, or becoming more active, after midlife still contribute to lowering dementia risk, especially in people who are overweight or obese at midlife (Tolppanen et al. 2015).

Even though the beneficial properties of physical activity and exercise on human health have been extensively reported in literature, the exact mechanism(s) behind this exercise-induced protective phenotype in the brain is still not well understood.

Nevertheless, it is consensual that among the myriad of cellular and subcellular adaptations that exercise can evoke, one of the most important relies on the modulation of mitochondrial network (Marques-Aleixo et al. 2012). As reported, exercise induces important brain mitochondrial adaptations in order to sustain increased metabolic demands (Dietrich et al. 2008). Among those, one could find an increased content and/or activity of several enzymes involved in aerobic energy production (Dietrich et al. 2008; Kirchner et al. 2008), increased activity of mitochondrial complexes I, III, and IV (Navarro et al. 2004), decreased expression/activation of several pro-apoptotic proteins (Um et al. 2008), increased mitochondrial biogenesis (Steiner et al. 2011) and antioxidant capacity (Camiletti-Moiron et al. 2013), as well as alterations in proteins involved in mitochondrial dynamics, apoptosis, and autophagic signaling (Marques-Aleixo et al. 2015). Previous data also confirm that endurance training attenuates neuronal cell apoptosis involved in the pathogenesis of AD by promoting reductions in brain cytochrome c, Bax, and caspases 3 and 9 levels (Cho et al. 2010; Um et al. 2008). Other important mitochondrial bioenergetics adaptations associated with voluntary exercise concerns an increase in the mitochondrial uncoupling protein 2 (UCP2) gene expression (Dietrich et al. 2008). Previously reported as exerting an important protection against Aβ toxicity and oxidative stress (Jun et al. 2015), UCPs, mainly UCP2, are often regarded as an effective strategy in the regulation of mitochondrial biogenesis by decreasing reactive oxygen species (ROS) overproduction, increasing ATP generation, and improving calcium homeostasis (Andrews et al. 2005).

So, exercise not only improves whole body energy metabolism but also restores brain glucose homeostasis. In this context, it represents an attractive strategy to reduce or reverse the diabesity-induced metabolic disturbances associated with neurodegeneration.

It is common sense to assume that one of the greatest factors contributing to the prevalence of obesity is the choice of diet, which presently contains large amounts of red meat, refined sugars, high fat foods, and refined grains in detriment of healthier aliments like fruits, vegetables, lean protein, and fiber (Fung et al. 2001). A choice that has as a consequence on ever-increasing rates of obesity, diabetes, and dementia (Winocur and Greenwood 2005; Kalmijn et al. 1997), thus stressing the importance of nutrition-related effects on brain function (Freeman et al. 2014). Within this context, defined as a moderate reduction in calorie intake of 20–40% in the absence of malnutrition, calorie restriction (CR) or dietary restriction (DR) is currently considered an experimental manipulation able to preserve metabolism during aging and extend the lifespan of a broad range of species, spanning from yeast to rodents and nonhuman primates (Amigo and Kowaltowski 2014; Srivastava and Haigis 2011; Ingram et al. 2006). An achievement attained through an improvement of mitochondrial function associated with the prevention of oxidative damage and mitochondrial ROS production, improvement of metabolic parameters, increased mitochondrial biogenesis through activation of the sirtuin1–peroxisome proliferator-activated receptor-gamma coactivator 1 alpha (PGC1α) pathway, and resistance to cellular stress (Desquiret-Dumas et al. 2013; Manzanero et al. 2011). As described by Lopez-Lluch et al. (2006), mitochondria under CR conditions show less oxygen consumption, reduced membrane potential, and generate less ROS than controls, but remarkably are able to maintain their critical ATP production. For instance, Yao et al. (2011), using the 3xTgAD mice fed with 2-deoxy-D-glucose (2-DG; a caloric restriction mimetic), observed that this dietary intervention was responsible for an improvement in brain mitochondrial bioenergetics paralleled with a reduction in oxidative stress. Also, dietary 2-DG promoted a reduction in Aβ generation and increased mechanisms of Aβ clearance, further suggesting DR as a disease-modifying intervention to delay progression of bioenergetics deficits in brain and associated amyloid burden (Yao et al. 2011). Moreover, others verified that CR exerts protection against HFD-induced cognitive decline via upregulation of brain-derived neurotrophic factor (BDNF)/ tropomyosin receptor kinase B (TrkB) through an antioxidant effect in the hippocampus of dietary-induced obese rats (Kishi et al. 2015). In a similar way, Yilmaz et al. (2011) reported positive effects of CR in modulating hippocampal NMDA receptors and preventing oxidative stress in an obese rat model.

So, research is increasingly focusing on understanding how interventions, such as improving nutritional status and modifying risk factors that may impact directly and/or indirectly brain functioning, can reduce the risk of neurocognitive impairment (Parrott and Greenwood 2007). As mentioned, measures that support or sustain insulin sensitivity, including exercise, caloric restriction, and loss of excess weight, will be beneficial in both diabesity and AD-related brain effects (de la Monte 2012).

5.5 Conclusions

There is no doubt that we are facing an unprecedented diabetogenic era. With the adoption of a Western diet and a sedentary life the number of obese diabetic people keeps risen every day. Hand in hand with the increase in those "metabolic plagues" is the increase in longevity and age-related neurodegenerative diseases such as AD. Gathering evidence has shown common pathogenic factors functioning in both conditions, namely altered brain energy and glucose metabolism and insulin resistance. Thereby, preventive and therapeutic agents and/or strategies focused on improving metabolic health could therefore help to maintain functional integrity in middle-aged and older adults and hence contribute to successful aging and brain health. Such challenges can be faced using lifestyle changes and pharmacological agents.

Acknowledgments The authors' work is supported by FEDER funds through the Operational Programme Competitiveness Factors—COMPETE and national funds by FCT—Foundation for Science and Technology under the project (PEst-C/SAU/LA0001/2013-2014) and strategic project UID/NEU/04539/2013. Susana Cardoso is recipient of a PostDoc fellowship from the Foundation for Science and Technology (FCT) (SFRH/BPD/95770/2013).

References

Aksenov MY, Tucker HM, Nair P, Aksenova MV, Butterfield DA, Estus S, Markesbery WR (1999) The expression of several mitochondrial and nuclear genes encoding the subunits of electron transport chain enzyme complexes, cytochrome c oxidase, and NADH dehydrogenase, in different brain regions in Alzheimer's disease. Neurochem Res 24(6):767–774

Amigo I, Kowaltowski AJ (2014) Dietary restriction in cerebral bioenergetics and redox state. Redox Biol 2:296–304

Andersen JV, Christensen SK, Aldana BI, Nissen JD, Tanila H, Waagepetersen HS (2016) Alterations in cerebral cortical glucose and glutamine metabolism precedes amyloid plaques in the APPswe/PSEN1dE9 Mouse Model of Alzheimer's Disease. Neurochem Res 42(6):1589–1598

Andrews ZB, Diano S, Horvath TL (2005) Mitochondrial uncoupling proteins in the CNS: in support of function and survival. Nat Rev Neurosci 6(11):829–840

Anstey KJ, Cherbuin N, Budge M, Young J (2011) Body mass index in midlife and late-life as a risk factor for dementia: a meta-analysis of prospective studies. Obes Rev 12(5):e426–e437

Apelt J, Mehlhorn G, Schliebs R (1999) Insulin-sensitive GLUT4 glucose transporters are colocalized with GLUT3-expressing cells and demonstrate a chemically distinct neuron-specific localization in rat brain. J Neurosci Res 57(5):693–705

Arnold SE, Lucki I, Brookshire BR, Carlson GC, Browne CA, Kazi H, Bang S, Choi BR, Chen Y, McMullen MF, Kim SF (2014) High fat diet produces brain insulin resistance, synaptodendritic abnormalities and altered behavior in mice. Neurobiol Dis 67:79–87

Atamna H, Frey WH 2nd (2007) Mechanisms of mitochondrial dysfunction and energy deficiency in Alzheimer's disease. Mitochondrion 7(5):297–310

Atkins CM, Selcher JC, Petraitis JJ, Trzaskos JM, Sweatt JD (1998) The MAPK cascade is required for mammalian associative learning. Nat Neurosci 1(7):602–609

Babri S, Badie HG, Khamenei S, Seyedlar MO (2007) Intrahippocampal insulin improves memory in a passive-avoidance task in male wistar rats. Brain Cogn 64(1):86–91

Baker LD, Cross DJ, Minoshima S, Belongia D, Watson GS, Craft S (2011) Insulin resistance and Alzheimer-like reductions in regional cerebral glucose metabolism for cognitively normal adults with prediabetes or early type 2 diabetes. Arch Neurol 68(1):51–57

Baskin DS, Browning JL, Pirozzolo FJ, Korporaal S, Baskin JA, Appel SH (1999) Brain choline acetyltransferase and mental function in Alzheimer disease. Arch Neurol 56(9):1121–1123

Benedict C, Hallschmid M, Hatke A, Schultes B, Fehm HL, Born J, Kern W (2004) Intranasal insulin improves memory in humans. Psychoneuroendocrinology 29(10):1326–1334

Benedict C, Hallschmid M, Schultes B, Born J, Kern W (2007) Intranasal insulin to improve memory function in humans. Neuroendocrinology 86(2):136–142

Benedict C, Brooks SJ, Kullberg J, Burgos J, Kempton MJ, Nordenskjold R, Nylander R, Kilander L, Craft S, Larsson EM, Johansson L, Ahlstrom H, Lind L, Schioth HB (2012) Impaired insulin sensitivity as indexed by the HOMA score is associated with deficits in verbal fluency and temporal lobe gray matter volume in the elderly. Diabetes Care 35(3):488–494

Biessels GJ, Kappelle AC, Utrecht Diabetic Encephalopathy Study G (2005) Increased risk of Alzheimer's disease in Type II diabetes: insulin resistance of the brain or insulin-induced amyloid pathology? Biochem Soc Trans 33(Pt 5):1041–1044

Biessels GJ, Kappelle AC, Bravenboer B, Erkelens DW, Gispen WH (1994) Cerebral function in diabetes mellitus. Diabetologia 37(7):643–650

Biessels GJ, van der Heide LP, Kamal A, Bleys RL, Gispen WH (2002) Ageing and diabetes: implications for brain function. Eur J Pharmacol 441(1–2):1–14

Bingham EM, Hopkins D, Smith D, Pernet A, Hallett W, Reed L, Marsden PK, Amiel SA (2002) The role of insulin in human brain glucose metabolism: an 18fluoro-deoxyglucose positron emission tomography study. Diabetes 51(12):3384–3390

Blazquez E, Velazquez E, Hurtado-Carneiro V, Ruiz-Albusac JM (2014) Insulin in the brain: its pathophysiological implications for States related with central insulin resistance, type 2 diabetes and Alzheimer's disease. Front Endocrinol 5:161

Bondy CA, Cheng CM (2004) Signaling by insulin-like growth factor 1 in brain. Eur J Pharmacol 490(1–3):25–31

Booth FW, Roberts CK, Laye MJ (2012) Lack of exercise is a major cause of chronic diseases. Compr Physiol 2(2):1143–1211

Bosco D, Fava A, Plastino M, Montalcini T, Pujia A (2011) Possible implications of insulin resistance and glucose metabolism in Alzheimer's disease pathogenesis. J Cell Mol Med 15(9):1807–1821

Boule NG, Haddad E, Kenny GP, Wells GA, Sigal RJ (2001) Effects of exercise on glycemic control and body mass in type 2 diabetes mellitus: a meta-analysis of controlled clinical trials. JAMA 286(10):1218–1227

Bubber P, Haroutunian V, Fisch G, Blass JP, Gibson GE (2005) Mitochondrial abnormalities in Alzheimer brain: mechanistic implications. Ann Neurol 57(5):695–703

Calvo-Ochoa E, Arias C (2015) Cellular and metabolic alterations in the hippocampus caused by insulin signalling dysfunction and its association with cognitive impairment during aging and Alzheimer's disease: studies in animal models. Diabetes Metab Res Rev 31(1):1–13

Camiletti-Moiron D, Aparicio VA, Aranda P, Radak Z (2013) Does exercise reduce brain oxidative stress? A systematic review. Scand J Med Sci Sports 23(4):e202–e212

Campbell RK (2011) Clarifying the role of incretin-based therapies in the treatment of type 2 diabetes mellitus. Clin Ther 33(5):511–527

Cao D, Lu H, Lewis TL, Li L (2007) Intake of sucrose-sweetened water induces insulin resistance and exacerbates memory deficits and amyloidosis in a transgenic mouse model of Alzheimer disease. J Biol Chem 282(50):36275–36282

Cardoso S, Correia S, Santos RX, Carvalho C, Santos MS, Oliveira CR, Perry G, Smith MA, Zhu X, Moreira PI (2009) Insulin is a two-edged knife on the brain. J Alzheimers Dis 18(3):483–507

Cardoso S, Carvalho C, Correia SC, Seica RM, Moreira PI (2016) Alzheimer's disease: from mitochondrial perturbations to mitochondrial medicine. Brain Pathol 26(5):632–647

Carvalho C, Cardoso S, Correia SC, Santos RX, Santos MS, Baldeiras I, Oliveira CR, Moreira PI (2012) Metabolic alterations induced by sucrose intake and Alzheimer's disease promote similar brain mitochondrial abnormalities. Diabetes 61(5):1234–1242

Cheke LG, Simons JS, Clayton NS (2016) Higher body mass index is associated with episodic memory deficits in young adults. Q J Exp Psychol 69(11):2305–2316

Chen Z, Zhong C (2013) Decoding Alzheimer's disease from perturbed cerebral glucose metabolism: implications for diagnostic and therapeutic strategies. Prog Neurobiol 108:21–43

Chen Z, Li L, Sun J, Ma L (2012a) Mapping the brain in type II diabetes: voxel-based morphometry using DARTEL. Eur J Radiol 81(8):1870–1876

Chen S, Liu AR, An FM, Yao WB, Gao XD (2012b) Amelioration of neurodegenerative changes in cellular and rat models of diabetes-related Alzheimer's disease by exendin-4. Age 34(5):1211–1224

Cheng Z, Tseng Y, White MF (2010) Insulin signaling meets mitochondria in metabolism. Trends Endocrinol Metab 21(10):589–598

Cho JY, Um HS, Kang EB, Cho IH, Kim CH, Cho JS, Hwang DY (2010) The combination of exercise training and alpha-lipoic acid treatment has therapeutic effects on the pathogenic phenotypes of Alzheimer's disease in NSE/APPsw-transgenic mice. Int J Mol Med 25(3):337–346

Chodzko-Zajko WJ, Moore KA (1994) Physical fitness and cognitive functioning in aging. Exerc Sport Sci Rev 22:195–220

Cole AR, Noble W, van Aalten L, Plattner F, Meimaridou R, Hogan D, Taylor M, LaFrancois J, Gunn-Moore F, Verkhratsky A, Oddo S, LaFerla F, Giese KP, Dineley KT, Duff K, Richardson JC, Yan SD, Hanger DP, Allan SM, Sutherland C (2007) Collapsin response mediator protein-2 hyperphosphorylation is an early event in Alzheimer's disease progression. J Neurochem 103(3):1132–1144

Combs CK, Johnson DE, Karlo JC, Cannady SB, Landreth GE (2000) Inflammatory mechanisms in Alzheimer's disease: inhibition of beta-amyloid-stimulated proinflammatory responses and neurotoxicity by PPARgamma agonists. J Neurosci 20(2):558–567

Correia SC, Santos RX, Carvalho C, Cardoso S, Candeias E, Santos MS, Oliveira CR, Moreira PI (2012) Insulin signaling, glucose metabolism and mitochondria: major players in Alzheimer's disease and diabetes interrelation. Brain Res 1441:64–78

Correia SC, Santos RX, Santos MS, Casadesus G, Lamanna JC, Perry G, Smith MA, Moreira PI (2013) Mitochondrial abnormalities in a streptozotocin-induced rat model of sporadic Alzheimer's disease. Curr Alzheimer Res 10(4):406–419

Cournot M, Marquie JC, Ansiau D, Martinaud C, Fonds H, Ferrieres J, Ruidavets JB (2006) Relation between body mass index and cognitive function in healthy middle-aged men and women. Neurology 67(7):1208–1214

Craft S, Watson GS (2004) Insulin and neurodegenerative disease: shared and specific mechanisms. Lancet Neurol 3(3):169–178

Craft S, Cholerton B, Baker LD (2013) Insulin and Alzheimer's disease: untangling the web. J Alzheimer's Dis 33(Suppl 1):S263–S275

Crane PK, Walker R, Larson EB (2013) Glucose levels and risk of dementia. N Engl J Med 369(19):1863–1864

Cunnane S, Nugent S, Roy M, Courchesne-Loyer A, Croteau E, Tremblay S, Castellano A, Pifferi F, Bocti C, Paquet N, Begdouri H, Bentourkia M, Turcotte E, Allard M, Barberger-Gateau P, Fulop T, Rapoport SI (2011) Brain fuel metabolism, aging, and Alzheimer's disease. Nutrition 27(1):3–20

D'Amico M, Di Filippo C, Marfella R, Abbatecola AM, Ferraraccio F, Rossi F, Paolisso G (2010) Long-term inhibition of dipeptidyl peptidase-4 in Alzheimer's prone mice. Exp Gerontol 45(3):202–207

De Felice FG, Vieira MN, Bomfim TR, Decker H, Velasco PT, Lambert MP, Viola KL, Zhao WQ, Ferreira ST, Klein WL (2009) Protection of synapses against Alzheimer's-linked toxins: insulin signaling prevents the pathogenic binding of Abeta oligomers. Proc Natl Acad Sci U S A 106(6):1971–1976

Deol H, Lekkakou L, Viswanath AK, Pappachan JM (2017) Combination therapy with GLP-1 analogues and SGLT-2 inhibitors in the management of diabesity: the real world experience. Endocrine 55(1):173–178

Desquiret-Dumas V, Gueguen N, Leman G, Baron S, Nivet-Antoine V, Chupin S, Chevrollier A, Vessieres E, Ayer A, Ferre M, Bonneau D, Henrion D, Reynier P, Procaccio V (2013) Resveratrol induces a mitochondrial complex I-dependent increase in NADH oxidation responsible for sirtuin activation in liver cells. J Biol Chem 288(51):36662–36675

Devaskar SU, Giddings SJ, Rajakumar PA, Carnaghi LR, Menon RK, Zahm DS (1994) Insulin gene expression and insulin synthesis in mammalian neuronal cells. J Biol Chem 269(11):8445–8454

Dhuria SV, Hanson LR, Frey WH 2nd (2010) Intranasal delivery to the central nervous system: mechanisms and experimental considerations. J Pharm Sci 99(4):1654–1673

Dietrich MO, Andrews ZB, Horvath TL (2008) Exercise-induced synaptogenesis in the hippocampus is dependent on UCP2-regulated mitochondrial adaptation. J Neurosci 28(42):10766–10771

Doyle P, Cusin I, Rohner-Jeanrenaud F, Jeanrenaud B (1995) Four-day hyperinsulinemia in euglycemic conditions alters local cerebral glucose utilization in specific brain nuclei of freely moving rats. Brain Res 684(1):47–55

Duarte AI, Santos MS, Oliveira CR, Rego AC (2005) Insulin neuroprotection against oxidative stress in cortical neurons--involvement of uric acid and glutathione antioxidant defenses. Free Radic Biol Med 39(7):876–889

Duarte AI, Santos P, Oliveira CR, Santos MS, Rego AC (2008) Insulin neuroprotection against oxidative stress is mediated by Akt and GSK-3beta signaling pathways and changes in protein expression. Biochim Biophys Acta 1783(6):994–1002

Duelli R, Maurer MH, Staudt R, Sokoloff L, Kuschinsky W (2001) Correlation between local glucose transporter densities and local 3-O-methylglucose transport in rat brain. Neurosci Lett 310(2–3):101–104

Elias MF, Elias PK, Sullivan LM, Wolf PA, D'Agostino RB (2005) Obesity, diabetes and cognitive deficit: the Framingham Heart Study. Neurobiol Aging 26(Suppl 1):11–16

Elias MF, Goodell AL, Waldstein SR (2012) Obesity, cognitive functioning and dementia: back to the future. J Alzheimer's Dis 30(Suppl 2):S113–S125

Feinstein DL (2003) Therapeutic potential of peroxisome proliferator-activated receptor agonists for neurological disease. Diabetes Technol Ther 5(1):67–73

Figlewicz DP, Patterson TA, Zavosh A, Brot MD, Roitman M, Szot P (1999) Neurotransmitter transporters: target for endocrine regulation. Horm Metab Res 31(5):335–339

Francis GJ, Martinez JA, Liu WQ, Xu K, Ayer A, Fine J, Tuor UI, Glazner G, Hanson LR, Frey WH 2nd, Toth C (2008) Intranasal insulin prevents cognitive decline, cerebral atrophy and white matter changes in murine type I diabetic encephalopathy. Brain 131(Pt 12):3311–3334

Freeman LR, Haley-Zitlin V, Rosenberger DS, Granholm AC (2014) Damaging effects of a high-fat diet to the brain and cognition: a review of proposed mechanisms. Nutr Neurosci 17(6):241–251

Freiherr J, Hallschmid M, Frey WH 2nd, Brunner YF, Chapman CD, Holscher C, Craft S, De Felice FG, Benedict C (2013) Intranasal insulin as a treatment for Alzheimer's disease: a review of basic research and clinical evidence. CNS Drugs 27(7):505–514

Freude S, Plum L, Schnitker J, Leeser U, Udelhoven M, Krone W, Bruning JC, Schubert M (2005) Peripheral hyperinsulinemia promotes tau phosphorylation in vivo. Diabetes 54(12):3343–3348

Frolich L, Blum-Degen D, Bernstein HG, Engelsberger S, Humrich J, Laufer S, Muschner D, Thalheimer A, Turk A, Hoyer S, Zochling R, Boissl KW, Jellinger K, Riederer P (1998) Brain insulin and insulin receptors in aging and sporadic Alzheimer's disease. J Neural Transm 105(4–5):423–438

Frolich L, Blum-Degen D, Riederer P, Hoyer S (1999) A disturbance in the neuronal insulin receptor signal transduction in sporadic Alzheimer's disease. Ann N Y Acad Sci 893:290–293

Fuenzalida K, Quintanilla R, Ramos P, Piderit D, Fuentealba RA, Martinez G, Inestrosa NC, Bronfman M (2007) Peroxisome proliferator-activated receptor gamma up-regulates the Bcl-2 anti-apoptotic protein in neurons and induces mitochondrial stabilization and protection against oxidative stress and apoptosis. J Biol Chem 282(51):37006–37015

Fukuyama H, Ogawa M, Yamauchi H, Yamaguchi S, Kimura J, Yonekura Y, Konishi J (1994) Altered cerebral energy metabolism in Alzheimer's disease: a PET study. J Nucl Med 35(1):1–6

Fulop T, Larbi A, Douziech N (2003) Insulin receptor and ageing. Pathol Biol 51(10):574–580

Fung TT, Rimm EB, Spiegelman D, Rifai N, Tofler GH, Willett WC, Hu FB (2001) Association between dietary patterns and plasma biomarkers of obesity and cardiovascular disease risk. Am J Clin Nutr 73(1):61–67

Garcia-Mesa Y, Lopez-Ramos JC, Gimenez-Llort L, Revilla S, Guerra R, Gruart A, Laferla FM, Cristofol R, Delgado-Garcia JM, Sanfeliu C (2011) Physical exercise protects against Alzheimer's disease in 3xTg-AD mice. J Alzheimers Dis 24(3):421–454

Gatta E, Lefebvre T, Gaetani S, dos Santos M, Marrocco J, Mir AM, Cassano T, Maccari S, Nicoletti F, Mairesse J (2016) Evidence for an imbalance between tau O-GlcNAcylation and phosphorylation in the hippocampus of a mouse model of Alzheimer's disease. Pharmacol Res 105:186–197

Gazdzinski S, Millin R, Kaiser LG, Durazzo TC, Mueller SG, Weiner MW, Meyerhoff DJ (2010) BMI and neuronal integrity in healthy, cognitively normal elderly: a proton magnetic resonance spectroscopy study. Obesity 18(4):743–748

Gerozissis K (2008) Brain insulin, energy and glucose homeostasis; genes, environment and metabolic pathologies. Eur J Pharmacol 585(1):38–49

Gold SM, Dziobek I, Sweat V, Tirsi A, Rogers K, Bruehl H, Tsui W, Richardson S, Javier E, Convit A (2007) Hippocampal damage and memory impairments as possible early brain complications of type 2 diabetes. Diabetologia 50(4):711–719

Gold M, Alderton C, Zvartau-Hind M, Egginton S, Saunders AM, Irizarry M, Craft S, Landreth G, Linnamagi U, Sawchak S (2010) Rosiglitazone monotherapy in mild-to-moderate Alzheimer's disease: results from a randomized, double-blind, placebo-controlled phase III study. Dement Geriatr Cogn Disord 30(2):131–146

Gong CX, Liu F, Grundke-Iqbal I, Iqbal K (2006) Impaired brain glucose metabolism leads to Alzheimer neurofibrillary degeneration through a decrease in tau O-GlcNAcylation. J Alzheimers Dis 9(1):1–12

Grillo CA, Piroli GG, Hendry RM, Reagan LP (2009) Insulin-stimulated translocation of GLUT4 to the plasma membrane in rat hippocampus is PI3-kinase dependent. Brain Res 1296:35–45

Grunblatt E, Salkovic-Petrisic M, Osmanovic J, Riederer P, Hoyer S (2007) Brain insulin system dysfunction in streptozotocin intracerebroventricularly treated rats generates hyperphosphorylated tau protein. J Neurochem 101(3):757–770

Gunstad J, Paul RH, Cohen RA, Tate DF, Gordon E (2006) Obesity is associated with memory deficits in young and middle-aged adults. Eat Weight Disord 11(1):e15–e19

Gunstad J, Paul RH, Cohen RA, Tate DF, Spitznagel MB, Gordon E (2007) Elevated body mass index is associated with executive dysfunction in otherwise healthy adults. Compr Psychiatry 48(1):57–61

Gustafson D, Rothenberg E, Blennow K, Steen B, Skoog I (2003) An 18-year follow-up of overweight and risk of Alzheimer disease. Arch Intern Med 163(13):1524–1528

Gustafson D, Lissner L, Bengtsson C, Bjorkelund C, Skoog I (2004) A 24-year follow-up of body mass index and cerebral atrophy. Neurology 63(10):1876–1881

Hallal PC, Andersen LB, Bull FC, Guthold R, Haskell W, Ekelund U, Lancet Physical Activity Series Working G (2012) Global physical activity levels: surveillance progress, pitfalls, and prospects. Lancet 380(9838):247–257

Hampel H, Wilcock G, Andrieu S, Aisen P, Blennow K, Broich K, Carrillo M, Fox NC, Frisoni GB, Isaac M, Lovestone S, Nordberg A, Prvulovic D, Sampaio C, Scheltens P, Weiner M, Winblad B, Coley N, Vellas B, Oxford Task Force G (2011) Biomarkers for Alzheimer's disease therapeutic trials. Prog Neurobiol 95(4):579–593

Hanson LR, Frey WH 2nd (2008) Intranasal delivery bypasses the blood-brain barrier to target therapeutic agents to the central nervous system and treat neurodegenerative disease. BMC Neurosci 9(Suppl 3):S5

Hassing LB, Dahl AK, Thorvaldsson V, Berg S, Gatz M, Pedersen NL, Johansson B (2009) Overweight in midlife and risk of dementia: a 40-year follow-up study. Int J Obes 33(8):893–898

Havrankova J, Roth J, Brownstein M (1978) Insulin receptors are widely distributed in the central nervous system of the rat. Nature 272(5656):827–829

Havrankova J, Roth J, Brownstein MJ (1979) Concentrations of insulin and insulin receptors in the brain are independent of peripheral insulin levels. Studies of obese and streptozotocin-treated rodents. J Clin Invest 64(2):636–642

Hendrickx H, McEwen BS, Ouderaa F (2005) Metabolism, mood and cognition in aging: the importance of lifestyle and dietary intervention. Neurobiol Aging 26(Suppl 1):1–5

Hildreth KL, Van Pelt RE, Schwartz RS (2012) Obesity, insulin resistance, and Alzheimer's disease. Obesity 20(8):1549–1557

Ho L, Qin W, Pompl PN, Xiang Z, Wang J, Zhao Z, Peng Y, Cambareri G, Rocher A, Mobbs CV, Hof PR, Pasinetti GM (2004) Diet-induced insulin resistance promotes amyloidosis in a transgenic mouse model of Alzheimer's disease. FASEB J 18(7):902–904

Ho AJ, Raji CA, Becker JT, Lopez OL, Kuller LH, Hua X, Lee S, Hibar D, Dinov ID, Stein JL, Jack CR Jr, Weiner MW, Toga AW, Thompson PM, Cardiovascular Health Study, ADNI (2010) Obesity is linked with lower brain volume in 700 AD and MCI patients. Neurobiol Aging 31(8):1326–1339

Ho AJ, Raji CA, Saharan P, DeGiorgio A, Madsen SK, Hibar DP, Stein JL, Becker JT, Lopez OL, Toga AW, Thompson PM, Alzheimer's Disease Neuroimaging I (2011) Hippocampal volume is related to body mass index in Alzheimer's disease. Neuroreport 22(1):10–14

Hogan CL, Mata J, Carstensen LL (2013) Exercise holds immediate benefits for affect and cognition in younger and older adults. Psychol Aging 28(2):587–594

Hokama M, Oka S, Leon J, Ninomiya T, Honda H, Sasaki K, Iwaki T, Ohara T, Sasaki T, LaFerla FM, Kiyohara Y, Nakabeppu Y (2014) Altered expression of diabetes-related genes in Alzheimer's disease brains: the Hisayama study. Cereb Cortex 24(9):2476–2488

Hopper I, Billah B, Skiba M, Krum H (2011) Prevention of diabetes and reduction in major cardiovascular events in studies of subjects with prediabetes: meta-analysis of randomised controlled clinical trials. Eur J Cardiovasc Prev Rehabil 18(6):813–823

Hossain P, Kawar B, El Nahas M (2007) Obesity and diabetes in the developing world--a growing challenge. N Engl J Med 356(3):213–215

Hoyer S (1991) Abnormalities of glucose metabolism in Alzheimer's disease. Ann N Y Acad Sci 640:53–58

Hoyer S (1998) Risk factors for Alzheimer's disease during aging. Impacts of glucose/energy metabolism. J Neural Transm Suppl 54:187–194

Hoyer S (2004) Causes and consequences of disturbances of cerebral glucose metabolism in sporadic Alzheimer disease: therapeutic implications. Adv Exp Med Biol 541:135–152

Hoyer S, Nitsch R (1989) Cerebral excess release of neurotransmitter amino acids subsequent to reduced cerebral glucose metabolism in early-onset dementia of Alzheimer type. J Neural Transm 75(3):227–232

Hoyer S, Henneberg N, Knapp S, Lannert H, Martin E (1996) Brain glucose metabolism is controlled by amplification and desensitization of the neuronal insulin receptor. Ann N Y Acad Sci 777:374–379

Inestrosa NC, Godoy JA, Quintanilla RA, Koenig CS, Bronfman M (2005) Peroxisome proliferator-activated receptor gamma is expressed in hippocampal neurons and its activation prevents beta-amyloid neurodegeneration: role of Wnt signaling. Exp Cell Res 304(1):91–104

Ingram DK, Roth GS, Lane MA, Ottinger MA, Zou S, de Cabo R, Mattison JA (2006) The potential for dietary restriction to increase longevity in humans: extrapolation from monkey studies. Biogerontology 7(3):143–148

Irie F, Fitzpatrick AL, Lopez OL, Kuller LH, Peila R, Newman AB, Launer LJ (2008) Enhanced risk for Alzheimer disease in persons with type 2 diabetes and APOE epsilon4: the Cardiovascular Health Study Cognition Study. Arch Neurol 65(1):89–93

Ito-Ishida A, Kakegawa W, Yuzaki M (2006) ERK1/2 but not p38 MAP kinase is essential for the long-term depression in mouse cerebellar slices. Eur J Neurosci 24(6):1617–1622

Jagust W, Harvey D, Mungas D, Haan M (2005) Central obesity and the aging brain. Arch Neurol 62(10):1545–1548

Janson J, Laedtke T, Parisi JE, O'Brien P, Petersen RC, Butler PC (2004) Increased risk of type 2 diabetes in Alzheimer disease. Diabetes 53(2):474–481

Jheng HF, Tsai PJ, Guo SM, Kuo LH, Chang CS, Su IJ, Chang CR, Tsai YS (2012) Mitochondrial fission contributes to mitochondrial dysfunction and insulin resistance in skeletal muscle. Mol Cell Biol 32(2):309–319

Joseph A, Antony S, Paulose CS (2008) Increased glutamate receptor gene expression in the cerebral cortex of insulin induced hypoglycemic and streptozotocin-induced diabetic rats. Neuroscience 156(2):298–304

Jun Z, Ibrahim MM, Dezheng G, Bo Y, Qiong W, Yuan Z (2015) UCP2 protects against amyloid beta toxicity and oxidative stress in primary neuronal culture. Biomed Pharmacother 74:211–214

Jurdak N, Lichtenstein AH, Kanarek RB (2008) Diet-induced obesity and spatial cognition in young male rats. Nutr Neurosci 11(2):48–54

Kalmijn S, Launer LJ, Ott A, Witteman JC, Hofman A, Breteler MM (1997) Dietary fat intake and the risk of incident dementia in the Rotterdam Study. Ann Neurol 42(5):776–782

Katsiardanis K, Diamantaras AA, Dessypris N, Michelakos T, Anastasiou A, Katsiardani KP, Kanavidis P, Papadopoulos FC, Stefanadis C, Panagiotakos DB, Petridou ET (2013) Cognitive impairment and dietary habits among elders: the Velestino Study. J Med Food 16(4):343–350

Kelley DE, Goodpaster BH (1999) Effects of physical activity on insulin action and glucose tolerance in obesity. Med Sci Sports Exerc 31(Suppl 11):S619–S623

Kelley DE, He J, Menshikova EV, Ritov VB (2002) Dysfunction of mitochondria in human skeletal muscle in type 2 diabetes. Diabetes 51(10):2944–2950

Kirchner L, Chen WQ, Afjehi-Sadat L, Viidik A, Skalicky M, Hoger H, Lubec G (2008) Hippocampal metabolic proteins are modulated in voluntary and treadmill exercise rats. Exp Neurol 212(1):145–151

Kishi T, Hirooka Y, Nagayama T, Isegawa K, Katsuki M, Takesue K, Sunagawa K (2015) Calorie restriction improves cognitive decline via up-regulation of brain-derived neurotrophic factor: tropomyosin-related kinase B in hippocampus ofobesity-induced hypertensive rats. Int Heart J 56(1):110–115

Kleinridders A, Ferris HA, Cai W, Kahn CR (2014) Insulin action in brain regulates systemic metabolism and brain function. Diabetes 63(7):2232–2243

Kohjima M, Sun Y, Chan L (2010) Increased food intake leads to obesity and insulin resistance in the tg2576 Alzheimer's disease mouse model. Endocrinology 151(4):1532–1540

Korf ES, van Straaten EC, de Leeuw FE, van der Flier WM, Barkhof F, Pantoni L, Basile AM, Inzitari D, Erkinjuntti T, Wahlund LO, Rostrup E, Schmidt R, Fazekas F, Scheltens P, Group LS (2007) Diabetes mellitus, hypertension and medial temporal lobe atrophy: the LADIS study. Diabet Med 24(2):166–171

Kothari V, Luo Y, Tornabene T, O'Neill AM, Greene MW, Geetha T, Babu JR (2017) High fat diet induces brain insulin resistance and cognitive impairment in mice. Biochim Biophys Acta 1863(2):499–508

Kummer MP, Heneka MT (2008) PPARs in Alzheimer's disease. PPAR Res 2008:403896

Kuusisto J, Koivisto K, Mykkanen L, Helkala EL, Vanhanen M, Hanninen T, Kervinen K, Kesaniemi YA, Riekkinen PJ, Laakso M (1997) Association between features of the insulin resistance syndrome and Alzheimer's disease independently of apolipoprotein E4 phenotype: cross sectional population based study. BMJ 315(7115):1045–1049

Landreth G, Jiang Q, Mandrekar S, Heneka M (2008) PPARgamma agonists as therapeutics for the treatment of Alzheimer's disease. Neurotherapeutics 5(3):481–489

Laurin D, Verreault R, Lindsay J, MacPherson K, Rockwood K (2001) Physical activity and risk of cognitive impairment and dementia in elderly persons. Arch Neurol 58(3):498–504

Lean ME (2010) Childhood obesity: time to shrink a parent. Int J Obes 34(1):1–3

Lee HK, Kumar P, Fu Q, Rosen KM, Querfurth HW (2009) The insulin/Akt signaling pathway is targeted by intracellular beta-amyloid. Mol Biol Cell 20(5):1533–1544

Leibson CL, Rocca WA, Hanson VA, Cha R, Kokmen E, O'Brien PC, Palumbo PJ (1997) The risk of dementia among persons with diabetes mellitus: a population-based cohort study. Ann N Y Acad Sci 826:422–427

Leroy K, Boutajangout A, Authelet M, Woodgett JR, Anderton BH, Brion JP (2002) The active form of glycogen synthase kinase-3beta is associated with granulovacuolar degeneration in neurons in Alzheimer's disease. Acta Neuropathol 103(2):91–99

Lester-Coll N, Rivera EJ, Soscia SJ, Doiron K, Wands JR, de la Monte SM (2006) Intracerebral streptozotocin model of type 3 diabetes: relevance to sporadic Alzheimer's disease. J Alzheimers Dis 9(1):13–33

Li ZG, Zhang W, Sima AA (2007) Alzheimer-like changes in rat models of spontaneous diabetes. Diabetes 56(7):1817–1824

Liu F, Iqbal K, Grundke-Iqbal I, Hart GW, Gong CX (2004) O-GlcNAcylation regulates phosphorylation of tau: a mechanism involved in Alzheimer's disease. Proc Natl Acad Sci U S A 101(29):10804–10809

Liu F, Shi J, Tanimukai H, Gu J, Gu J, Grundke-Iqbal I, Iqbal K, Gong CX (2009) Reduced O-GlcNAcylation links lower brain glucose metabolism and tau pathology in Alzheimer's disease. Brain 132(Pt 7):1820–1832

Liu Y, Liu F, Grundke-Iqbal I, Iqbal K, Gong CX (2011) Deficient brain insulin signalling pathway in Alzheimer's disease and diabetes. J Pathol 225(1):54–62

Liu Z, Patil IY, Jiang T, Sancheti H, Walsh JP, Stiles BL, Yin F, Cadenas E (2015) High-fat diet induces hepatic insulin resistance and impairment of synaptic plasticity. PLoS One 10(5):e0128274

Lizcano JM, Alessi DR (2002) The insulin signalling pathway. Curr Biol 12(7):R236–R238

Long-Smith CM, Manning S, McClean PL, Coakley MF, O'Halloran DJ, Holscher C, O'Neill C (2013) The diabetes drug liraglutide ameliorates aberrant insulin receptor localisation and signalling in parallel with decreasing both amyloid-beta plaque and glial pathology in a mouse model of Alzheimer's disease. NeuroMolecular Med 15(1):102–114

Lopez-Lluch G, Hunt N, Jones B, Zhu M, Jamieson H, Hilmer S, Cascajo MV, Allard J, Ingram DK, Navas P, de Cabo R (2006) Calorie restriction induces mitochondrial biogenesis and bio-energetic efficiency. Proc Natl Acad Sci U S A 103(6):1768–1773

Lucignani G, Namba H, Nehlig A, Porrino LJ, Kennedy C, Sokoloff L (1987) Effects of insulin on local cerebral glucose utilization in the rat. J Cereb Blood Flow Metab 7(3):309–314

Ma XH, Zhong P, Gu Z, Feng J, Yan Z (2003) Muscarinic potentiation of GABA(A) receptor currents is gated by insulin signaling in the prefrontal cortex. J Neurosci 23(4):1159–1168

Maccioni RB, Farias G, Morales I, Navarrete L (2010) The revitalized tau hypothesis on Alzheimer's disease. Arch Med Res 41(3):226–231

MacPherson RE, Baumeister P, Peppler WT, Wright DC, Little JP (2015) Reduced cortical BACE1 content with one bout of exercise is accompanied by declines in AMPK, Akt, and MAPK signaling in obese, glucose-intolerant mice. J Appl Physiol 119(10):1097–1104

Manczak M, Reddy PH (2012) Abnormal interaction of VDAC1 with amyloid beta and phosphorylated tau causes mitochondrial dysfunction in Alzheimer's disease. Hum Mol Genet 21(23):5131–5146

Manczak M, Park BS, Jung Y, Reddy PH (2004) Differential expression of oxidative phosphorylation genes in patients with Alzheimer's disease: implications for early mitochondrial dysfunction and oxidative damage. NeuroMolecular Med 5(2):147–162

Manzanero S, Gelderblom M, Magnus T, Arumugam TV (2011) Calorie restriction and stroke. Exp Transl Stroke Med 3:8

Marques-Aleixo I, Oliveira PJ, Moreira PI, Magalhaes J, Ascensao A (2012) Physical exercise as a possible strategy for brain protection: evidence from mitochondrial-mediated mechanisms. Prog Neurobiol 99(2):149–162

Marques-Aleixo I, Santos-Alves E, Balca MM, Rizo-Roca D, Moreira PI, Oliveira PJ, Magalhaes J, Ascensao A (2015) Physical exercise improves brain cortex and cerebellum mitochondrial bioenergetics and alters apoptotic, dynamic and auto(mito)phagy markers. Neuroscience 301:480–495

Martin D, Salinas M, Lopez-Valdaliso R, Serrano E, Recuero M, Cuadrado A (2001) Effect of the Alzheimer amyloid fragment Abeta(25-35) on Akt/PKB kinase and survival of PC12 cells. J Neurochem 78(5):1000–1008

Mattson MP, Gary DS, Chan SL, Duan W (2001) Perturbed endoplasmic reticulum function, synaptic apoptosis and the pathogenesis of Alzheimer's disease. Biochem Soc Symp 67:151–162

McEwen BS, Reagan LP (2004) Glucose transporter expression in the central nervous system: relationship to synaptic function. Eur J Pharmacol 490(1–3):13–24

McNay EC, Ong CT, McCrimmon RJ, Cresswell J, Bogan JS, Sherwin RS (2010) Hippocampal memory processes are modulated by insulin and high-fat-induced insulin resistance. Neurobiol Learn Mem 93(4):546–553

Messier C, Teutenberg K (2005) The role of insulin, insulin growth factor, and insulin-degrading enzyme in brain aging and Alzheimer's disease. Neural Plast 12(4):311–328

Mielke JG, Taghibiglou C, Liu L, Zhang Y, Jia Z, Adeli K, Wang YT (2005) A biochemical and functional characterization of diet-induced brain insulin resistance. J Neurochem 93(6):1568–1578

Miles W, Root HF (1922) Psychologic tests applied in diabetic patients. Arch Intern Med 30:767–777

Minoshima S, Giordani B, Berent S, Frey KA, Foster NL, Kuhl DE (1997) Metabolic reduction in the posterior cingulate cortex in very early Alzheimer's disease. Ann Neurol 42(1):85–94

Mistur R, Mosconi L, Santi SD, Guzman M, Li Y, Tsui W, de Leon MJ (2009) Current challenges for the early detection of Alzheimer's disease: brain imaging and CSF studies. J Clin Neurol 5(4):153–166

Mittal K, Mani RJ, Katare DP (2016) Type 3 diabetes: cross talk between differentially regulated proteins of type 2 diabetes mellitus and Alzheimer's disease. Sci Rep 6:25589

Moloney AM, Griffin RJ, Timmons S, O'Connor R, Ravid R, O'Neill C (2010) Defects in IGF-1 receptor, insulin receptor and IRS-1/2 in Alzheimer's disease indicate possible resistance to IGF-1 and insulin signalling. Neurobiol Aging 31(2):224–243

de la Monte SM (2009) Insulin resistance and Alzheimer's disease. BMB Rep 42(8):475–481

de la Monte SM (2012) Therapeutic targets of brain insulin resistance in sporadic Alzheimer's disease. Front Biosci 4:1582–1605

de la Monte SM (2014) Type 3 diabetes is sporadic Alzheimer's disease: mini-review. Eur Neuropsychopharmacol 24(12):1954–1960

de la Monte SM, Tong M, Lester-Coll N, Plater M Jr, Wands JR (2006) Therapeutic rescue of neurodegeneration in experimental type 3 diabetes: relevance to Alzheimer's disease. J Alzheimer's Dis 10(1):89–109

Moreira PI, Santos MS, Moreno AM, Seica R, Oliveira CR (2003) Increased vulnerability of brain mitochondria in diabetic (Goto-Kakizaki) rats with aging and amyloid-beta exposure. Diabetes 52(6):1449–1456

Moreira PI, Santos MS, Oliveira CR (2007) Alzheimer's disease: a lesson from mitochondrial dysfunction. Antioxid Redox Signal 9(10):1621–1630

Moroz N, Tong M, Longato L, Xu H, de la Monte SM (2008) Limited Alzheimer-type neurodegeneration in experimental obesity and type 2 diabetes mellitus. J Alzheimers Dis 15(1):29–44

Morris JK, Vidoni ED, Honea RA, Burns JM (2014) Impaired glycemia and Alzheimer's disease. Neurobiol Aging 35(10):e23

Navarro A, Gomez C, Lopez-Cepero JM, Boveris A (2004) Beneficial effects of moderate exercise on mice aging: survival, behavior, oxidative stress, and mitochondrial electron transfer. Am J Physiol Regul Integr Comp Physiol 286(3):R505–R511

Nepal B, Brown LJ, Anstey KJ (2014) Rising midlife obesity will worsen future prevalence of dementia. PLoS One 9(9):e99305

Neumann KF, Rojo L, Navarrete LP, Farias G, Reyes P, Maccioni RB (2008) Insulin resistance and Alzheimer's disease: molecular links and clinical implications. Curr Alzheimer Res 5(5):438–447

Nuzzo D, Picone P, Baldassano S, Caruana L, Messina E, Marino Gammazza A, Cappello F, Mule F, Di Carlo M (2015) Insulin resistance as common molecular denominator linking obesity to Alzheimer's disease. Curr Alzheimer Res 12(8):723–735

Ott A, Stolk RP, van Harskamp F, Pols HA, Hofman A, Breteler MM (1999) Diabetes mellitus and the risk of dementia: the Rotterdam Study. Neurology 53(9):1937–1942

Ott V, Benedict C, Schultes B, Born J, Hallschmid M (2012) Intranasal administration of insulin to the brain impacts cognitive function and peripheral metabolism. Diabetes Obes Metab 14(3):214–221

Palleschi L, Vetta F, De Gennaro E, Idone G, Sottosanti G, Gianni W, Marigliano V (1996) Effect of aerobic training on the cognitive performance of elderly patients with senile dementia of Alzheimer type. Arch Gerontol Geriatr 22(Suppl 1):47–50

Pannacciulli N, Del Parigi A, Chen K, Le DS, Reiman EM, Tataranni PA (2006) Brain abnormalities in human obesity: a voxel-based morphometric study. NeuroImage 31(4):1419–1425

Park CR, Seeley RJ, Craft S, Woods SC (2000) Intracerebroventricular insulin enhances memory in a passive-avoidance task. Physiol Behav 68(4):509–514

Parrott MD, Greenwood CE (2007) Dietary influences on cognitive function with aging: from high-fat diets to healthful eating. Ann N Y Acad Sci 1114:389–397

Patrone C, Eriksson O, Lindholm D (2014) Diabetes drugs and neurological disorders: new views and therapeutic possibilities. Lancet Diabetes Endocrinol 2(3):256–262

Paulus GF, de Vaan LE, Verdam FJ, Bouvy ND, Ambergen TA, van Heurn LW (2015) Bariatric surgery in morbidly obese adolescents: a systematic review and meta-analysis. Obes Surg 25(5):860–878

Peila R, Rodriguez BL, Launer LJ, Honolulu-Asia Aging S (2002) Type 2 diabetes, APOE gene, and the risk for dementia and related pathologies: The Honolulu-Asia Aging Study. Diabetes 51(4):1256–1262

Peters SA, Huxley RR, Woodward M (2014) Diabetes as a risk factor for stroke in women compared with men: a systematic review and meta-analysis of 64 cohorts, including 775,385 individuals and 12,539 strokes. Lancet 383(9933):1973–1980

Petrov D, Pedros I, Artiach G, Sureda FX, Barroso E, Pallas M, Casadesus G, Beas-Zarate C, Carro E, Ferrer I, Vazquez-Carrera M, Folch J, Camins A (2015) High-fat diet-induced deregulation of hippocampal insulin signaling and mitochondrial homeostasis deficiencies contribute to Alzheimer disease pathology in rodents. Biochim Biophys Acta 1852(9):1687–1699

Pintana H, Apaijai N, Chattipakorn N, Chattipakorn SC (2013) DPP-4 inhibitors improve cognition and brain mitochondrial function of insulin-resistant rats. J Endocrinol 218(1):1–11

Pope SK, Shue VM, Beck C (2003) Will a healthy lifestyle help prevent Alzheimer's disease? Annu Rev Public Health 24:111–132

Popovic M, Biessels GJ, Isaacson RL, Gispen WH (2001) Learning and memory in streptozotocin-induced diabetic rats in a novel spatial/object discrimination task. Behav Brain Res 122(2):201–207

Prickett C, Brennan L, Stolwyk R (2015) Examining the relationship between obesity and cognitive function: a systematic literature review. Obes Res Clin Pract 9(2):93–113

Profenno LA, Porsteinsson AP, Faraone SV (2010) Meta-analysis of Alzheimer's disease risk with obesity, diabetes, and related disorders. Biol Psychiatry 67(6):505–512

Pugazhenthi S, Qin L, Reddy PH (2016) Common neurodegenerative pathways in obesity, diabetes, and Alzheimer's disease. Biochim Biophys Acta. doi:10.1016/j.bbadis.2016.04.017

Qizilbash N, Gregson J, Johnson ME, Pearce N, Douglas I, Wing K, Evans SJ, Pocock SJ (2015) BMI and risk of dementia in two million people over two decades: a retrospective cohort study. Lancet Diabetes Endocrinol 3(6):431–436

Radak Z, Hart N, Sarga L, Koltai E, Atalay M, Ohno H, Boldogh I (2010) Exercise plays a preventive role against Alzheimer's disease. J Alzheimer's Dis 20(3):777–783

Raji CA, Ho AJ, Parikshak NN, Becker JT, Lopez OL, Kuller LH, Hua X, Leow AD, Toga AW, Thompson PM (2010) Brain structure and obesity. Hum Brain Mapp 31(3):353–364

Rasgon NL, Kenna HA, Wroolie TE, Kelley R, Silverman D, Brooks J, Williams KE, Powers BN, Hallmayer J, Reiss A (2011) Insulin resistance and hippocampal volume in women at risk for Alzheimer's disease. Neurobiol Aging 32(11):1942–1948

Reagan LP (2005) Neuronal insulin signal transduction mechanisms in diabetes phenotypes. Neurobiol Aging 26(Suppl 1):56–59

Reinert KR, Po'e EK, Barkin SL (2013) The relationship between executive function and obesity in children and adolescents: a systematic literature review. J Obes 2013:820956

Risner ME, Saunders AM, Altman JF, Ormandy GC, Craft S, Foley IM, Zvartau-Hind ME, Hosford DA, Roses AD, Rosiglitazone in Alzheimer's Disease Study G (2006) Efficacy of rosiglitazone in a genetically defined population with mild-to-moderate Alzheimer's disease. Pharmacogenomics J 6(4):246–254

Rivera EJ, Goldin A, Fulmer N, Tavares R, Wands JR, de la Monte SM (2005) Insulin and insulin-like growth factor expression and function deteriorate with progression of Alzheimer's disease: link to brain reductions in acetylcholine. J Alzheimers Dis 8(3):247–268

Ryan SM, Kelly AM (2016) Exercise as a pro-cognitive, pro-neurogenic and anti-inflammatory intervention in transgenic mouse models of Alzheimer's disease. Ageing Res Rev 27:77–92

Salkovic-Petrisic M, Hoyer S (2007) Central insulin resistance as a trigger for sporadic Alzheimer-like pathology: an experimental approach. J Neural Transm Suppl 72:217–233

Salkovic-Petrisic M, Tribl F, Schmidt M, Hoyer S, Riederer P (2006) Alzheimer-like changes in protein kinase B and glycogen synthase kinase-3 in rat frontal cortex and hippocampus after damage to the insulin signalling pathway. J Neurochem 96(4):1005–1015

Santos RX, Correia SC, Alves MG, Oliveira PF, Cardoso S, Carvalho C, Seica R, Santos MS, Moreira PI (2014) Mitochondrial quality control systems sustain brain mitochondrial bioenergetics in early stages of type 2 diabetes. Mol Cell Biochem 394(1–2):13–22

Saroka RM, Kane MP, Busch RS, Watsky J, Hamilton RA (2015) Sglt-2 Inhibitor therapy added to Glp-1 agonist therapy in the management of T2dm. Endocr Pract 21(12):1315–1322

Schauer PR, Bhatt DL, Kashyap SR (2014) Bariatric surgery versus intensive medical therapy for diabetes. N Engl J Med 371(7):682

Schechter R, Yanovitch T, Abboud M, Johnson G 3rd, Gaskins J (1998) Effects of brain endogenous insulin on neurofilament and MAPK in fetal rat neuron cell cultures. Brain Res 808(2):270–278

Scheen AJ (2010) Central nervous system: a conductor orchestrating metabolic regulations harmed by both hyperglycaemia and hypoglycaemia. Diabetes Metab 36(Suppl 3):S31–S38

Schipper HM (2011) Apolipoprotein E: implications for AD neurobiology, epidemiology and risk assessment. Neurobiol Aging 32(5):778–790

Sebastiao I, Candeias E, Santos MS, de Oliveira CR, Moreira PI, Duarte AI (2014) Insulin as a bridge between Type 2 diabetes and Alzheimer disease—how anti-diabetics could be a solution for dementia. Front Endocrinol 5:110

Segev Y, Livne A, Mints M, Rosenblum K (2016) Concurrence of high fat diet and APOE gene induces allele specific metabolic and mental stress changes in a mouse model of Alzheimer's disease. Front Behav Neurosci 10:170

Selcher JC, Atkins CM, Trzaskos JM, Paylor R, Sweatt JD (1999) A necessity for MAP kinase activation in mammalian spatial learning. Learn Mem 6(5):478–490

Shulman GI (2000) Cellular mechanisms of insulin resistance. J Clin Invest 106(2):171–176

Simpson IA, Chundu KR, Davies-Hill T, Honer WG, Davies P (1994a) Decreased concentrations of GLUT1 and GLUT3 glucose transporters in the brains of patients with Alzheimer's disease. Ann Neurol 35(5):546–551

Simpson IA, Vannucci SJ, Maher F (1994b) Glucose transporters in mammalian brain. Biochem Soc Trans 22(3):671–675

Sims EA, Danforth E Jr, Horton ES, Bray GA, Glennon JA, Salans LB (1973) Endocrine and metabolic effects of experimental obesity in man. Recent Prog Horm Res 29:457–496

Sims NR, Bowen DM, Allen SJ, Smith CC, Neary D, Thomas DJ, Davison AN (1983) Presynaptic cholinergic dysfunction in patients with dementia. J Neurochem 40(2):503–509

Skeberdis VA, Lan J, Zheng X, Zukin RS, Bennett MV (2001) Insulin promotes rapid delivery of N-methyl-D-aspartate receptors to the cell surface by exocytosis. Proc Natl Acad Sci U S A 98(6):3561–3566

Smith GE (2016) Healthy cognitive aging and dementia prevention. Am Psychol 71(4):268–275

Srivastava S, Haigis MC (2011) Role of sirtuins and calorie restriction in neuroprotection: implications in Alzheimer's and Parkinson's diseases. Curr Pharm Des 17(31):3418–3433

Steiner JL, Murphy EA, McClellan JL, Carmichael MD, Davis JM (2011) Exercise training increases mitochondrial biogenesis in the brain. J Appl Physiol 111(4):1066–1071

Subramanian S, John M (2012) Intranasal administration of insulin lowers amyloid-beta levels in rat model of diabetes. Indian J Exp Biol 50(1):41–44

Talbot K, Wang HY, Kazi H, Han LY, Bakshi KP, Stucky A, Fuino RL, Kawaguchi KR, Samoyedny AJ, Wilson RS, Arvanitakis Z, Schneider JA, Wolf BA, Bennett DA, Trojanowski JQ, Arnold SE (2012) Demonstrated brain insulin resistance in Alzheimer's disease patients is associated with IGF-1 resistance, IRS-1 dysregulation, and cognitive decline. J Clin Invest 122(4):1316–1338

Tan ZS, Beiser AS, Fox CS, Au R, Himali JJ, Debette S, Decarli C, Vasan RS, Wolf PA, Seshadri S (2011) Association of metabolic dysregulation with volumetric brain magnetic resonance imaging and cognitive markers of subclinical brain aging in middle-aged adults: the Framingham Offspring Study. Diabetes Care 34(8):1766–1770

Thomas DE, Elliott EJ, Naughton GA (2006) Exercise for type 2 diabetes mellitus. Cochrane Database Syst Rev 3:CD002968

Tolppanen AM, Solomon A, Kulmala J, Kareholt I, Ngandu T, Rusanen M, Laatikainen T, Soininen H, Kivipelto M (2015) Leisure-time physical activity from mid- to late life, body mass index, and risk of dementia. Alzheimer's Dement 11(4):434–443. e436

Tomlinson DR, Gardiner NJ (2008) Diabetic neuropathies: components of etiology. J Peripher Nerv Syst 13(2):112–121

Toyoda H, Zhao MG, Xu H, Wu LJ, Ren M, Zhuo M (2007) Requirement of extracellular signal-regulated kinase/mitogen-activated protein kinase for long-term potentiation in adult mouse anterior cingulate cortex. Mol Pain 3:36

Ueno H, Mizuta M, Shiiya T, Tsuchimochi W, Noma K, Nakashima N, Fujihara M, Nakazato M (2014) Exploratory trial of intranasal administration of glucagon-like peptide-1 in Japanese patients with type 2 diabetes. Diabetes Care 37(7):2024–2027

Um HS, Kang EB, Leem YH, Cho IH, Yang CH, Chae KR, Hwang DY, Cho JY (2008) Exercise training acts as a therapeutic strategy for reduction of the pathogenic phenotypes for Alzheimer's disease in an NSE/APPsw-transgenic model. Int J Mol Med 22(4):529–539

Verma S, Hussain ME (2017) Obesity and diabetes: an update. Diabetes Metab Syndr 11(1):73–79

Verstynen TD, Weinstein AM, Schneider WW, Jakicic JM, Rofey DL, Erickson KI (2012) Increased body mass index is associated with a global and distributed decrease in white matter microstructural integrity. Psychosom Med 74(7):682–690

Vincent AM, Edwards JL, McLean LL, Hong Y, Cerri F, Lopez I, Quattrini A, Feldman EL (2010) Mitochondrial biogenesis and fission in axons in cell culture and animal models of diabetic neuropathy. Acta Neuropathol 120(4):477–489

Wang B, Yang L, Wang Z, Zheng H (2007) Amyolid precursor protein mediates presynaptic localization and activity of the high-affinity choline transporter. Proc Natl Acad Sci U S A 104(35):14140–14145

Ward MA, Carlsson CM, Trivedi MA, Sager MA, Johnson SC (2005) The effect of body mass index on global brain volume in middle-aged adults: a cross sectional study. BMC Neurol 5:23

Watson GS, Cholerton BA, Reger MA, Baker LD, Plymate SR, Asthana S, Fishel MA, Kulstad JJ, Green PS, Cook DG, Kahn SE, Keeling ML, Craft S (2005) Preserved cognition in patients with early Alzheimer disease and amnestic mild cognitive impairment during treatment with rosiglitazone: a preliminary study. Am J Geriatr Psychiatry 13(11):950–958

Webber L, Divajeva D, Marsh T, McPherson K, Brown M, Galea G, Breda J (2014) The future burden of obesity-related diseases in the 53 WHO European-Region countries and the impact of effective interventions: a modelling study. BMJ Open 4(7):e004787

Whitmer RA, Gunderson EP, Barrett-Connor E, Quesenberry CP Jr, Yaffe K (2005) Obesity in middle age and future risk of dementia: a 27 year longitudinal population based study. BMJ 330(7504):1360

Whitmer RA, Gustafson DR, Barrett-Connor E, Haan MN, Gunderson EP, Yaffe K (2008) Central obesity and increased risk of dementia more than three decades later. Neurology 71(14):1057–1064

Wilcox G (2005) Insulin and insulin resistance. Clin Biochem Rev 26(2):19–39

Wild S, Roglic G, Green A, Sicree R, King H (2004) Global prevalence of diabetes: estimates for the year 2000 and projections for 2030. Diabetes Care 27(5):1047–1053

Willette AA, Xu G, Johnson SC, Birdsill AC, Jonaitis EM, Sager MA, Hermann BP, La Rue A, Asthana S, Bendlin BB (2013) Insulin resistance, brain atrophy, and cognitive performance in late middle-aged adults. Diabetes Care 36(2):443–449

Willette AA, Bendlin BB, Starks EJ, Birdsill AC, Johnson SC, Christian BT, Okonkwo OC, La Rue A, Hermann BP, Koscik RL, Jonaitis EM, Sager MA, Asthana S (2015a) Association of insulin resistance with cerebral glucose uptake in late middle-aged adults at risk for Alzheimer disease. JAMA Neurol 72(9):1013–1020

Willette AA, Johnson SC, Birdsill AC, Sager MA, Christian B, Baker LD, Craft S, Oh J, Statz E, Hermann BP, Jonaitis EM, Koscik RL, La Rue A, Asthana S, Bendlin BB (2015b) Insulin resistance predicts brain amyloid deposition in late middle-aged adults. Alzheimer's Dement 11(5):504–510. e501

Williamson R, McNeilly A, Sutherland C (2012) Insulin resistance in the brain: an old-age or new-age problem? Biochem Pharmacol 84(6):737–745

Winocur G, Greenwood CE (2005) Studies of the effects of high fat diets on cognitive function in a rat model. Neurobiol Aging 26(Suppl 1):46–49

Winocur G, Greenwood CE, Piroli GG, Grillo CA, Reznikov LR, Reagan LP, McEwen BS (2005) Memory impairment in obese Zucker rats: an investigation of cognitive function in an animal model of insulin resistance and obesity. Behav Neurosci 119(5):1389–1395

Wolf PA, Beiser A, Elias MF, Au R, Vasan RS, Seshadri S (2007) Relation of obesity to cognitive function: importance of central obesity and synergistic influence of concomitant hypertension. The Framingham Heart Study. Curr Alzheimer Res 4(2):111–116

Yan J, Liu XH, Han MZ, Wang YM, Sun XL, Yu N, Li T, Su B, Chen ZY (2015) Blockage of GSK3beta-mediated Drp1 phosphorylation provides neuroprotection in neuronal and mouse models of Alzheimer's disease. Neurobiol Aging 36(1):211–227

Yang Y, Ma D, Wang Y, Jiang T, Hu S, Zhang M, Yu X, Gong CX (2013) Intranasal insulin ameliorates tau hyperphosphorylation in a rat model of type 2 diabetes. J Alzheimers Dis 33(2):329–338

Yao J, Chen S, Mao Z, Cadenas E, Brinton RD (2011) 2-Deoxy-D-glucose treatment induces ketogenesis, sustains mitochondrial function, and reduces pathology in female mouse model of Alzheimer's disease. PLoS One 6(7):e21788

Yilmaz N, Vural H, Yilmaz M, Sutcu R, Sirmali R, Hicyilmaz H, Delibas N (2011) Calorie restriction modulates hippocampal NMDA receptors in diet-induced obese rats. J Recept Signal Transduct Res 31(3):214–219

Yin F, Boveris A, Cadenas E (2014) Mitochondrial energy metabolism and redox signaling in brain aging and neurodegeneration. Antioxid Redox Signal 20(2):353–371

Zhang L, Trushin S, Christensen TA, Bachmeier BV, Gateno B, Schroeder A, Yao J, Itoh K, Sesaki H, Poon WW, Gylys KH, Patterson ER, Parisi JE, Diaz Brinton R, Salisbury JL, Trushina E (2016) Altered brain energetics induces mitochondrial fission arrest in Alzheimer's Disease. Sci Rep 6:18725

Zhao WQ, Townsend M (2009) Insulin resistance and amyloidogenesis as common molecular foundation for type 2 diabetes and Alzheimer's disease. Biochim Biophys Acta 1792(5):482–496

Zhao W, Chen H, Xu H, Moore E, Meiri N, Quon MJ, Alkon DL (1999) Brain insulin receptors and spatial memory. Correlated changes in gene expression, tyrosine phosphorylation, and signaling molecules in the hippocampus of water maze trained rats. J Biol Chem 274(49):34893–34902

Zhao WQ, Chen H, Quon MJ, Alkon DL (2004a) Insulin and the insulin receptor in experimental models of learning and memory. Eur J Pharmacol 490(1–3):71–81

Zhao L, Teter B, Morihara T, Lim GP, Ambegaokar SS, Ubeda OJ, Frautschy SA, Cole GM (2004b) Insulin-degrading enzyme as a downstream target of insulin receptor signaling cascade: implications for Alzheimer's disease intervention. J Neurosci 24(49):11120–11126

Zheng WH, Quirion R (2009) Glutamate acting on N-methyl-D-aspartate receptors attenuates insulin-like growth factor-1 receptor tyrosine phosphorylation and its survival signaling properties in rat hippocampal neurons. J Biol Chem 284(2):855–861

Zhong N, Weisgraber KH (2009) Understanding the association of apolipoprotein E4 with Alzheimer disease: clues from its structure. J Biol Chem 284(10):6027–6031

Zimmet P, Alberti KG, Shaw J (2001) Global and societal implications of the diabetes epidemic. Nature 414(6865):782–787

Chapter 6
The Influence of Adipose Tissue on Brain Development, Cognition, and Risk of Neurodegenerative Disorders

Liliana Letra and Isabel Santana

Abstract The brain is a highly metabolic organ and thus especially vulnerable to changes in peripheral metabolism, including those induced by obesity-associated adipose tissue dysfunction. In this context, it is likely that the development and maturation of neurocognitive circuits may also be affected and modulated by metabolic environmental factors, beginning in utero. It is currently recognized that maternal obesity, either pre-gestational or gestational, negatively influences fetal brain development and elevates the risk of cognitive impairment and neuropsychiatric disorders in the offspring. During infancy and adolescence, obesity remains a limiting factor for healthy neurodevelopment, especially affecting executive functions but also attention, visuospatial ability, and motor skills. In middle age, obesity seems to induce an accelerated brain aging and thus may increase the risk of age-related neurodegenerative diseases such as Alzheimer's disease. In this chapter we review and discuss experimental and clinical evidence focusing on the influence of adipose tissue dysfunction on neurodevelopment and cognition across lifespan, as well as some possible mechanistic links, namely the role of the most well studied adipokines.

Keywords Obesity • Neurodevelopment • Cognition • Neurodegenerative disorders • Lifespan • Alzheimer's disease

L. Letra (✉)
Institute of Physiology, Institute for Biomedical Imaging and Life Sciences—IBILI, Faculty of Medicine, University of Coimbra, Coimbra, Portugal

Neurology Department, Centro Hospitalar do Baixo Vouga, Aveiro, Portugal
e-mail: lrletra@fmed.uc.pt

I. Santana
Neurology Department, Centro Hospitalar e Universitário de Coimbra, Coimbra, Portugal

Faculty of Medicine, University of Coimbra, Coimbra, Portugal

CNC—Center for Neuroscience and Cell Biology, University of Coimbra, Coimbra, Portugal

© Springer International Publishing AG 2017
L. Letra, R. Seiça (eds.), *Obesity and Brain Function*, Advances in Neurobiology 19, DOI 10.1007/978-3-319-63260-5_6

6.1 Introduction

In the past few decades, obesity has been associated with serious negative brain health outcomes such as developmental cognitive deficits, neurodegenerative diseases, and neuropsychiatric disorders. The growing interest of several researchers in identifying modifiable and preventable conditions that can reduce the risk of such diseases has contributed to major advancements in the understanding on how peripheral-derived hormonal signals modulate central nervous system beyond energy homeostasis. In this chapter, we address how the development and maturation of neurocognitive circuits may be affected and modulated by obesity throughout lifespan, beginning in utero, and may constitute an aging accelerator, determining the risk of developing late-onset neurodegenerative diseases (Fig. 6.1).

6.2 The Influence of Adipose Tissue on Neurodevelopment and Cognition Across Lifespan

6.2.1 Maternal Obesity

Several large epidemiologic studies have described lower cognitive capabilities, developmental delay, increased incidence of autism spectrum and attention deficit hyperactivity disorders as well as psychiatric diseases in the offspring of obese individuals. When comparing relative parental contribution, maternal obesity seems to be the most determinant to these adverse outcomes, being associated with a 1.3- to 3.6-fold increase in the risk of cognitive impairment and neuropsychiatric disorders in their offspring (Edlow et al. 2014; Edlow 2017). Therefore, it is highly suggestive that in utero environment has substantial impact in

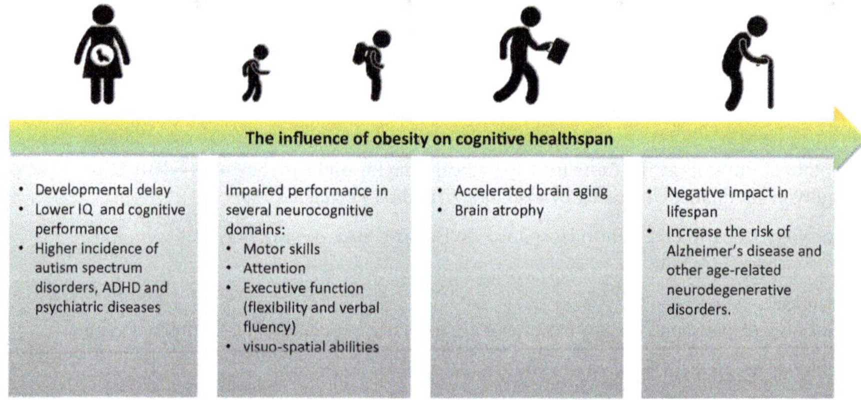

Fig. 6.1 Obesity, neurodevelopment, and cognition across lifespan

neurodevelopment. In fact, data from epidemiological and experimental studies show that alterations in maternal physiology can negatively influence the fetoplacental unit. Animal models of maternal diet-induced obesity have provided important mechanistic insights that improved our comprehension on the association between maternal obesity and adverse neurodevelopment outcomes. They describe impaired neurogenesis and neuronal arborization (mainly, though not exclusively in the hippocampus) with subsequent cognitive impairment and behavioral abnormalities namely anxiety and depressive-like behaviors as well as eating disorders. These alterations may arise as a consequence of disrupted neuronal fetal programming which, in turn, results from an association between altered placental permeability and increased oxidative and inflammatory burden, impaired metabolic signaling (characterized by insulin and leptin resistance), impaired serotoninergic and mesolimbic dopaminergic signaling, and impaired brain-derived neurotrophic factor (BDNF)-mediated synaptic plasticity (Edlow 2017). Moreover, women that in addition have elevated gestational weight gain have a threefold increased risk of IQ (Intelligence Quotient) deficit in the offspring (Huang et al. 2014). Overall, not only is important to promote prenatal dietary and lifestyle changes, but is also essential to minimize gestational weight gain in order to prevent adverse fetal neurodevelopment outcomes.

6.2.2 Obesity During Infancy and Adolescence

When considering infancy and adolescence, the existing literature supports that obesity is negatively linked to functioning in several neurocognitive domains such as motor skills, attention, executive functions, and visuospatial abilities (Liang et al. 2014). Notoriously, in this broad area of executive performance, a more consistent inverse relationship is found with mental flexibility and verbal fluency (Cserjesi et al. 2007; Verdejo-García et al. 2010; Delgado-Rico et al. 2012) relatively to impulsivity, planning, and decision-making. In addition, overweight children display less effective inhibition strategies (Lokken et al. 2009; Nederkoorn et al. 2012) and more difficulties in dealing with delayed gratification (Bruce et al. 2011). Results from studies on the association between obesity and general cognition (verbal and nonverbal domains), language, learning, and memory are, otherwise, less congruent (Liang et al. 2014). Findings from research in this area also demonstrate a two-way relationship between early-life obesity and cognition once mental deficits also influence decision-making, impulsivity, and overall behaviors (food-related, physical activity) that increase, by themselves, the risk of future obesity (Seeyave et al. 2009). Moreover, physical activity, rather than BMI, has been proposed as the most important differentiator of global academic scores (London and Castrechini 2011). Furthermore we cannot overlook socioeconomic factors, which determine the amount and quality of care and stimulation in key developmental stages (Barrigas and Fragoso 2012).

6.2.3 Middle-Age Obesity as a Model of Accelerated Aging

Aging is associated with significant changes on adipose tissue composition and function contributing to age-related phenotypes and disorders (Tchkonia et al. 2010). Around middle or early-old ages a cascade of events is triggered in adipose tissue independently of BMI: body fat percentage increases, distribution shifts from subcutaneous to visceral depots, adipose pro-inflammatory signaling pathways are upregulated, progenitor cell function decline as well as adipose tissue miRNA processing, fatty acid storage capacity decreases with consequent ectopic lipid deposition, and adipose tissue browning declines (Palmer and Kirkland 2016). In summary, with aging, adipose tissue becomes dysfunctional with consequent dysregulation of adipose-derived hormone production and signaling. In overweight or obese individuals, these changes are accelerated and may play a critical role in the etiology of obesity-related complications, including in the increased risk of age-related neurological diseases.

On the other hand, caloric restriction, which also causes profound changes in adipose tissue, increases lifespan in humans and other species and seems to prevent or delay age-related pathologies (Van Cauwenberghe et al. 2016). Multiple interactive pathways and molecular mechanisms are involved in the beneficial effects of caloric restriction on neurons, such as insulin-dependent pathways, FoxO transcription factors, sirtuins, and peroxisome proliferator-activated receptors (Martin et al. 2006 for review). Research that focus specifically on the impact of reducing visceral fat deposits (either through diet or using surgery) indicate a positive effect mediated by an improvement of age-related insulin resistance related to adipokine secretion. In fact, animal models, which characteristically present extended lifespan, also present reduced adiposity and high plasma adiponectin levels suggesting that upregulation of adiponectin and insulin sensitivity may represent a key mediator of longevity-regulatory pathways. Similar results were reported in studies conducted on centenarians and addressing this subject (Arai et al. 2011).

In the field of neurodegenerative diseases, Alzheimer's disease (AD) has been, by far, the pathology most frequently associated with obesity. This association is less robust or inconsistently supported in other degenerative diseases, and contrary to the general idea expressed throughout this chapter, obesity does not always convey a higher risk for neurological disease. A large prospective study runned by O'Reilly et al. (2013) demonstrated that young obese adults had a 30–40% lower risk of developing amyotrophic lateral sclerosis (ALS) compared to normal weighted individuals. Other groups had previously described a strong link between low BMI and decreased survival in patients with ALS (Paganoni et al. 2011; Shimizu et al. 2012), which have motivated the inclusion of dietary supplementation in the treatment of these patients. Increased expression of adiponectin, IL-6, IL-8, lipocalin-2, PAI-1, and TNFα was described in ALS patients, contrarily to leptin, although no relation was found with functional scales or disease duration. In addition, riluzole, the only drug approved in the treatment of ALS, has no effect on circulating adipokine levels (Ngo et al. 2015).

Parkinson's disease (PD) is the second most common neurodegenerative disease after AD and although neuropathological studies indicate that striatal dopamine D2 receptor availability is lower in obese when compared to lean individuals (Wang et al. 2001), epidemiological links with obesity are lacking. In fact, BMI or abdominal obesity does not seem to significantly alter the risk of developing the disease (Palacios et al. 2011). In addition, the few studies on adipokine levels in PD report that serum levels of leptin, adiponectin, and resistin are comparable between patients and controls (matched for age, gender and fat mass) and there is no correlation with clinical parameters (Lorefält et al. 2009; Aziz et al. 2011; Rocha et al. 2014). Furthermore, the levels of these adipokines do not suffer any variation with subthalamic deep brain stimulation, as described by Novakova et al. (2011).

Considering that Alzheimer's disease is the most prevalent neurodegenerative disorder and the most frequent cause of dementia and also because AD has been more extensively associated with obesity and its associated metabolic effects, we review and discuss below how changes in the adipokinome can influence the risk of developing this disease.

6.3 Adipokines Implicated in Alzheimer's Disease

6.3.1 *Leptin*

Leptin is a pleiotropic adipokine with actions on appetite regulation, immune response, bone homeostasis, reproduction, and also on cognitive functions. When directly administered on the dentate gyrus or CA1 hippocampal regions, leptin is able to enhance long-term potentiation (Shanley et al. 2001; Wayner et al. 2004), and influence hippocampal synaptic plasticity by enhancing N-methyl-D-aspartate (NMDA) receptors (Durakoglugil et al. 2005) thus promoting beneficial effects on cognition. Data from cell cultures and AD animal models demonstrate the role of leptin in amyloid precursor protein (APP) metabolism, namely a reduction of amyloid-beta (Aβ) peptide production by blocking β-secretase activity and increasing Apolipoprotein E (ApoE) dependent Aβ uptake (Fewlass et al. 2004). Moreover, leptin promotes phosphorylation (and deactivation) of glycogen synthase kinase beta (GSK-3β), which is the main responsible for abnormal tau phosphorylation (Greco et al. 2009). Additionally, Liu et al. (2012) demonstrated that obese leptin-resistant-mice (*db/db*) had multiple AD-like brain changes including increased phosphorylated tau and Aβ as well as decreased synaptic proteins, with consequent impairment of their performance in cognitive tasks.

Despite extensive animal research on leptin, studies in human populations are limited. A subgroup of the Framingham cohort and the Health ABC Study have demonstrated that higher leptin levels were associated with higher brain volume and a reduction in the risk of dementia, 4 and 7. 7 years after leptin was assayed, respectively (Lieb et al. 2009; Holden et al. 2009). On the other hand, Rajagopalan et al.

(2013) reported that obese elderly, with higher leptin levels, presented with greater global brain atrophy, highlighting the deleterious effect of central leptin insufficiency associated with obesity. In fact, despite peripheral hyperproduction of leptin in obese individuals, the brain becomes insensitive to its action. To our knowledge, only two groups investigated the association between CSF leptin and AD pathology. While Bonda and his group (Bonda et al. 2014) demonstrated an upregulation of leptin in AD patients' CSF and hippocampus, Maioli group (Maioli et al. 2015) reports no significant differences between controls, MCI, and AD patients. Interestingly, Bonda et al. (2014) also observed a positive correlation between CSF leptin and Braak staging (plaque and tangle distribution at different stages of Alzheimer's disease progression) as well as co-localization of ObR (leptin's receptor) with neurofibrillary tangles.

6.3.2 Adiponectin

Although there is no solid evidence of adiponectin intracerebral production, its receptors - AdipoRs - are widely expressed in the brain, namely in the nucleus basalis of Meynert and in the hippocampus (Thundyil et al. 2012), two key structures affected in Alzheimer's disease. Moreover, T-cadherin, which is thought to represent an adiponectin-binding protein (Denzel et al. 2010), is also present in the hippocampus and has been assigned with an important role in cognitive circuitries (Rivero et al. 2015). The neuroprotective role of adiponectin was first proven by Jeon and Qiu (Jeon et al. 2009; Qiu et al. 2011) in excitotoxic conditions. The effect is mediated by adenosine monophosphate-activated protein kinase (AMPK) activation, which is able to minimize neural insult acting as a regulatory factor between neuronal growth and death in response to stress (Erol 2008). Recent data have also highlighted the role of sphingolipid metabolism in the pleiotropic effects of adiponectin once activation of AdipoRs stimulates intracellular catabolism of ceramide, increasing the levels of its anti-apoptotic metabolite sphingosin-1-phosphate (S1p) (Holland et al. 2011; Turer and Scherer 2012). Additionally, Chan and his group (Chan et al. 2012) raised the possibility that NF-κB (nuclear factor kappa-light-chain-enhancer of activated B cells) suppression may also contribute to neuroprotection. Most experimental studies are, indeed, consistent in assigning adiponectin a neuroprotective role in Aβ-induced neurotoxicity. A couple of studies have reported diminished hippocampal neurogenesis in adiponectin-haploinsufficient and adiponectin-deficient mice, which are reversed with intracerebroventricular administration (Zhang et al. 2016; Yu et al. 2014) and more recently, Ng and his group have described for the first time that chronic adiponectin deficiency in aged adiponectin-knockout mice leads to AD-like pathology and cognitive deficits (Ng et al. 2016). Involvement of adiponectin in AD including in the prodromal stage of disease (Mild Cognitive Impairment—MCI) has also been investigated. The J-SHIPP Study (Kamogawa et al. 2010) was the first to demonstrate that higher plasma adiponectin levels had a protective effect against the development of MCI (although only in men) and Une et al. 2011 showed higher concentrations of plasma and CSF adiponectin in MCI compared to normal controls. On the other hand, Teixeira et al. (2013) demonstrated that circulating adiponectin levels were reduced in MCI and AD patients and did not predict progression to dementia, while others have failed to demonstrate any correlation between plasmatic levels of this adipokine and AD at all (Bigalke et al. 2011; Warren et al. 2012). The only meta-analysis assessing this correlation favors the studies in which higher peripheral levels of Adpn were found in AD patients when compared to controls (Ma et al. 2016). Not surprisingly, when analyzing the few observational studies published, obvious methodological caveats are present and may explain the incongruency of results. To date, no study have demonstrated a clear association between baseline adiponec-tinemia and the risk of MCI or progression to dementia, despite growing evidence showing that adiponectin may be related to both CSF and imaging AD biomarkers (Waragai et al. 2016).

6.3.3 Resistin

Resistin was firstly proposed as a molecule of interest for AD in 2010 (Hu et al. 2010), but it was Liu et al. (2013) that confirmed a protective effect of this adipokine against Aβ-induced neurotoxicity. Since no changes were detected in the production of Aβ, the author pointed out a mediation through basic mechanisms of neurodegen-eration, namely improvement of mitochondrial function, decreased toxicity of reactive oxygen, or prevention of apoptosis. Others (Leung et al. 2015) also reported higher levels of this adipokine in subjects with lower Aβ42 levels, suggesting that it might constitute an amyloid-related CSF biomarkers in AD. Moreover, recent research describes higher serum resistin levels in AD patients when compared to normal controls, highlighting a possible role of this molecule in the insulin resis-tance and pro-inflammatory environment typically found in the AD brain (Kizilarslanoğlu et al. 2015; Demirci et al. 2017).

6.3.4 Visfatin

Visfatin or nicotinamide phosphoribosyltransferase (NAMPT) is a key enzyme in the synthesis of nicotinamideadenine dinucleotide (NAD), ubiquitously expressed and widely distributed in visceral adipose tissue and brain, including hippocampus and cerebral cortex. Its biological role is not entirely understood but it seems to be insulin-sensitizer and a pro-inflammatory cytokine (Al-Suhaimi and Shehzad 2013). Elevated serum levels of this protein have been reported in obese and elderly while its brain levels are reduced, compromising NAD biosynthesis and thus able to pro-mote neurodegeneration in these individuals (Adams 2008; Liu et al. 2012). A direct effect of this protein on the pathophysiology of AD remains unclear, but interestingly, some genetic alterations in genes implicated in NAD metabolism have been correlated with AD, suggesting that it can influence the production of amyloid beta and its deposition in senile plaques (Donmez et al. 2010; Villela et al. 2014).

6.4 Conclusions

Despite some contradictory and unresolved observations, the influence of adipose tissue on the brain structure and function during lifespan is unquestionable. Depending on the stage of development, adipose tissue differently affects cognition, beginning in utero a highly dynamic crosstalk with the brain. We hope to encourage further research on the subject, as obesity is already an epidemic metabolic disorder and is essential to understand its impact on brain health, especially on the incidence of neurodegenerative diseases.

References

Adams JD Jr (2008) Alzheimer's disease, ceramide, visfatin and NAD. CNS Neurol Disord Drug Targets 7(6):492–498

Al-Suhaimi EA, Shehzad A (2013) Leptin, resistin and visfatin: the missing link between endocrine metabolic disorders and immunity. Eur J Med Res 18:12

Arai Y, Takayama M, Abe Y, Hirose N (2011) Adipokines and aging. J Atheroscler Thromb 18(7):545–550

Aziz NA, Pijl H, Frölich M, Roelfsema F, Roos RA (2011) Leptin, adiponectin, and resistin secretion and diurnal rhythmicity are unaltered in Parkinson's disease. Mov Disord 26(4):760–761

Barrigas C, Fragoso I (2012) Obesity, academic performance and reasoning ability in Portuguese students between 6 and 12 years old. J Biosoc Sci 44:165–179

Bigalke B, Schreitmüller B, Sopova K, Paul A, Stransky E, Gawaz M, Stellos K, Laske C (2011) Adipocytokines and CD34 progenitor cells in Alzheimer's disease. PLoS One 6(5):e20286

Bonda DJ, Stone JG, Torres SL, Siedlak SL, Perry G, Kryscio R, Jicha G, Casadesus G, Smith MA, Zhu X, Lee HG (2014) Dysregulation of leptin signaling in Alzheimer disease: evidence for neuronal leptin resistance. J Neurochem 128(1):162–172

Bruce AS, Black WR, Bruce JM, Daldalian M, Martin LE, Davis AM (2011) Ability to delay gratification and BMI in preadolescence. Obesity (Silver Spring) 19:1101–1102

Chan KH, Lam KS, Cheng OY, Kwan JS, Ho PW, Cheng KK, Chung SK, Ho JW, Guo VY, Xu A (2012) Adiponectin is protective against oxidative stress induced cytotoxicity in amyloid-beta neurotoxicity. PLoS One 7(12):e52354

Cserjesi R, Molnar D, Luminet O, Lenard L (2007) Is there any relationship between obesity and mental flexibility in children? Appetite 49:675–678

Delgado-Rico E, Rio-Valle JS, Gonzalez-Jimenez E, Campoy C, Verdejo-Garcia A (2012) BMI predicts emotion-driven impulsivity and cognitive inflexibility in adolescents with excess weight. Obesity (Silver Spring) 20:1604–1610

Demirci S, Aynalı A, Demirci K, Demirci S, Arıdoğan BC (2017) The serum levels of resistin and its relationship with other proinflammatory cytokines in patients with Alzheimer's disease. Clin Psychopharmacol Neurosci 15(1):59–63

Denzel MS, Scimia MC, Zumstein PM, Walsh K, Ruiz-Lozano P, Ranscht B (2010) T-cadherin is critical for adiponectin-mediated cardioprotection in mice. J Clin Invest 120:4342–4352

Donmez G, Wang D, Cohen DE, Guarente L (2010) SIRT1 suppresses beta-amyloid production by activating the alpha-secretase gene ADAM10. Cell 142:320–332

Durakoglugil M, Irving AJ, Harvey J (2005) Leptin induces a novel form of NMDA receptor-dependent long- term depression. J Neurochem 95(2):396–405

Edlow AG, Vora NL, Hui L, Wick HC, Cowan JM, Bianchi DW (2014) Maternal obesity affects fetal neurodevelopmental and metabolic gene expression: a pilot study. PLoS One 9(2):e88661

Edlow AG (2017) Maternal obesity and neurodevelopmental and psychiatric disorders in offspring. Prenat Diagn 37(1):95–110

Erol A (2008) An integrated and unifying hypotesis for the metabolic basis of sporadic Alzheimer's disease. JAD 13(3):241–253

Fewlass DC, Noboa K, Pi-Sunyer FX, Johnston JM, Yan SD, Tezapsidis N (2004) Obesity-related leptin regulates Alzheimer's Abeta. FASEB J 18(15):1870–1878

Greco SJ, Sarkar S, Casadesus G, Zhu X, Smith MA, Ashford JW, Johnston JM, Tezapsidis N (2009) Leptin inhibits glycogen synthase kinase-3beta to prevent tau phosphorylation in neuronal cells. Neurosci Lett 455(3):191–194

Holden KF, Lindquist K, Tylavsky FA, Rosano C, Harris TB, Yaffe K (2009) Serum leptin level and cognition in the elderly: findings from the health ABC study. Neurobiol Aging 30(9):1483–1489

Holland WL, Miller RA, Wang ZV et al (2011) Receptor-mediated activation of ceramidase activity initiates the pleiotropic actions of adiponectin. Nat Med 17(1):55–63

Hu WT, Chen-Plotkin A, Arnold SE, Grossman M, Clark CM, Shaw LM, Pickering E, Kuhn M, Chen Y, McCluskey L, Elman L, Karlawish J, Hurtig HI, Siderowf A, Lee VM, Soares H, Trojanowski JQ (2010) Novel CSF biomarkers for Alzheimer's disease and mild cognitive impairment. Acta Neuropathol 119(6):669–678

Huang L, Yu X, Keim S, Li L, Zhang L, Zhang J (2014) Maternal prepregnancy obesity and child neurodevelopment in the collaborative perinatal project. Int J Epidemiol 43(3):783–792

Jeon BT, Shin HJ, Kim JB et al (2009) Adiponectin protects hippocampal neurons against kainic acid-induced excitotoxicity. Brain Res Rev 61(2):81–88

Kamogawa K, Kohara K, Tabara Y, Uetani E, Nagai T, Yamamoto M, Igase M, Miki T (2010) Abdominal fat, adipose-derived hormones and mild cognitive impairment: the J-SHIPP study. Dement Geriatr Cogn Disord 30(5):432–439

Kizilarslanoğlu MC, Kara Ö, Yeşil Y, Kuyumcu ME, Öztürk ZA, Cankurtaran M, Rahatli S, Pakaştiçali N, Çinar E, Halil MG, Sener B, Cankurtaran ES, Arioğul S (2015) Alzheimer disease, inflammation, and novel inflammatory marker: resistin. Turk J Med Sci 45(5):1040–1046

Leung YY, Toledo JB, Nefedov A, Polikar R, Raghavan N, Xie SX, Farnum M, Schultz T, Baek Y, Deerlin VV, WT H, Holtzman DM, Fagan AM, Perrin RJ, Grossman M, Soares HD, Kling MA, Mailman M, Arnold SE, Narayan VA, Lee VM, Shaw LM, Baker D, Wittenberg GM, Trojanowski JQ, Wang LS (2015) Identifying amyloid pathology-related cerebrospinal fluid biomarkers for Alzheimer's disease in a multicohort study. Alzheimers Dement (Amst) 1(3):339–348

Liang J, Matheson BE, Kaye WH, Boutelle KN (2014) Neurocognitive correlates of obesity and obesity-related behaviors in children and adolescents. Int J Obes (Lond) 38(4):494–506

Lieb W, Beiser AS, Vasan RS et al (2009) Association of plasma leptin levels with incident Alzheimer's disease and MRI measures of brain aging: the Framingham study. JAMA 302(23):2565–2572

Liu LY, Wang F, Zhang XY, Huang P, YB L, Wei EQ, Zhang WP (2012) Nicotinamide phosphoribosyltransferase may be involved in age-related brain diseases. PLoS One 7(10).e44933

Liu J, Chi N, Chen H, Zhang J, Bian Y, Cui G, Xiu C (2013) Resistin protection against endogenous Aβ neuronal cytotoxicity from mi- tochondrial pathway. Brain Res 1523:77–84

Lokken KL, Boeka AG, Austin HM, Gunstad J, Harmon CM (2009) Evidence of executive dysfunction in extremely obese adolescents: a pilot study. Surg Obes Relat Dis 5:547–552

London RA, Castrechini S (2011) A longitudinal examination of the link between youth physical fitness and academic achievement. J Sch Health 81:400–408

Lorefält B, Toss G, Granérus AK (2009) Weight loss, body fat mass, and leptin in Parkinson's disease. Mov Disord 24(6):885–890

Ma J, Zhang W, Wang H et al (2016) Peripheral blood adipokines and insulin levels in patients with Alzheimer's disease: a replication study and meta-analysis. Curr Alzheimer Res 13:1–11

Maioli S, Lodeiro M, Merino-Serrais P, Falahati F, Khan W, Puerta E, Codita A, Rimondini R, Ramirez MJ, Simmons A, Gil-Bea F, Westman E, Initiative C-MA A's DN (2015) Alterations in brain leptin signalling in spite of unchanged CSF leptin levels in Alzheimer's disease. Aging Cell 14(1):122–129

Martin B, Mattson MP, Maudsley S (2006) Caloric restriction and intermittent fasting: two potential diets for successful brain aging. Ageing Res Rev 5(3):332–353

Nederkoorn C, Coelho JS, Guerrieri R, Houben K, Jansen A (2012) Specificity of the failure to inhibit responses in overweight children. Appetite 59:409–413

Ng RC, Cheng OY, Kwan JSC et al (2016) Chronic adiponectin deficiency leads to Alzheimer's disease-like cognitive impairments through AMPK inactivation and cerebral insulin resistance in aged mice. Mol Neurodegener 11:71

Ngo ST, Steyn FJ, Huang L, Mantovani S, Pfluger CM, Woodruff TM, O'Sullivan JD, Henderson RD, McCombe PA (2015) Altered expression of metabolic proteins and adipokines in patients with amyotrophic lateral sclerosis. J Neurol Sci 357(1–2):22–27

Novakova L, Haluzik M, Jech R, Urgosik D, Ruzicka F, Ruzicka E (2011) Hormonal regulators of food intake and weight gain in Parkinson's disease after subthalamic nucleus stimulation. Neuro Endocrinol Lett 32(4):437–441

O'Reilly ÉJ, Wang H, Weisskopf MG, Fitzgerald KC, Falcone G, McCullough ML et al (2013) Premorbid body mass index and risk of amyotrophic lateral sclerosis. Amyotroph Lateral Scler Frontotemporal Degener 14:205–211

Paganoni S, Deng J, Jaffa M, Cudkowicz ME, Wills A-M (2011) Body mass index, not dyslipidemia, is an independent predictor of survival in amyotrophic lateral sclerosis. Muscle Nerve 44:20–24

Palacios N, Gao X, McCullough ML, Jacobs EJ, Patel AV, Mayo T, Schwarzschild MA, Ascherio A (2011) Obesity, diabetes, and risk of Parkinson's disease. Mov Disord 26(12):2253–2259

Qiu G, Wan R, Hu J, Mattson MP, Spangler E, Liu S, Yau SY, Lee TM, Gleichmann M, Ingram DK, So KF, Zou S (2011) Adiponectin protects rat hippocampal neurons against excitotoxicity. Age (Dordr) 33(2):155–165

Rajagopalan P, Toga AW, Jack CR, Weiner MW, Thompson PM (2013) Fat-mass-related hormone, plasma leptin, predicts brain volumes in the elderly. Neuroreport 24(2):58–62

Rivero O, Selten MM, Sich S et al (2015) Cadherin-13, a risk gene for ADHD and comorbid disorders, impacts GABAergic function in hippocampus and cognition. Transl Psychiatry 5:e655

Rocha NP, Scalzo PL, Barbosa IG, de Sousa MS, Morato IB, Vieira EL, Christo PP, Reis HJ, Teixeira AL (2014) Circulating levels of adipokines in Parkinson's disease. J Neurol Sci 339(1–2):64–68

Seeyave DM, Coleman S, Appugliese D, Corwyn RF, Bradley RH, Davidson NS et al (2009) Ability to delay gratification at age 4 years and risk of overweight at age 11 years. Arch Pediatr Adolesc Med 163:303–308

Shanley LJ, Irving AJ, Harvey J (2001) Leptin enhances NMDA receptor function and modulates hippocampal synaptic plasticity. J Neurosci 21(24):RC186

Shimizu T, Nagaoka U, Nakayama Y, Kawata A, Kugimoto C, Kuroiwa Y, Kawai MT, Nishizawa M, Mihara B, Arahata H, Fujii N, Namba R, Ito H, Imai T, Nobukuni K, Kondo K, Ogino M, Nakajima T, Komori T (2012) Reduction rate of body mass index predicts prognosis for survival in amyotrophic lateral sclerosis: a multicenter study in Japan. Amyotroph Lateral Scler 13:363–366

Tchkonia T, Morbeck DE, Von Zglinicki T, Van Deursen J, Lustgarten J, Scrable H, Khosla S, Jensen MD, Kirkland JL (2010) Fat tissue, aging, and cellular senescence. Aging Cell 9:667–684

Teixeira AL, Diniz BS, Campos AC, Miranda AS, Rocha NP, Talib LL, Gattaz WF, Forlenza OV (2013) Decreased levels of circulating adiponectin in mild cognitive impairment and Alzheimer's disease. NeuroMolecular Med 15(1):115–121

Thundyil J, Pavlovski D, Sobey CG, Arumugam T V (2012) Adiponectin receptor signalling in the brain. Br J Pharmacol 165:313–327

Turer AT, Scherer PE (2012) Adiponectin: mechanistic insights and clinical implications. Diabetologia 55(9):2319–2326

Une K, Takei YA, Tomita N, Asamura T, Ohrui T, Furukawa K, Arai H (2011) Adiponectin in plasma and cerebrospinal fluid in MCI and Alzheimer's disease. Eur J Neurol 18(7):1006–1009

Van Cauwenberghe C, Vandendriessche C, Libert C, Vandenbroucke RE (2016) Caloric restriction: beneficial effects on brain aging and Alzheimer's disease. Mamm Genome 27(7–8):300–319

Verdejo-García A, Pérez-Expósito M, Schmidt-Río-Valle J, Fernández-Serrano MJ, Cruz F, Pérez-García M, López-Belmonte G, Martín-Matillas M, Martín-Lagos JA, Marcos A, Campoy C (2010) Selective alterations within executive functions in adolescents with excess weight. Obesity (Silver Spring) 18:1572–1578

Villela D, Schlesinger D, Suemoto CK, Grinberg LT, Rosenberg C (2014) A microdeletion in Alzheimer's disease disrupts NAMPT gene. J Genet 93(2):535–537

Wang GJ, Volkow ND, Logan J, Pappas NR, Wong CT, Zhu W, Netusil N, Fowler JS (2001) Brain dopamine and obesity. Lancet 357(9253):354–357

Waragai M, Adame A, Trinh I et al (2016) Possible involvement of adiponectin, the anti-diabetes molecule, in the pathogenesis of Alzheimer's disease. J Alzheimers Dis 52:1453–1459

Warren MW, Hynan LS, Weiner MF (2012) Lipids and adipokines as risk factors for Alzheimer's disease. J Alzheimers Dis 29(1):151–157

Wayner MJ, Armstrong DL, Phelix CF, Oomura Y (2004) Orexin-A (Hypocretin-1) and leptin enhance LTP in the dentate gyrus of rats in vivo. Peptides 25(6):991–996

Yu S, Li A, Hoo RLC et al (2014) Physical exercise-induced hippocampal neurogenesis and anti-depressant effects are mediated by the adipocyte hormone adiponectin. Proc Natl Acad Sci U S A 111:15810–15815

Zhang D, Wang X, Adiponectin Exerts LX (2016) Neurotrophic effects on dendritic dentate gyrus of male mice. Endocrinology 157:2853–2869

Chapter 7
Cerebrovascular Disease: Consequences of Obesity-Induced Endothelial Dysfunction

Liliana Letra and Cristina Sena

Abstract Despite the well-known global impact of overweight and obesity in the incidence of cerebrovascular disease, many aspects of this association are still inconsistently defined. In this chapter we aim to present a critical review on the links between obesity and both ischemic and hemorrhagic stroke and discuss its influence on functional outcomes, survival, and current treatments to acute and chronic stroke. The role of cerebrovascular endothelial function and respective modulation is also described as well as its laboratory and clinical assessment. In this context, the major contributing mechanisms underlying obesity-induced cerebral endothelial function (adipokine secretion, insulin resistance, inflammation, and hypertension) are discussed. A special emphasis is given to the participation of adipokines in the pathophysiology of stroke, namely adiponectin, leptin, resistin, apelin, and visfatin.

Keywords Obesity • Endothelial dysfunction • Cerebrovascular disease • Stroke

L. Letra (✉)
Institute of Physiology, Institute for Biomedical Imaging and Life Sciences—IBILI,
Faculty of Medicine, University of Coimbra, Coimbra, Portugal

Neurology Department, Centro Hospitalar do Baixo Vouga, Aveiro, Portugal
e-mail: lrletra@fmed.uc.pt

C. Sena
Institute of Physiology, Institute for Biomedical Imaging and Life Sciences—IBILI,
Faculty of Medicine, University of Coimbra, Coimbra, Portugal

© Springer International Publishing AG 2017
L. Letra, R. Seiça (eds.), *Obesity and Brain Function*, Advances in
Neurobiology 19, DOI 10.1007/978-3-319-63260-5_7

7.1   Epidemiology of Obesity and Stroke

7.1.1   Obesity and Ischemic Stroke

Obesity is a highly prevalent risk factor for ischemic stroke (Strazzullo et al. 2010), representing a logical target, though rather difficult and complex, for both primary and secondary prevention strategies. Some dozens of studies have examined the association between obesity and cerebrovascular disease. Overall they used body mass index (BMI) as a marker of adiposity, though some authors have preferred measures of central obesity such as waist circumference, waist-to-hip ratio, or waist-to-height ratio. Although the prevalence of obesity among patients with established cerebrovascular disease has not been extensively explored, it is estimated that it may range from 18% to 44%, including all studies independently of the anthropometric measure used. In fact, some authors estimate that for every 1-unit increase in BMI, the risk for ischemic stroke increases around 5%, and this applies also to BMI under 30 (Kernan et al. 2013). Noteworthy, measures of central obesity are better predictors of stroke in the majority of the epidemiological studies (Suk et al. 2003; Winter et al. 2008; Bodenant et al. 2011). This comes in line with the belief that the detrimental effects of abdominal obesity are due to visceral adipose tissue (VAT) dysfunction, which is strongly correlated with traditional vascular disease risk factors, such as insulin resistance, systemic inflammation, dyslipidemia, and arterial hypertension (Hajer et al. 2008). All of these are morbidities that often coexist in the obese patient and can, independently, increase the risk of ischemic cerebrovascular disease, fact that must be accounted when interpreting results from research on this field. Atrial fibrillation and obstructive sleep apnea are other important risk factors for ischemic stroke that are, at the same time, more prevalent in obese individuals (Zhang et al. 2015). The relative risk for stroke associated with obesity seems also somehow dependent on age once it seems to be higher in middle-aged compared with elderly individuals (Whitlock et al. 2009; Wormser et al. 2011). Interestingly, some studies report that overweight and obese patients who survive the first stroke event tend to have improved subsequent cerebrovascular disease and mortality—the so-called "obesity paradox" (Towfighi and Ovbiagele 2009; Vemmos et al. 2011; Ovbiagele et al. 2011). This could lead to a change in the nutritional recommendations included in secondary prevention guidelines. However, these results should be looked with caution while important selection bias are found in some studies, such as the exclusion of patients with severely disabling stroke, as well as the difficulty on gathering the same cluster of nonadipose stroke risk factors (which also predispose to subsequent vascular events) within the groups studied. Additionally, obese patients may tend to have more lacunar type of stroke, which is generally characterized by faster recovery and more favorable prognosis. More, obese patients may have stricter follow-up, with consequent better control of their comorbidities and better clinical outcomes (Katsnelson and Rundek 2011; Oesch et al. 2017). On the other hand, there are a considerable number of factors that can contribute for the decreased post-stroke long-term survival observed in patients with normal and underweight

BMI. It is difficult to discriminate if low/normal BMI is a result of healthy habits, medical illness including depression, malnutrition, neglect or lower socioeconomic status. Besides, patients recovering from an acute cerebrovascular event can experience a catabolic state: fever, sympathetic activation, endothelial and insulin sensitivity dysfunction which can overall lead to muscle wasting and weight loss (Scherbakov et al. 2011; Katsnelson and Rundek 2011). This can be aggravated if the patient is immobilized, and thus looses bone mineral density and muscle area, or has dysphagia with consequent decrease in food intake. Future randomized and prospective clinical trials are needed to better define the interactions between obesity and stroke, taking into account that BMI has significant limitations and more accurate tools should be used to assess the relation between obesity or adipose tissue dysfunction and the risk of ischemic stroke.

7.1.2 *Obesity and Hemorrhagic Stroke*

Although the association between obesity and cerebrovascular disease has been more consistently demonstrated for ischemic stroke, there is also evidence of the link with intracerebral hemorrhage (ICH). It is already recognized that obese individuals tend to present higher levels of factors VII, VIII, IX, and XII and von Willebrand factor (vWF), higher levels of fibrinogen and plasminogen activator inhibitor-1 (PAI-1), and low tissue plasminogen activator (tPA) activity and activated protein C (APC) (Rosito et al. 2004), what would presumably make these patients less prone to bleed. However, several studies enrolling a significant number of patients have demonstrated conflicting results. Some indicate that BMI is associated with ICH risk (Bazzano et al. 2010; Song et al. 2004; Silventoinen et al. 2009) but others did not find any association (Strazzullo et al. 2010; Wang et al. 2013). Interestingly, extremes of BMI, both low and high, predict increased risk for ICH. Low cholesterol levels could be responsible for increased ICH risk in individuals with low BMI (Gostynski et al. 2004; Woo et al. 2005), while very high BMI-related conditions, namely hypertension, contribute to cerebral small vessel disease (Meissner 2016). An important limitation found in some of the published studies on this topic is the fact that some do not account for the likely differential effect of adiposity on the pathogenic subtype of ICH. Factors that damage small penetrating arteries leading to deep subcortical hemorrhage are probably different from those favoring lobar hemorrhages related to cerebral amyloid angiopathy (Bith et al. 2011). Overall, obesity seems to be associated with increased risk of deep ICH, although with only a modest direct effect once hypertensive vasculopathy is the major contributor, whereas weak or no association has been found between BMI and risk of lobar ICH.

7.1.3   Influence of Obesity on Stroke Treatment

In the last decade we have assisted to considerable advances in acute stroke care. Thrombolysis with alteplase (intravenous rtPA) and endovascular trombectomy have significantly improved survival and functional outcomes of acute ischemic stroke. However, little is known about the impact of body weight on the outcomes after these treatments, even though alteplase dosis calculation is made according to the patient's weight (0.9 × total body weight) while the maximum dosis is limited to 90 mg. This means that patients weighting more than 100 kg may not be given sufficient trombolytic dosis to promote revascularization. Some small studies have compared clinical outcomes and safety after intravenous thrombolysis between obese and lean patients but reached discordant results (Sarikaya et al. 2011; Seo et al. 2013; Seet et al. 2014). To date, the only large prospective observational multicenter study (896 patients included) conducted by Branscheidt et al. (2016) has demonstrated that early neurological improvement (measured by NIHSS at 24 h after thrombolysis, as predictor of recanalization) as well as mortality and clinical outcome at 3 months were similar among different BMI groups, suggesting that the actual dosage scheme of alteplase is appropriate. In addition, the majority of studies seem to agree that obesity does not influence the risk of symptomatic intracranial hemorrhage after rtPA treatment.

Other relevant question regarding the treatment of patients with stroke is whether obesity influences the efficacy of anticoagulant drugs, which are indicated in the secondary prevention of cardioembolic ischemic cerebrovascular events. Current recommendations for direct oral anticoagulants (DOACs) imply the use of a fixed dose irrespective of high BMI. Analyzing the results from the most relevant trials, plasmatic levels of these drugs show a great diversity according to the total body weight, though statistical significance was never described (Güler et al. 2015). In fact, several case reports have called attention to the occurrence of stroke or systemic embolism in obese patients on DOACs raising concerns about the adequate dosage in these patients. These cases refer mainly to patients taking dabigatran, presumably due to increased creatinine clearance that may occur in obese patients. Therefore, it may be rational to use, in these patients, DOACs with lower renal clearance (RC), namely Apixaban (RC = 27%) and Edoxaban (RC = 35%) (Bounameaux and John Camm 2014). Obese patients taking warfarin require significantly longer median time and higher daily dose to achieve therapeutic INR (Wallace et al. 2013) when compared to normal weighted patients, but the association of obesity with higher risk of bleeding in these patients is controversial (Ogunsua et al. 2015; Hart et al. 2017). Furthermore, obesity has also been described as an independent predictor of anticoagulation reversal failure with prothrombin complex concentrate, what is relevant in those patients presenting with acute intracranial hemorrhage (Chu et al. 2016). Interestingly, the effects of intentional weight loss on stroke risk (ischemic and hemorrhagic) are much less explored, though some studies report no clear benefit of intensive lifestyle modifications in the risk of stroke (Wing et al. 2013).

7.2 Cerebrovascular Endothelial Function: Physiology, Modulation, Laboratory, and Clinical Assessment

7.2.1 Physiology and Modulation of Cerebrovascular Endothelial Function

The vascular endothelium is the main regulator of vascular homeostasis due to its interaction with both the circulating cells and those present in the vascular wall (Pantoni 2010; Sena et al. 2013). Endothelial cells (ECs) are capable of sensing changes in the hemodynamic forces and blood factors, and respond by releasing different substances. The normal paracrine and autocrine functions of ECs include the synthesis of a series of substances that moderate vascular tone, mediation of local inflammation, depression of leukocyte migration and control of permeability, regulation of proliferation and migration of smooth muscle cells, control of platelet adhesion and aggregation, antioxidant activity, and anticoagulant and profibrinolytic effects (Wardlaw et al. 2013; Fig. 7.1). Similarly, cerebral endothelial cells also have effects on vascular tone regulating cerebral blood flow (CBF) by releasing vasodilators (nitric oxide, prostacyclin, bradykinin, endothelium-derived hyperpolarizing factors, etc.) and vasoconstrictors (endothelin-1 [ET-1], endothelium-derived constrictor factor, etc.) (Pescini and Abbate 2014). The regulation of vascular tone is obtained through the balanced production of vasodilators and vasoconstrictors in response to a variety of stimuli (Michiels 2003; Pearson 1999; Fig. 7.1). The most abundant mediator of normal vascular function is nitric oxide (NO), generated from L-arginine through endothelial NO synthase. Another endothelium-derived vasodilator is prostacyclin (PGI_2), generated by cyclooxygenase (COX) and arachidonic acid metabolism. Endothelium-derived vasoactive factors participate in the maintenance of resting CBF and may play a role in coordinating the vasodilatation of intraparenchymal arterioles with that of upstream pial arteries, and in local adjustments of flow in response to mechanical forces (Murphey 2012). Brain perfusion and cerebral vascular reactivity are essential for maintaining energy-dependent processes, generating sufficient metabolic substrate and also clearing the products produced by neuronal activity (Drake and Iadecola 2007). In addition, ECs participate in the regulation of the blood–brain barrier (BBB). Cerebral ECs have a low vesicular transport and are linked to each other by tight junctions, which prevent the entry of hydrophilic substances into the brain (Burger and Touyz 2012). Specialized transport proteins on the EC membrane regulate the bidirectional transfer of substances into and from the brain parenchyma (Burger and Touyz 2012). The integrity of the BBB is vital to maintain the homeostasis of the cerebral microenvironments crucial for normal brain function.

ECs have a central role in maintenance of a nonthrombogenic surface by providing activators and inhibitors of the coagulation and fibrinolysis systems. NO and PGI_2 have important antiplatelet effects, limiting aggregation and permeability. ECs maintain blood fluidity by promoting the activity of numerous anticoagulant pathways. Once activated, ECs express adhesion molecules on their surfaces thus allow-

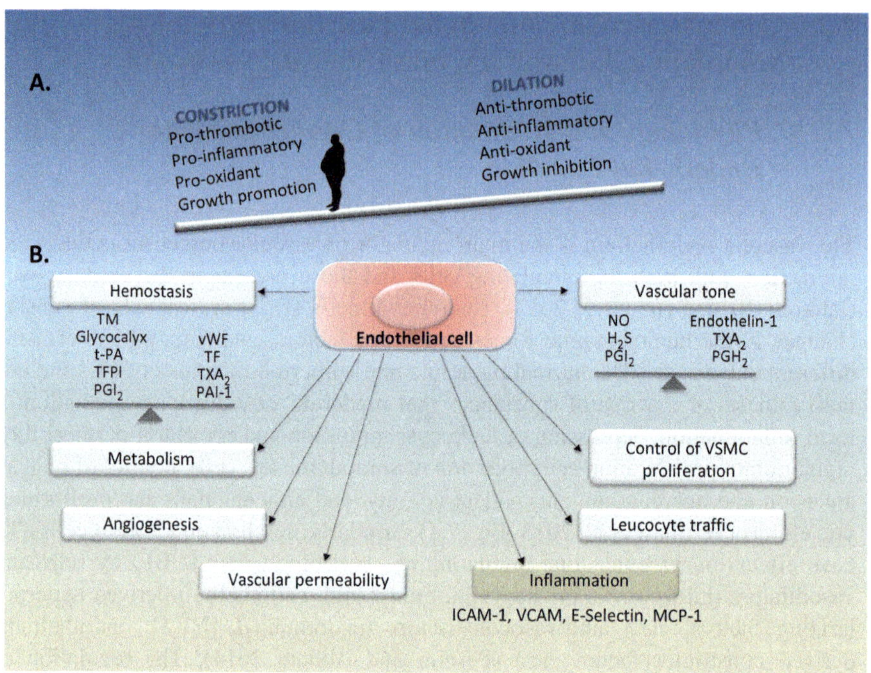

Fig. 7.1 Vascular unbalance between dilation and constriction under pathological obese conditions. (**A**) Obesity leads to endothelial activation and dysfunction promoting a major unbalance towards constriction, triggering thrombotic, inflammatory, oxidant and growth promotion events. Below (**B**) are represented the major functions of endothelial cells: regulation of vascular tone (vasodilators: NO, H_2S, PGI_2; vasoconstrictors: ET_1, TXA_2, PGH_2), control of VSMC proliferation, leukocyte traffic, inflammation, permeability, angiogenesis, metabolism, and hemostasis (anticoagulant: TM, glycocalyx, t-PA, TFPI, PGI_2; prothrombotic: vWF, TF, TXA_2, PAI-1). *ET1* endothelin-1, H_2S hydrogen sulfide, *ICAM* intercellular adhesion molecule, *MCP-1* monocyte chemoattractant protein-1, *NO* nitric oxide, *PAI-1* plasminogen activator inhibitor-1, PGH_2 prostaglandin H_2, PGI_2 prostacyclin, *TF* tissue factor, *TFPI* tissue factor pathway inhibitor, *TM* thrombomodulin, *t-PA* tissue plasminogen activator, TXA_2 thromboxane A_2, *VCAM* vascular cell adhesion molecule, *control of VSMC* vascular smooth muscle cells, *vWF* von Willebrand factor

ing the binding of circulating leukocytes. Chemotactic factors attract leukocytes along a chemical gradient to the site of injury. Adhesion molecules involved in leukocyte recruitment include selectins, addressins, integrins, and immunoglobulins. On the other hand, ECs are also involved in angiogenesis and endothelial repair. Growth factors, like vascular endothelial growth factor (VEGF), are mediators of angiogenesis. VEGF is the most specific for endothelium and along with cytokines, chemokines, matrix metalloproteinases (MMPs), and extracellular matrix macromolecules may be involved in endothelial proliferation directly or indirectly by means of the stimulation and the production of other angiogenic factors. Circulating endothelial progenitor cells are hematopoietic stem cells with limited pluripotent potential, mainly involved in formation of new vessels after ischemia and in repairing damaged endothelium.

Endothelial dysfunction is usually described as endothelial-based abnormalities that promote vasoconstriction, inflammation, increased permeability, atherosclerosis, and thrombosis (Fig. 7.1). Endothelial dysfunction is generally considered to be the earliest event in the initiation of vascular disease and also plays a key role throughout the disease process. Both genetic and clinical data have established that the vasomotor component of endothelial dysfunction is predictive of cardiovascular events (Volpe et al. 1996; Lind et al. 2011; Green et al. 2014). In the brain, the consequences of endothelial dysfunction include atherosclerosis, large and small vessel disease, loss of CBF autoregulation, inability to adequately adjust blood supply, stroke, dementia, and blood–brain barrier abnormalities (Abbott and Friedman 2012). Vascular dysfunction in cerebral arteries has also been associated with cognitive impairments with advancing age and disease. Endothelial dysfunction is a reversible process that needs to be monitored in order to prevent vascular disease and reduce morbidity and mortality of patients.

7.2.2 Laboratory and Clinical Assessment of Cerebrovascular Endothelial Function

Several circulating factors are important vascular risk factors due to their harmful effects on ECs. For instance, homocysteine, C-reactive protein (CRP), and asymmetric dimethylarginine (ADMA) may all be considered as endothelial aggressors (Cheng et al. 2009; Szmitko et al. 2003; Franceschelli et al. 2013). Dysfunctional endothelium releases ET-1, which along with decreased NO release, leads to vasoconstriction and further reduction of CBF. Vascular injury could also lead to the degradation of the basal lamina by MMPs and the release of cellular-fibronectin. Subsequent clot formation from damaged vessels could also activate the fibrinolytic and coagulation pathways. Thrombosis in vessels leads to the activation of both coagulation (PAI-1) and fibrinolytic system (tPA). Increased vWF is an indicator of damaged endothelium. Thus, MMPs, cellular-fibronectin, ADMA, ET-1, NO, tPA, PAI-1, and vWF are potential biomarkers of vascular injury.

In addition, other markers of inflammation (hsCRP, tumor necrosis factor-α, interleukin-6, lipoprotein-associated phospholipase A2, S110B proteins, cellular adhesion molecules), oxidative stress (superoxide dismutases, glutathiones), and metabolism (microalbuminuria, urinary albumin/creatinine ratio, hyperhomocysteinemia) could be considered but, to date, most biomarkers lack sensitivity or specificity to be of clinical use.

Circulating ECs have also been proposed as diagnostic markers of endothelial damage caused by systemic vascular disease (Makin et al. 2004; Lee et al. 2005; Boos et al. 2006, 2007; Woywodt et al. 2012; Deb et al. 2013; Fadini et al. 2015). It was recently described, in newborn piglets, a relationship between cerebral vascular disorders and circulating ECs of brain vessel origin, the brain-derived ECs, suggesting that they serve as a sensitive indicator of cerebral vascular endothelial damage and long-term cerebral blood flow dysregulation caused by epileptic seizures (Pourcyrous et al. 2015).

The majority of the existing studies evaluate biomarkers measured from blood samples that reflect systemic endothelial function, which does not necessarily correspond to what happens at the level of brain endothelium.

The study of endothelial circulating biomarkers needs to be strictly correlated with other more specific measurements of brain endothelial function, for instance, advanced magnetic resonance imaging (MRI) technologies with intravenous gadolinium contrast agent combined with perfusion imaging or cerebrovascular reactivity (Poggesi et al. 2015).

The noninvasive methods for early detection of cerebrovascular damage in clinical settings (Woywodt et al. 2012) are limited. Cerebral endothelium is a cell monolayer on the inner wall of the vessels difficult to image. Intraluminal ultrasound techniques enable imaging of morphological changes. However, these techniques do not enable an evaluation of cerebral endothelial function. In the past, cerebrovascular reactivity (CVR) to L-arginine by means of transcranial Doppler (TCD) has emerged as a reliable marker for evaluation of cerebral endothelial function (Micieli et al. 1997, 1999; Okamoto et al. 2001; Zvan et al. 2002; Zimmermann et al. 2004). The reduction of the CVR translates in an altered capacity for vasodilation of the cerebral arteries, which has been associated with the future development of cerebrovascular disease (Wijnhoud et al. 2011). TCD is a relatively simple, noninvasive, and low-cost bedside test that is widely available. It offers a unique real-time velocity measurement of intracranial vessels (McDonnell et al. 2013). TCD can also be utilized to assess CVR, i.e., the vasodilation of cerebral arterioles in response to a different physiological stimulus such as an increase in the partial pressure of arterial CO_2 (hypercapnia), infusion of acetazolamide or L-arginine (Ainslie and Duffin 2009; Willie et al. 2011; Fierstra et al. 2013).

The CVR to L-arginine is reported to be a reliable marker for cerebral endothelial function (Perko et al. 2011; John et al. 2015). Following a strict standardized protocol the method enables reproducible measurements. Its diagnostic utility for evaluation of endothelial impairment has been compared to flow-mediated dilatation as a gold standard for systemic endothelial function and intima-media thickness as a marker for morphological changes.

Additionally, other techniques have been used to assess cerebral endothelial function, including the vascular response to hypercapnia or infusion of acetazolamide, expressing the response as a percentage increase in mean arterial blood velocity in the middle cerebral artery or basilar artery. Change in blood oxygen level dependent (BOLD) signal during hypercapnia detected using functional MRI (Hund-Georgiadis et al. 2003), percent increase in regional cerebral blood flow in response to hypercapnia using stable Xenon CT (Mochizuki et al. 1997) and "dynamic cerebral autoregulation" (the ability to restore cerebral blood flow following sudden changes in perfusion pressure; Immink et al. 2005) are also used. Measurements of the hemodynamic response to hypercapnia with both functional MRI and functional Near-InfraRed Spectroscopy (fNIRS) are practical, noninvasive methods to assess CVR in humans. These results support the use of both functional MRI and fNIRS as biomarkers for evaluation of cerebral endothelial dysfunction.

7.3 Mechanisms of Obesity-Induced Cerebrovascular Endothelial Dysfunction: Altered Adipokine Secretion, Hypertension, and Insulin Resistance

Obesity is a major determinant of vascular diseases. It leads to several physiologic changes, including hyperinsulinemia and insulin resistance, increased activation of the sympathetic nervous system, sodium retention, and increased oxidative stress (Dearborn et al. 2015; Rao et al. 2015; dos Santos Moreira et al. 2015; Thorp and Schlaich 2015). Obesity also promotes progressive endothelial dysfunction, not only in large arteries, but also at the level of microcirculation (Bagi et al. 2012; Barton et al. 2012; Campia et al. 2012). The etiology of cerebrovascular endothelial dysfunction in obesity is complex with metabolic, inflammatory, and hemodynamic factors playing a role (Fig. 7.2).

7.3.1 Adipokines

Adipokines were suggested to be a link between obesity and vascular diseases. Several studies have reported the impact of adipokines on the vascular wall describing changes in the release of endothelium-derived vasoactive factors and also influencing local inflammation, growth, and remodeling (Lee et al. 2009; Miao and Li 2012; Hui et al. 2012; Jamaluddin et al. 2012). Adipokines like leptin can counteract systemic and central nervous system molecular alterations associated with Alzheimer's disease (Greco et al. 2009; Chakrabarti et al. 2015). These findings provide some support for the possibility that secretions from adipose tissue may impact brain inflammation and subsequently alter the risk of endothelial dysfunction in obesity. It is also important to profile the secretome based on the source of adipose tissue (lean vs. obese mouse models or human subjects), which may determine positive or negative adipose-vascular-brain crosstalk.

Direct experimental evidence for adipose-brain crosstalk in vivo presents many technical challenges, particularly in humans. It was recently suggested, using an in vitro model with adipose tissue organ culture conditioned media applied to human SH-SY5Y neuronal cells, that adipokines can regulate adipose-brain crosstalk and can play a role in neuroprotection or neurodegeneration depending on the adiposity status of the individual (Wan et al. 2015). The results suggest that lean adipose tissue may secrete certain adipokines that are protective towards neurons whereas in obesity, adipose tissue secretome changes and loses this protective

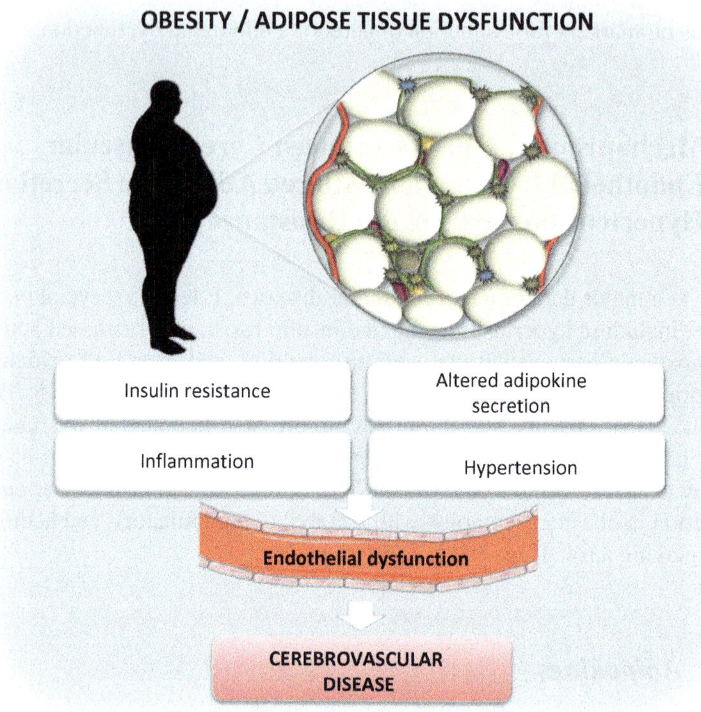

Fig. 7.2 Mechanisms of obesity-induced cerebrovascular endothelial dysfunction. Insulin resistance, inflammation, altered adipokine secretion (reduced adiponectin and enhanced leptin circulating levels), and hypertension are systemic consequences of a dysfunctional adipose tissue. They promote progressive endothelial dysfunction in large and small arteries, and may contribute to the pathophysiological events leading to ischemic or hemorrhagic stroke

potential. It remains to be determined which adipokine(s) may possesses these neuroprotective properties (Little and Safdar 2015).

Adipokines are surely important links between obesity and cerebrovascular disease. Obesity leads to adipose tissue dysfunction, triggering the release of proinflammatory adipokines that can directly act on vascular tissues to promote disease. The adipokine imbalance can also affect the function of metabolically important tissues promoting insulin resistance and inflammation and indirectly contributing to vascular disease.

The influence of adipokines at cerebrovascular function is increasingly suggested but very few studies address this topic and provide evidence to support the adipokine-cerebrovascular vessel crosstalk.

Several studies have demonstrated that lower adiponectin levels are closely related to endothelial dysfunction measured by forearm blood flow (a systemic evaluation of endothelial function) and an increased risk of coronary artery disease (Barseghian et al. 2011). Hypoadiponectinemia associated with obesity enhances endothelial dysfunction and predicts future cerebro- and cardiovascular diseases (Shimabukuro et al. 2003). Accordingly, obesity-dependent hypoadiponectinemia was associated with increased common carotid intima-media thickness in young and middle-aged women (Bang et al. 2007). Noteworthy, adiponectin administration in vivo reduced sepsis-induced microvascular dysfunction leading to BBB dysfunction in the mouse brain through a mechanism that involves E-selectin expression (Vachharajani et al. 2012). Thus, adiponectin seems to protect BBB integrity by reducing BBB permeability, microvascular MMP-9 expression, and parenchymal leukocyte accumulation (Cheng et al. 2009). Furthermore, adiponectin induces eNOS activation and consequentially increases CBF during ischemia (Nishimura et al. 2008). In addition, adiponectin has been shown to decrease expression of pro-inflammatory cytokines, by inhibiting NF-kB (Cheng et al. 2009; Spranger et al. 2006). A similar decrease in inflammatory cytokine expression was found in cultured brain ECs treated with adiponectin (Spranger et al. 2006). Adiponectin has been extensively shown to protect the vascular endothelium in the periphery (Sena et al. 2017). Therefore, a loss of these protective effects may explain increases in stroke damage and microvascular complications in obese mice where adiponectin levels are chronically decreased. In agreement, recent data described that obesity exacerbates experimental ischemia by increasing apoptosis of adiponectin-expressing neurons (Wu et al. 2016).

Resistin can also increase the risk of stroke by promoting systemic inflammation and endothelial dysfunction, both playing a significant role in atherosclerosis (Reilly et al. 2005). It was previously found that the association between resistin and insulin resistance remained significant after adjustment for obesity as well as markers for inflammation and endothelial dysfunction (Rajpathak et al. 2011). The participation of resistin in systemic endothelial dysfunction in insulin-resistant patients related to its direct effect on ECs promoting the release of ET-1 (Verma et al. 2003). It was recently documented that high resistin levels were associated with cerebrovascular symptomatology and low chemerin levels were associated with carotid disease severity, suggesting that these adipokines may act as potential markers for plaque instability and stroke risk (Gasbarrino et al. 2016). Further studies are needed to corroborate that adipokines may be a link between excessive adipose tissue accumulation and cerebrovascular dysfunction.

7.3.2 Hypertension

Hypertension is also associated with obesity and is known to modify functional hyperemia and endothelial function (Iadecola and Davisson 2008). Endothelial function, crucial for endothelium-dependent dilation and regulation of myogenic reactivity, is impaired in hypertension. The mechanism are related to vasodilator alterations involving NO, epoxyeicosatrienoic acids, and ion channels, including calcium-activated potassium channels and transient receptor potential vanilloid channel 4 (Pires et al. 2013). Reductions in cerebrovascular responses to endothelium-dependent vasodilators have been described in animal models of chronic hypertension (Didion et al. 2000; Capone et al. 2011, 2012a). Functional hyperemia is attenuated in mice treated with slow-pressor angiotensin II (Ang II) and in spontaneously hypertensive rats (Capone et al. 2012a; Calcinaghi et al. 2013). The cerebrovascular dysfunction precedes the hypertension induced by Ang II infusion and persists beyond the elevation in blood pressure (Capone et al. 2011). Accordingly, systemic endothelial dysfunction precedes hypertension in diet-induced animal models of insulin resistance (Katakam et al. 1998). In addition, diet-induced obesity causes cerebral vessel remodeling and increases the damage caused by ischemic stroke (Deutsch et al. 2009; Osmond et al. 2009).

In humans, the association of chronically or acutely elevated blood pressure with markers of inflammation has also been documented. Circulating levels of sICAM-1, sVCAM-1, sE-selectin, and MCP-1 are increased in patients with essential hypertension (Madej et al. 2005; Palomo et al. 2003). Increasing levels of adhesion molecules and chemoattractant molecules could induce monocyte adhesion on the vascular surface and migration into subendothelial lesions in both aorta and brain vessels.

Hypertension also impairs neurovascular coupling (Kazama et al. 2003) and CVR to CO_2, a measure of brain endothelial function (Serrador et al. 2005). Hypertension exposes the cerebral microvasculature to pulsatile pressure and flow that cause vascular endothelium and smooth muscle cell tearing (O'Rourke and Safar 2005).

Confirming studies in animal models, patients with chronic hypertension have CBF diminished and the cerebrovascular dysfunction correlated with cognitive deficits (Jennings et al. 2005). Direct evidence of altered cerebrovascular endothelial responses in humans with hypertension is lacking. But the NO synthase inhibitor NG-nitro-L-arginine methyl ester does not reduce retinal blood flow in patients with hypertension, a finding consistent with NO-dependent endothelial dysfunction (Delles et al. 2004). Previous studies have suggested that hypertension disrupts key cerebrovascular control mechanisms aimed at maintaining the energy homeostasis of the brain, which act in concert with the structural alterations of cerebral blood vessels described earlier to produce brain dysfunction and damage (Faraco and Iadecola 2013).

Several lines of evidence suggest that reactive oxygen species (ROS) are key mediators of the cerebrovascular damage produced by hypertension. Indeed, oxidative stress has also been implicated in other vascular effects of hypertension, including vascular remodeling and inflammation (De Silva and Faraci 2012). Hypertension promotes ROS production in cerebral blood vessels and ROS scavengers counteract the effects of hypertension on functional hyperemia and endothelial dysfunction, including alterations of the BBB (Faraco and Iadecola 2013). Cerebrovascular ET-1, via ET type A receptor and ROS, is also involved in the cerebrovascular effects of hypertension induced by chronic intermittent hypoxia (Capone et al. 2012b).

Importantly, hypertension can promote several cellular damages years before symptoms develop, through an event cascade that thickens and stiffens artery walls throughout the body. The increase in arterial stiffness was closely associated with the onset of hypertension, indicative of the well-established relationship between elevated arterial pressure and vascular remodeling (Pires et al. 2013).

Studies of experimental models of other cerebral vascular risk factors such as hypercholesterolemia (Kitayama et al. 2007; Miller et al. 2010), diabetes (Mayhan et al. 1991, 2006; Didion et al. 2005, 2007; Arrick et al. 2007; Ergul et al. 2009), and obesity (Lynch et al. 2013; Brooks et al. 2015) indicate that all are associated with functional abnormalities and oxidative stress. Genetic deletion of Nox2 in hypercholesterolemic apolipoprotein E-deficient mice prevents elevations in ROS production and impaired vasodilator capacity of cerebral vessels (Miller et al. 2010). Similarly, Nox2-deficient mice are protected against obesity-induced dysfunction of cerebral arterioles (Lynch et al. 2013). Moreover, activation of poly(ADP)-ribose polymerase (Arrick et al. 2007), Rho kinase (Didion et al. 2005), and Ang II type 1 receptors (Arrick et al. 2007) have also been shown to be involved in cerebral microvascular dysfunction. Thus, as in models of hypertension and aging, the generation of ROS by NADPH oxidases and the activation of key signaling pathways such as poly(ADP)-ribose polymerase and Rho kinase might be a common underlying mechanism of cerebral vascular dysfunction caused by many of the known risk factors for cerebrovascular disease.

A recent study found that myogenic tone and reactivity are different in collateral versus noncollateral arterioles (Chan et al. 2016). During hypertension, for example, collaterals have increased tone and impaired vasodilation, which may result in lower levels of collateral dependent blood flow, contributing to greater infarct size in response to ischemia (Chan et al. 2016).

7.3.3 Insulin Resistance

Insulin resistance is fundamental to the metabolic syndrome and drives the adverse effects of obesity on the brain, likely due to the associated endothelial dysfunction and abnormalities in peripheral vascular reactivity. It constitutes a vital risk factor that causes oxidative stress and endothelial dysfunction in systemic vessels of obese animal models and humans (Katakam et al. 1998; Prakash et al. 2016).

The link between insulin resistance and cerebrovascular dysfunction in obesity has not been clearly established. Data regarding the impact of insulin resistance in the CVR is limited and contradictory. It was recently reported a correlation between CVR and insulin resistance in type 2 diabetes patients, although Rodríguez-Flores et al. (2014) found a significant negative linear association between homeostatic model assessment of insulin resistance (HOMA-IR) and breath-holding index (a method to assess CVR) in obese individuals. There contradictory results are probably explained by the different parameters used for evaluating CVR and also due to different groups of patients. Prakash et al. (2016) reported significant correlations between HOMA-IR and Delta cerebrovascular conductance index (CVCi), but not between HOMA-IR and other CVR parameters. According to the authors, this supports the view that calculation of CVR as delta CVCi is more physiologically plausible than other CVR parameters.

Overall, careful investigation of the role of endothelial function and cerebral vascular reactivity is warranted to understand the adverse effects of obesity-associated metabolic syndrome on brain integrity.

7.4 Adipokines and the Risk of Stroke

Despite the well-known global impact of overweight and obesity in the incidence of cerebrovascular disease, many aspects of this association are still not consistently defined. In the last decade several groups have focused their research on adipokines, namely on their capacity to predict stroke incidence and its outcomes. Even though their participation in the pathophysiology of stroke has been reported by independent studies, there are still incongruences and contradictory results, that may relate with their susceptibility to other factors such as gender, age, systemic inflammation, renal function, and circadian rhythms (Kantorová et al. 2015). Moreover, it is still controversial their role as cerebrovascular risk factors independently of the presence of obesity or adipose tissue dysfunction. Below we resume the state of current knowledge on the association between the most extensively studied adipokines and stroke.

7.4.1 Adiponectin

Growing interest on the association between adiponectin and the risk of stroke is based on the fundamental role played by this adipokine in vascular homeostasis and protection against early atherosclerotic events (Nishimura et al. 2008; Vaiopoulos et al. 2012), including its effects on the cerebrovascular endothelium (Ouchi et al. 2011). It has been shown that adiponectin is not present in intact vessel walls but accumulates at sites of damaged endothelium, for example, after cerebral ischemia-reperfusion injury, although the exact mechanisms involved are still elusive (Nishimura et al. 2008). Therefore, it is easy to accept and understand results that converge into the idea that adiponectin is protective against ischemic cerebrovascular disease (Kantorova et al. 2011; Bang et al. 2007), is associated with carotid intima-media thickness (IMT) (Bang et al. 2007; Lo et al. 2006), and constitute an independent predictor of mortality after ischemic stroke (Efstathiou et al. 2005). On the other hand, some studies report no impact of adiponectin levels on ischemic stroke incidence, while others refer that it has a deleterious effect (Ogorodnikova et al. 2010; Rajpathak et al. 2011; Wannamethee et al. 2013). The fact that adiponectin can be increased in heart failure and positively associated with aging must be taken into account when interpreting these results, although some authors advocate that these conditions may stimulate an overproduction of adiponectin to counterbalance atherosclerosis (Stott et al. 2009). Several studies have otherwise tried to correlate stroke subtype with adiponectin levels, but besides a relative consensus on the association between low adiponectinemia and atherotrombotic stroke (Kantorova et al. 2011; Bang et al. 2007; Kim et al. 2012), it seems that the levels of this adipokine are weakly correlated with the etiology of the ischemic event (Kantorová et al. 2015). A recent meta-analysis published by Gorgui et al. (2017) indicates that high circulating adiponectin levels showed an 8% increase in the risk of ischemic stroke, independently of HDL levels. Currently, it remains unclear whether specific molecular-weight fractions of adiponectin may influence risk of stroke, once all studies published to date measure total or HMW adiponectin.

The benefits of adiponectin have also been explored in the context of hemorrhagic stroke. Osuka et al. (2012) have detected a transient but significant (200-fold) increase in adiponectin concentration in CSF within 24 h after subarachnoid hemorrhage, which was interpreted as a strategy to counteract vasospasm via activation of the AMPK/eNOS signaling pathway. These results are, in part, corroborated by the study of Takeuchi et al. (2014) which describes an association between low plasma adiponectin concentration and the development of vasospasm-induced delayed cerebral ischemia. Additionally, after ICH there seems to be an increase on the plasma adiponectin levels within the first 24 h which remain elevated for a period of 7 days when compared to controls. In these patients, plasma adiponectin concentration was an independent predictor for short-term mortality (Wang et al. 2011).

7.4.2 Leptin

The pathological obesity-associated burden includes a leptin resistance state characterized by decreased action of this adipokine on brain parenchyma and vasculature, despite normal or even high plasmatic and CSF levels. Although high circulating leptin levels are strongly correlated with several vascular risk factors, hyperleptinemia has been inconsistently associated with increased risk of ischemic stroke and some authors describe a discordant gender specificity (Bouziana et al. 2016). Higher levels of leptin have also been associated with the acute-phase of ischemic stroke (Söderberg et al. 2003). The vast majority of the research on this field lies on cross-sectional studies although this may not be the optimal study design to address this association, once leptin has been pointed as an acute stress responder. Notwithstanding, there are prospective studies assessing the risk of first-ever cerebral ischemic event based on leptin levels measured at baseline, some of them with 10 years of follow-up. A meta-analysis including some of these prospective studies did not find any significant association between leptin levels and risk of stroke (Saber et al. 2015). Interestingly, the same authors describe an inverse association between leptin and the risk of stroke for elderly subjects in the top waist-to-hip ratio quartile. They hypothezise that in states of relative leptin resistance, such as obesity, low leptin levels may increase the risk of vascular events based on the evidence that hyperleptinemia can protect nonadipose tissues from lipotoxicity and oxidative damage and play an important role in NO regulation and production. Regarding the potential therapeutical application of leptin, experimental data has not been supported by clinical results. For example, although leptin administration to a mice model of transient focal cerebral ischemia is able to decrease total infarct volume (Zhang et al. 2013), the study conducted by Calleja et al. (2013) in ischemic stroke patients treated with rtPA describes an association between higher leptin levels and larger infarct volumes. Leptin has also been associated with increased risk of hemorrhagic stroke (Söderberg et al. 1999, 2003) and elevated levels were associated with poor outcomes after ICH (Zhao et al. 2012). Moreover, leptin may play a critical role in secondary brain injury around the hematoma via the STAT3 signaling pathway in microglia, which transduce a pro-inflammatory harmful signal. Interestingly, the inhibition of STAT3 in the absence of hyperleptinemia did not result in perihematomal edema and inflammation after ICH (Kim et al. 2013), suggesting that its role as modulator in these cases is controversial.

7.4.3 Resistin

Contrarily to what is observed with both adipokines previously described, resistin is mainly produced by macrophages and monocytes and to a smaller extent by adipocytes. It can, in theory, increase the risk of ischemic stroke by promoting atherosclerosis and systemic and vascular inflammation as described previously in this chapter. Moreover, resistin mRNA has been shown to increase in the hypoxic/ischemic cortex (Wiesner et al. 2006) and it has been pointed as a potential biomarker for carotid plaque instability (Gasbarrino et al. 2016). Some large epidemiological studies have tried to evaluate the capacity of resistin to influence and predict the risk of ischemic stroke such as the PRIME study and the Women's Health Initiative study, that considered resistin an independent predictor of ischemic stroke (Prugger et al. 2012; Rajpathak et al. 2011). On the other hand, other groups did not find any correlation, although reported that higher levels of resistin were associated with higher risk of myocardial infarction (Weikert et al. 2008). To our knowledge, only two studies have addressed the association between resistin and stroke severity and outcomes. Efstathiou et al. (2007) reported that higher plasma levels of resistin (measured within 24 h after atherothrombotic ischemic stroke) were independently associated with increased risk of mortality or dependency at 5 years of follow-up. Another small case-control study described a positive correlation between stroke severity and peripheral levels of resistin in women with acute ischemic stroke (Kochanowski et al. 2012). However other inflammatory cytokines, such as TNF-α, are upregulated in these conditions, as demonstrated in the study mentioned before, and may also influence stroke outcomes. Interestingly, results from the Hisayama Study have shown that elevated resistinemia was an independent risk factor for ischemic cerebrovascular disease in a Japanese population, especially lacunar and atherothrombotic stroke, while no relation was observed for hemorrhagic stroke (Osawa et al.

2009). Other groups described higher plasma levels of resistin in patients with spontaneous intracerebral hemorrhage and suggest that it could be involved in the inflammatory component of hemorrhagic stroke, and thus associated with poor outcome (Dong et al. 2010, 2012).

7.4.4 Apelin

The neuroprotective effect of apelin has been explored in the last decade both in vitro and in vivo studies. This adipokine soon became a potential target in cerebrovascular disease since it is an analogue of the angiotensin II receptor, and in addition, its receptor, AGTRL1, was associated with ischemic stroke in genome-wide association studies (Hata et al. 2011). The potential benefits and neuroprotective effects of apelin-13, the most active isoform, have been demonstrated by the improvement of neurological deficits in ischemia-reperfusion injury models. Apelin can protect cells from programmed death by preventing oxidative and ER stress, and possibly by suppressing anti-inflammatory cascades and autophagy, mediated by PI3K/AKT signaling pathway. Apelin can facilitate angiogenesis and reperfusion, which seem to be dependent on an increase in vascular endothelial growth factor-vascular endothelial growth factor receptor 2 (VEGF-VEGFR2) signaling (Wu et al. 2017). The only clinical study to date investigating the role of apelin in acute ischemic stroke failed to demonstrate an association between the event and apelin circulating levels within the first 24 h, although no data exist on its levels on follow-up. However, patients with carotid atherosclerosis tend to have lower serum apelin concentration when compared to those without carotid atherosclerosis (Kadoglou et al. 2014). The participation of apelin in the pathophysiology of ICH was also investigated by Bao et al. (2016). They described the capacity of this adipokine in reducing brain edema and apoptosis following ICH in a mouse model, through MMP-9 downregulation and by concomitant caspase-3 downregulation and Bcl-2 upregulation, respectively. These results show that the neuroprotective effects of this adipokine are not limited to ischemic injury, although the research in hemorrhagic stroke and its relation with adipokines is overlooked.

7.4.5 Visfatin

Accumulating data from clinical studies have provided evidence on the association between ischemic and hemorrhagic stroke and the elevation of visfatin plasmatic levels. Visfatin, also called nicotinamide phosphoribosyltransferase (NAMPT) and secreted mainly by visceral adipose tissue, is a rate-limiting enzyme that participates in metabolic signaling pathways involved in cell survival, and thus a promising therapeutic target in cerebrovascular disease (Wang and Miao 2015; Chen et al. 2015). Its neuroprotective, proneuritogenic, and proangiogenic effects have been recently demonstrated (Wang and Miao 2015; Chen et al. 2015), although its potential role in neuroinflammation and the fact that it seems to be a pro-inflammatory mediator in carotid atherosclerotic plaques (Dahl et al. 2007) still needs clarification.

7.5 Conclusions

Recent advances in knowledge surrounding dysfunctional adipose tissue and obesity have provided new insights into the mechanisms of obesity-induced cerebrovascular endothelial dysfunction: insulin resistance, inflammation, altered adipokine secretion (reduced adiponectin and enhanced leptin circulating levels), and hypertension. This review highlighted the contributions of all these mechanisms that promote progressive endothelial dysfunction in large and small cerebral arteries, and may contribute to the pathophysiological events leading to ischemic or hemorrhagic stroke. Understanding these mechanisms in detail will enable to find novel biomarkers and new therapeutic strategies to improve stroke incidence as well as its functional outcomes and survival

References

Abbott NJ, Friedman A (2012) Overview and introduction: the blood-brain barrier in health and disease. Epilepsia 53(Suppl 6):1–6

Ainslie PN, Duffin J (2009) Integration of cerebrovascular CO_2 reactivity and chemoreflex control of breathing: mechanisms of regulation, measurement, and interpretation. Am J Physiol Regul Integr Comp Physiol 296:R1473–R1495

Arrick DM, Sharpe GM, Sun H, Mayhan WG (2007) Diabetes-induced cerebrovascular dysfunction: role of poly(ADP-ribose) polymerase. Microvasc Res 73:1–6

Bagi Z, Feher A, Cassuto J (2012) Microvascular responsiveness in obesity: implications for therapeutic intervention. Br J Pharmacol 165:544–560

Bang OY, Saver JL, Ovbiagele B, Choi YJ, Yoon SR, Lee KH (2007) Adiponectin levels in patients with intracranial atherosclerosis. Neurology 68:1931–1937

Bao H, Yang X, Huang Y, Qiu H, Huang G, Xiao H, Kuai J (2016) The neuroprotective effect of apelin-13 in a mouse model of intracerebral hemorrhage. Neurosci Lett 628:219–224

Barseghian A, Gawande D, Bajaj M (2011) Adiponectin and vulnerable atherosclerotic plaques. J Am Coll Cardiol 57:761–770

Barton M, Baretella O, Meyer MR (2012) Obesity and risk of vascular disease: importance of endothelium-dependent vasoconstriction. Br J Pharmacol 165:591–602

Bazzano LA, Gu D, Whelton MR, Wu X, Chen CS, Duan X et al (2010) Body mass index and risk of stroke among Chinese men and women. Ann Neurol 67:11–20

Biffi A, Cortellini L, Nearnberg CM, Ayres AM, Schwab K, Gilson AJ, Rost NS, Goldstein JN, Viswanathan A, Greenberg SM, Rosand J (2011) Body mass index and etiology of intracerebral hemorrhage. Stroke 42(9):2526–2530

Bodenant M, Kuulasmaa K, Wagner A, Kee F, Palmieri L, Ferrario MM, Montaye M, Amouyel P, Dallongeville J, MORGAM Project (2011) Measures of abdominal adiposity and the risk of stroke: the MOnica Risk, Genetics, Archiving and Monograph (MORGAM) study. Stroke 42(10):2872–2877

Boos CJ, Lip GY, Blann AD (2006) Circulating endothelial cells in cardiovascular disease. J Am Coll Cardiol 48:1538–1547

Boos CJ, Jaumdally RJ, RJ MF, Varma C, Lip GY (2007) Circulating endothelial cells and von Willebrand factor as indices of endothelial damage/dysfunction in coronary artery disease: a comparison of central vs. peripheral levels and effects of coronary angioplasty. J Thromb Haemost 5:630–632

Bounameaux H, John Camm A (2014) Edoxaban: an update on the new oral direct factor Xa inhibitor. Drugs 74(11):1209–1231

Bouziana S, Tziomalos K, Goulas A, Hatzitolios AI (2016) The role of adipokines in ischemic stroke risk stratification. Int J Stroke 11 (4):389–398

Branscheidt M, Schneider J, Michel P, Eskioglou E, Kaegi G, Stark R, Fischer U, Jung S, Arnold M, Wertli M, Held U, Wegener S, Luft A, Sarikaya H (2016) No impact of body mass index on outcome in stroke patients treated with IV thrombolysis BMI and IV thrombolysis outcome. PLoS One 11(10):e0164413

Brooks SD, DeVallance E, d'Audiffret AC, Frisbee SJ, Tabone LE, Shrader CD, Frisbee JC, Chantler PD (2015) Metabolic syndrome impairs reactivity and wall mechanics of cerebral resistance arteries in obese Zucker rats. Am J Physiol Heart Circ Physiol 309:H1846–H1859

Burger D, Touyz RM (2012) Cellular biomarkers of endothelial health: microparticles, endothelial progenitor cells, and circulating endothelial cells. J Am Soc Hypertens 6:85–99

Calcinaghi N, Wyss MT, Jolivet R, Singh A, Keller AL, Winnik S, Fritschy JM, Buck A, Matter CM, Weber B (2013) Multimodal imaging in rats reveals impaired neurovascular coupling in sustained hypertension. Stroke 44:1957–1964

Calleja AI, Cortijo E, García-Bermejo P, Reyes J, Bermejo JF, Muñoz MF, Fernandez-Herranz R, Arenillas JF (2013) Blood biomarkers of insulin resistance in acute stroke patients treated with intravenous thrombolysis: Temporal profile and prognostic value. J Diab Res Clin Metab 2(1):2

Campia U, Tesauro M, Cardillo C (2012) Human obesity and endothelium-dependent responsiveness. Br J Pharmacol 165:561–573

Capone C, Faraco G, Park L, Cao X, Davisson RL, Iadecola C (2011) The cerebrovascular dysfunction induced by slow pressor doses of angiotensin II precedes the development of hypertension. Am J Physiol Heart Circ Physiol 300:H397–H407

Capone C, Faraco G, Peterson JR, Coleman C, Anrather J, Milner TA, Pickel VM, Davisson RL, Iadecola C (2012a) Central cardiovascular circuits contribute to the neurovascular dysfunction in angiotensin II hypertension. J Neurosci 32:4878–4886

Capone C, Faraco G, Coleman C, Young CN, Pickel VM, Anrather J, Davisson RL, Iadecola C (2012b) Endothelin 1-dependent neurovascular dysfunction in chronic intermittent hypoxia. Hypertension 60:106–113

Chakrabarti S, Khemka VK, Banerjee A, Chatterjee G, Ganguly A, Biswas A (2015) Metabolic risk factors of sporadic Alzheimer's disease: implications in the pathology, pathogenesis and treatment. Aging Dis 6:282–299

Chan SL, Sweet JG, Bishop N, Cipolla MJ (2016) Pial collateral reactivity during hypertension and aging: understanding the function of collaterals for stroke therapy. Stroke 47:1618–1625

Chen X, Zhao S, Song Y, Shi Y, Leak RK, Cao G (2015) The role of nicotinamide phosphoribosyltransferase in cerebral ischemia. Curr Top Med Chem 15(21):2211–2221

Cheng Z, Yang X, Wang H (2009) Hyperhomocysteinemia and endothelial dysfunction. Curr Hypertens Rev 5:158–165

Chu C, Tokumaru S, Izumi K, Nakagawa K (2016) Obesity increases risk of anticoagulation reversal failure with prothrombin complex concentrate in those with intracranial hemorrhage. Int J Neurosci 126(1):62–66

Dahl TB, Yndestad A, Skjelland M, Øie E, Dahl A, Michelsen A, Damås JK, Tunheim SH, Ueland T, Smith C, Bendz B, Tonstad S, Gullestad L, Frøland SS, Krohg-Sørensen K, Russell D, Aukrust P, Halvorsen B (2007) Increased expression of visfatin in macrophages of human unstable carotid and coronary atherosclerosis: possible role in inflammation and plaque destabilization. Circulation 115(8):972–980

De Silva TM, Faraci FM (2012) Effects of angiotensin II on the cerebral circulation: role of oxidative stress. Front Physiol 3:484

Dearborn JL, Schneider AL, Sharrett AR, Mosley TH, Bezerra DC, Knopman DS, Selvin E, Jack CR, Coker LH, Alonso A, Wagenknecht LE, Windham BG, Gottesman RF (2015) Obesity, insulin resistance, and incident small vessel disease on magnetic resonance imaging: atherosclerosis Risk in Communities Study. Stroke 46:3131–3136

Deb M, Gerdes S, Heeren M, Lambrecht J, Worthmann H, Goldbecker A, Tryc AB, Lovric S, Schulz-Schaeffer W, Brandis A, Dengler R, Weissenborn K, Haubitz M (2013) Circulating endothelial cells as potential diagnostic biomarkers in primary central nervous system vasculitis. J Neurol Neurosurg Psychiatry 84:732–734

Delles C, Michelson G, Harazny J, Oehmer S, Hilgers KF, Schmieder RE (2004) Impaired endotelial function of the retinal vasculature in hypertensive patients. Stroke 35:1289–1293

Deutsch C, Portik-Dobos V, Smith AD, Ergul A, Dorrance AM (2009) Diet-induced obesity causes cerebral vessel remodeling and increases the damage caused by ischemic stroke. Microvasc Res 78:100–106

Didion SP, Sigmund CD, Faraci FM (2000) Impaired endothelial function in transgenic mice expressing both human renin and human angiotensinogen. Stroke 31:760–764

Didion SP, Lynch CM, Baumbach GL, Faraci FM (2005) Impaired endothelium-dependent responses and enhanced influence of Rho-kinase in cerebral arterioles in type II diabetes. Stroke 36:342–347

Didion SP, Lynch CM, Faraci FM (2007) Cerebral vascular dysfunction in TallyHo mice: a new model of Type II diabetes. Am J Physiol Heart Circ Physiol 292:H1579–H1583

Dong XQ, Hu YY, Yu WH, Zhang ZY (2010) High concentrations of resistin in the peripheral blood of patients with acute basal ganglia hemorrhage are associated with poor outcome. J Crit Care 25(2):243–247

Dong XQ, Du Q, Yu WH, Zhang ZY, Zhu Q, Che ZH, Wang H, Chen J, Yang SB, Wen JF (2012) Plasma resistin, associated with single nucleotide polymorphism −420, is correlated with C-reactive protein in Chinese Han patients with spontaneous basal ganglia hemorrhage. Genet Mol Res 11(3):1841–1850

dos Santos Moreira MC, de Jesus Pinto IS, Mourão AA, Fajemiroye JO, Colombari E, da Silva Reis ÂA, Freiria-Oliveira AH, Ferreira-Neto ML, Pedrino GR (2015) Does the sympathetic nervous system contribute to the pathophysiology of metabolic syndrome? Front Physiol 6:234

Drake CT, Iadecola C (2007) The role of neuronal signaling in controlling cerebral blood flow. Brain Lang 102:141–152

Efstathiou SP, Tsioulos DI, Tsiakou AG, Gratsias YE, Pefanis AV, Mountokalakis TD (2005) Plasma adiponectin levels and five-year survival after first-ever ischemic stroke. Stroke 36(9):1915–1919

Efstathiou SP, Tsiakou AG, Tsioulos DI, Panagiotou TN, Pefanis AV, Achimastos AD, Mountokalakis TD (2007) Prognostic significance of plasma resistin levels in patients with atherothrombotic ischemic stroke. Clin Chim Acta 378(1–2):78–85

Ergul A, Li W, Elgebaly MM, Bruno A, Fagan SC (2009) Hyperglycemia, diabetes and stroke: focus on the cerebrovasculature. Vasc Pharmacol 51:44–49

Fadini GP, Bonora BM, Marcuzzo G, Marescotti MC, Cappellari R, Pantano G, Sanzari MC, Duran X, Vendrell J, Plebani M, Avogaro A (2015) Circulating stem cells associate with adiposity and future metabolic deterioration in healthy subjects. J Clin Endocrinol Metab 100:4570–4578

Faraco G, Iadecola C (2013) Hypertension: a harbinger of stroke and dementia. Hypertension 62:810–817

Fierstra J, Sobczyk O, Battisti-Charbonney A, Mandell DM, Poublanc J, Crawley AP, Mikulis DJ, Duffin J, Fisher JA (2013) Measuring cerebrovascular reactivity: what stimulus to use? J Physiol 591:5809–5821

Franceschelli S, Ferrone A, Pesce M, Riccioni G, Speranza L (2013) Biological functional relevance of asymmetric dimethylarginine (ADMA) in cardiovascular disease. Int J Mol Sci 14:24412–24421

Gasbarrino K, Mantzoros C, Gorgui J, Veinot JP, Lai C, Daskalopoulou SS (2016) Circulating chemerin is associated with carotid plaque instability, whereas resistin is related to cerebrovascular symptomatology. Arterioscler Thromb Vasc Biol 36:1670–1678

Gorgui J, Gasbarrino K, Georgakis MK, Karalexi MA, Nauche B, Petridou ET, Daskalopoulou SS (2017) Circulating adiponectin levels in relation to carotid atherosclerotic plaque presence, ischemic stroke risk, and mortality: A systematic review and meta-analyses. Metabolism 69:51–66

Gostynski M, Gutzwiller F, Kuulasmaa K, Döring A, Ferrario M, Grafnetter D, Pajak A et al (2004) Analysis of the relationship between total cholesterol, age, body mass index among males and females in the WHO MONICA Project. Int J Obes Relat Metab Disord 28:1082–1090

Greco SJ, Bryan KJ, Sarkar S, Zhu X, Smith MA, Ashford JW, Johnston JM, Tezapsidis N, Casadesus G (2009) Leptin reduces pathology and improves memory in a transgenic mouse model of Alzheimer's disease. J Alzheimers Dis 19:1155–1167

Green DJ, Dawson EA, Groenewoud HM, Jones H, Thijssen DH (2014) Is flow-mediated dilation nitric oxide mediated? A meta-analysis. Hypertension 63:376–382

Güler E, Babur Güler G, Demir GG, Hatipoğlu S (2015) A review of the fixed dose use of new oral anticoagulants in obese patients: is it really enough? Anatol J Cardiol 15(12):1020–1029

Hajer GR, van Haeften TW, Visseren FL (2008) Adipose tissue dysfunction in obesity, diabetes, and vascular diseases. Eur Heart J 29:2959–2971

Hart R, Veenstra DL, Boudreau DM, Roth JA (2017) Impact of Body Mass Index and Genetics on Warfarin Major Bleeding Outcomes in a Community Setting. Am J Med 130(2):222–228

Hata J, Kubo M, Kiyohara Y (2011) Genome-wide association study for ischemic stroke based on the Hisayama study. Nihon Eiseigaku Zasshi 66(1):47–52

Hui X, Lam KS, Vanhoutte PM, Xu A (2012) Adiponectin and cardiovascular health: an update. Br J Pharmacol 165:574–590

Hund-Georgiadis M, Zysset S, Naganawa S, Norris DG, Yves von Cramon D (2003) Determination of cerebrovascular reactivity by means of fMRI signal changes in cerebral microangiopathy: a correlation with morphological abnormalities. Cerebrovasc Dis 16:158–165

Iadecola C, Davisson RL (2008) Hypertension and cerebrovascular dysfunction. Cell Metab 7:476–484

Immink RV, van Montfrans GA, Stam J, Karemaker JM, Diamant M, van Lieshout JJ (2005) Dynamic cerebral autoregulation in acute lacunar and middle cerebral artery territory ischemic stroke. Stroke 36:2595–2600

Jamaluddin MDS, Weakley S, Yao Q, Chen C (2012) Resistin: functional roles and therapeutic considerations for cardiovascular disease. Br J Pharmacol 165:622–632

Jennings JR, Muldoon MF, Ryan C, Price JC, Greer P, Sutton-Tyrrell K, van der Veen FM, Meltzer CC (2005) Reduced cerebral blood flow response and compensation among patients with untreated hypertension. Neurology 64:1358–1365

John S, Hajj-Ali RA, Min D, Calabrese LH, Cerejo R, Uchino K (2015) Reversible cerebral vasoconstriction syndrome: Is it more than just cerebral vasoconstriction? Cephalalgia 35:631–634

Kadoglou NP, Fotiadis G, Lambadiari V, Maratou E, Dimitriadis G, Liapis CD (2014) Serum levels of novel adipokines in patients with acute ischemic stroke: potential contribution to diagnosis and prognosis. Peptides 57:12–16

Kantorova E, Chomova M, Kurca E, Sivak S, Zelenak K, Kučera P, Galajda P (2011) Leptin, adiponectin and ghrelin, new potential mediators of ischemic stroke. Neuro Endocrinol Lett 32(5):716–721

Kantorová E, Jesenská Ľ, Čierny D, Zeleňák K, Sivák Š, Stančík M, Galajda P, Nosáľ V, Kurča E (2015) The intricate network of adipokines and stroke. Int J Endocrinol 2015:967698

Katakam PV, Ujhelyi MR, Hoenig ME, Miller AW (1998) Endothelial dysfunction precedes hypertension in diet-induced insulin resistance. Am J Physiol 275:R788–R792

Katsnelson M, Rundek T (2011) Obesity paradox and stroke: noticing the (fat) man behind the curtain. Stroke 42(12):3331–3332

Kazama K, Wang G, Frys K, Anrather J, Iadecola C (2003) Angiotensin II attenuates functional hyperemia in the mouse somatosensory cortex. Am J Physiol Heart Circ Physiol 285:H1890–H1899

Kernan WN, Inzucchi SE, Sawan C, Macko RF, Furie KL (2013) Obesity: a stubbornly obvious target for stroke prevention. Stroke 44(1):278–286

Kim BJ, Lee SH, Ryu WS, Kim CK, Yoon BW (2012) Adipocytokines and ischemic stroke: differential associations between stroke subtypes. J Neurol Sci 312(1–2):117–122

Kim CK, Ryu WS, Choi IY, Kim YJ, Rim D, Kim BJ, Jang H, Yoon BW, Lee SH (2013) Detrimental effects of leptin on intracerebral hemorrhage via the STAT3 signal pathway. J Cereb Blood Flow Metab 33:944–953

Kitayama J, Faraci FM, Lentz SR, Heistad DD (2007) Cerebral vascular dysfunction during hypercholesterolemia. Stroke 38:2136–2141

Kochanowski J, Grudniak M, Baranowska-Bik A, Wolinska-Witort E, Kalisz M, Baranowska B, Bik W (2012) Resistin levels in women with ischemic stroke. Neuro Endocrinol Lett 33(6):603–607

Lee KW, Blann AD, Lip GY (2005) Plasma markers of endothelial damage/dysfunction, inflammation and thrombogenesis in relation to TIMI risk stratification in acute coronary syndromes. Thromb Haemost 94:1077–1083

Lee DE, Kehlenbrink S, Lee H, Hawkins M, Yudkin JS (2009) Getting the message across: mechanisms of physiological cross talk by adipose tissue. Am J Physiol Endocrinol Metab 296:E1210–E1229

Lind L, Berglund L, Larsson A, Sundstrom J (2011) Endothelial function in resistance and conduit arteries and 5-year risk of cardiovascular disease. Circulation 123:1545–1551

Little JP, Safdar A (2015) Adipose-brain crosstalk: do adipokines have a role in neuroprotection? Neural Regen Res 10:1381–1382

Lo J, Dolan SE, Kanter JR, Hemphill LC, Connelly JM, Lees RS, Grinspoon SK (2006) Effects of obesity, body composition, and adiponectin on carotid intima-media thickness in healthy women. J Clin Endocrinol Metab 91(5):1677–1682

Lynch CM, Kinzenbaw DA, Chen X, Zhan S, Mezzetti E, Filosa J, Ergul A, Faulkner JL, Faraci FM, Didion SP (2013) Nox2-derived superoxide contributes to cerebral vascular dysfunction in diet-induced obesity. Stroke 44:3195–3201

Madej A, Okopień B, Kowalski J, Haberka M, Herman ZS (2005) Plasma concentrations of adhesion molecules and chemokines in patients with essential hypertension. Pharmacol Rep 57:878–881

Makin AJ, Blann AD, Chung NA, Silverman SH, Lip GY (2004) Assessment of endothelial damage in atherosclerotic vascular disease by quantification of circulating endothelial cells. Relationship with von Willebrand factor and tissue factor. Eur Heart J 25:371–376

Mayhan WG, Simmons LK, Sharpe GM (1991) Mechanism of impaired responses of cerebral arterioles during diabetes mellitus. Am J Phys 260(Pt 2):H319–H326

Mayhan WG, Arrick DM, Sharpe GM, Patel KP, Sun H (2006) Inhibition of NAD(P)H oxidase alleviates impaired NOS-dependent responses of pial arterioles in type 1 diabetes mellitus. Microcirculation 13:567–575

McDonnell MN, Berry NM, Cutting MA, Keage HA, Buckley JD, Howe PR (2013) Transcranial Doppler ultrasound to assess cerebrovascular reactivity: reliability, reproducibility and effect of posture. Peer J 1:e65

Meissner A (2016) Hypertension and the brain: a risk factor for more than heart disease. Cerebrovasc Dis 42(3–4):255–262

Miao CY, Li ZY (2012) The role of perivascular adipose tissue in vascular smooth muscle cell growth. Br J Pharmacol 165:643–658

Michiels C (2003) Endothelial cell functions. J Cell Physiol 196:430–443

Micieli G, Bosone D, Costa A, Cavallini A, Marcheselli S, Pompeo F, Nappi G (1997) Opposite effects of L-arginine and nitroglycerin on cerebral blood velocity: nitric oxide precursors and cerebral blood velocity. J Neurol Sci 150:71–75

Micieli G, Bosone D, Zappoli F, Marcheselli S, Argenteri A, Nappi G (1999) Vasomotor response to CO_2 and L-Arginine in patients with severe internal carotid artery stenosis; pre- and post-surgical evaluation with transcranial Doppler. J Neurol Sci 163:153–158

Miller AA, De Silva TM, Judkins CP, Diep H, Drummond GR, Sobey CG (2010) Augmented superoxide production by Nox2-containing NADPH oxidase causes cerebral artery dysfunction during hypercholesterolemia. Stroke 41:784–789

Mochizuki Y, Oishi M, Takasu T (1997) Cerebral blood flow in single and multiple lacunar infarctions. Stroke 28:1458–1460

Moreira MC, Pinto IS, Mourão AA, Fajemiroye JO, Colombari E, Reis ÂA, Freiria-Oliveira AH, Ferreira-Neto ML, Pedrino GR (2015) Does the sympathetic nervous system contribute to the pathophysiology of metabolic syndrome? Front Physiol 6:234

Murphey HS (2012) Inflammation. In: Streyer R (ed) Rubin's pathology: clinicopathologic foundations of medicine. Lippincott Williams and Wilkins, Philadelphia, PA, pp 23–44

Nishimura M, Izumiya Y, Higuchi A, Shibata R, Qiu J, Kudo C, Shin HK, Moskowitz MA, Ouchi N (2008) Adiponectin prevents cerebral ischemic injury through endothelial nitric oxide synthase dependent mechanisms. Circulation 117:216–223

O'Rourke MF, Safar ME (2005) Relationship between aortic stiffening and microvascular disease in brain and kidney: cause and logic of therapy. Hypertension 46:200–204

Oesch L, Tatlisumak T, Arnold M, Sarikaya H (2017) Obesity paradox in stroke—Myth or reality? A systematic review. PLoS One 12(3):e0171334

Ogorodnikova AD, Wassertheil-Smoller S, Mancuso P, Sowers MR, Rajpathak SN, Allison MA, Baird AE, Rodriguez B, Wildman RP (2010) High-molecular-weight adiponectin and incident ischemic stroke in postmenopausal women: a Women's Health Initiative Study. Stroke 41(7):1376–1381

Ogunsua AA, Touray S, Lui JK, Ip T, Escobar JV, Gore J (2015) Body mass index predicts major bleeding risks in patients on warfarin. J Thromb Thrombolysis 40(4):494–498

Okamoto M, Etani H, Yagita Y, Kinoshita N, Nukada T-A (2001) Diminished reserve for cerebral vasomotor response to L-arginine in the elderly: evaluation by transcranial Doppler sonography. Gerontology 47:131–135

Osawa H, Doi Y, Makino H, Ninomiya T, Yonemoto K, Kawamura R, Hata J, Tanizaki Y, Iida M, Kiyohara Y (2009) Diabetes and hypertension markedly increased the risk of ischemic stroke associated with high serum resistin concentration in a general Japanese population: the Hisayama Study. Cardiovasc Diabetol 8:60

Osmond JM, Mintz JD, Dalton B, Stepp DW (2009) Obesity increases blood pressure, cerebral vascular remodeling, and severity of stroke in the Zucker rat. Hypertension 53:381–386

Osuka K, Watanabe Y, Yasuda M, Takayasu M (2012) Adiponectin activates endothelial nitric oxide synthase through AMPK signaling after subarachnoid hemorrhage. Neurosci Lett 514(1):2–5

Ouchi N, Parker JL, Lugus JJ, Walsh K (2011) Adipokines in inflammation and metabolic disease. Nat Rev Immunol 11(2):85–97

Ovbiagele B, Bath PM, Cotton D, Vinisko R, Diener HC (2011) Obesity and recurrent vascular risk after a recent ischemic stroke. Stroke 42(12):3397–3402

Palomo I, Marín P, Alarcón M, Gubelin G, Viñambre X, Mora E, Icaza G (2003) Patients with essential hypertension present higher levels of sE-selectin and sVCAM-1 than normotensive volunteers. Clin Exp Hypertens 25:517–523

Pantoni L (2010) Cerebral small vessel disease: from pathogenesis and clinical characteristics to therapeutic challenges. Lancet Neurol 9:689–701

Pearson JD (1999) Endothelial cell function and thrombosis. Baillieres Best Pract Res Clin Haematol 12:329–341

Perko D, Pretnar-Oblak J, Sabovic M, Zaletel M, Zvan B (2011) Associations between cerebral and systemic endothelial function in migraine patients: a post-hoc study. BMC Neurol 11:146

Pescini F, Abbate R (2014) Markers of endothelial dysfunction, oxidative stress, and inflammation in cerebral small vessel disease. In: Gorelick P (ed) Cerebral small vessel disease. Cambridge University Press, Cambridge, pp 192–199

Pires PW, Dams Ramos CM, Matin N, Dorrance AM (2013) The effects of hypertension on the cerebral circulation. Am J Physiol Heart Circ Physiol 304:H1598–H1614

Poggesi A, Pasi M, Pescini F, Pantoni L, Inzitari D (2015) Circulating biologic markers of endothelial dysfunction in cerebral small vessel disease: a review. J Cereb Blood Flow Metab 116:1–15

Pourcyrous M, Basuroy S, Tcheranova D, Arheart KL, Elabiad MT, Leffler CW, Parfenova H (2015) Brain-derived circulating endothelial cells in peripheral blood of newborn infants with seizures: a potential biomarker for cerebrovascular injury. Physiol Rep 3. doi:10.14814/phy2.12345. pii: e12345

Prakash K, Chandran DS, Khadgawat R, Jaryal AK, Deepak KK (2016) Correlations between endothelial function in the systemic and cerebral circulation and insulin resistance in type 2 diabetes mellitus. Diab Vasc Dis Res 13:49–55

Prugger C, Luc G, Haas B, Arveiler D, Machez E, Ferrieres J, Ruidavets JB, Bingham A, Montaye M, Amouyel P, Yarnell J, Kee F, Ducimetiere P, Empana JP, PRIME Study Group (2012) Adipocytokines and the risk of ischemic stroke: the PRIME Study. Ann Neurol 71(4):478–486

Rajpathak SN, Kaplan RC, Wassertheil-Smoller S, Cushman M, Rohan TE, McGinn AP, Wang T, Strickler HD, Scherer PE, Mackey R, Curb D, Ho GY (2011) Resistin, but not adiponectin and leptin, is associated with the risk of ischemic stroke among postmenopausal women: results from the Women's Health Initiative. Stroke 42(7):1813–1820

Rao A, Pandya V, Whaley-Connell A (2015) Obesity and insulin resistance in resistant hypertension: implications for the kidney. Adv Chronic Kidney Dis 22:211–217

Reilly MP, Lehrke M, Wolfe ML, Rohatgi A, Lazar MA, Rader DJ (2005) Resistin is an inflammatory marker of atherosclerosis in humans. Circulation 111:932–939

Rodríguez-Flores M, García-García E, Cano-Nigenda CV, Cantú-Brito C (2014) Relationship of obesity and insulin resistance with the cerebrovascular reactivity: a case control study. Cardiovasc Diabetol 13:2

Rosito GA, D'Agostino RB, Massaro J, Lipinska I, Mittleman MA, Sutherland P, Wilson PW, Levy D, Muller JE, Tofler GH (2004) Association between obesity and a prothrombotic state: the Framingham Offspring Study. Thromb Haemost 91:683–689

Saber H, Himali JJ, Shoamanesh A, Beiser A, Pikula A, Harris TB, Roubenoff R, Romero JR, Kase CS, Vasan RS, Seshadri S (2015) Serum leptin levels and the risk of stroke: the Framingham Study. Stroke 46(10):2881–2885

Sarikaya H, Elmas F, Arnold M, Georgiadis D, Baumgartner RW (2011) Impact of obesity on stroke outcome after intravenous thrombolysis. Stroke 42(8):2330–2332

Scherbakov N, Dirnagl U, Doehner W (2011) Body weight after stroke: lessons from the obesity paradox. Stroke 42(12):3646–3650

Seet RC, Zhang Y, Wijdicks EF, Rabinstein AA (2014) Thrombolysis outcomes among obese and overweight stroke patients: an age- and National Institutes of Health Stroke Scale-matched comparison. J Stroke Cerebrovasc Dis 23(1):1–6

Sena CM, Pereira AM, Seiça R (2013) Endothelial dysfunction—a major mediator of diabetic vascular disease. Biochim Biophys Acta 1832:2216–2231

Sena CM, Pereira A, Fernandes R, Letra L, Seiça RM (2017) Adiponectin improves endothelial function in mesenteric arteries of high-fat fed wistar rats: role of perivascular adipose tissue. Br J Pharmacol. doi:10.1111/bph.13756. In press

Seo JH, Jeong HY, Noh S, Kim E, Ji K-H, Bae JS et al (2013) Relationship of body mass index and mortality for acute ischemic stroke patients after thrombolysis therapy. J Neurocrit Care 6(2):92–96

Serrador JM, Sorond FA, Vyas M, Gagnon M, Iloputaife ID, Lipsitz LA (2005) Cerebral pressure-flow relations in hypertensive elderly humans: transfer gain in different frequency domains. J Appl Physiol 98:151–159

Shimabukuro M, Higa N, Asahi T, Oshiro Y, Takasu N, Tagawa T, Ueda S, Shimomura I, Funahashi T, Matsuzawa Y (2003) Hypoadiponectinemia is closely linked to endothelial dysfunction in man. J Clin Endocr Metab 88:3236–3240

Silventoinen K, Magnusson PK, Tynelius P, Batty GD, Rasmussen F (2009) Association of body size and muscle strength with incidence of coronary heart disease and cerebrovascular diseases: a population-based cohort study of one million Swedish men. Int J Epidemiol 38:110–118

Söderberg S, Ahrén B, Stegmayr B, Johnson O, Wiklund PG, Weinehall L, Hallmans G, Olsson T (1999) Leptin is a risk marker for first-ever hemorrhagic stroke in a population-based cohort. Stroke 30(2):328–337

Söderberg S, Stegmayr B, Ahlbeck-Glader C, Slunga-Birgander L, Ahrén B, Olsson T (2003) High leptin levels are associated with stroke. Cerebrovasc Dis 15(1–2):63–69

Song YM, Sung J, Davey Smith G, Ebrahim S (2004) Body mass index and ischemic and hemorrhagic stroke: a prospective study in Korean men. Stroke 35:831–836

Spranger J, Verma S, Göhring I, Bobbert T, Seifert J, Sindler AL, Pfeiffer A, Hileman SM, Tschöp M, Banks WA (2006) Adiponectin does not cross the blood-brain barrier but modifies cytokine expression of brain endothelial cells. Diabetes 55:141–147

Stott DJ, Welsh P, Rumley A, Robertson M, Ford I, Sattar N, Westendorp RG, Jukema JW, Cobbe SM, Lowe GD (2009) Adipocytokines and risk of stroke in older people: a nested case-control study. Int J Epidemiol 38(1):253–261

Strazzullo P, D'Elia L, Cairella G, Garbagnati F, Cappuccio FP, Scalfi L (2010) Excess body weight and incidence of stroke: meta-analysis of prospective studies with 2 million participants. Stroke 41(5):e418–e426

Suk SH, Sacco RL, Boden-Albala B, Cheun JF, Pittman JG, Elkind MS et al (2003) Abdominal obesity and risk of ischemic stroke: the Northern Manhattan Stroke Study. Stroke 34:1586–1592

Szmitko PE, Wang CH, Weisel RD, de Almeida JR, Anderson TJ, Verma S (2003) New markers of inflammation and endothelial cell activation: Part I. Circulation 108:1917–1923

Takeuchi S, Wada K, Otani N, Osada H, Nagatani K, Mori K (2014) Temporal profile of plasma adiponectin level and delayed cerebral ischemia in patients with subarachnoid hemorrhage. J Clin Neurosci 21(6):1007–1010

Thorp A, Schlaich MP (2015) Relevance of sympathetic nervous system activation in obesity and metabolic syndrome. J Diabetes Res 2015:341583

Towfighi A, Ovbiagele B (2009) The impact of body mass index on mortality after stroke. Stroke 40:2704–2708

Vachharajani V, Cunningham C, Yoza B, Carson J Jr, Vachharajani TJ, McCall C (2012) Adiponectin-deficiency exaggerates sepsis-induced microvascular dysfunction in the mouse brain. Obesity (Silver Spring) 20:498–504

Vaiopoulos AG, Marinou K, Christodoulides C, Koutsilieris M (2012) The role of adiponectin in human vascular physiology. Int J Cardiol 155(2):188–193

Vemmos K, Ntaios G, Spengos K, Savvari P, Vemmou A, Pappa T et al (2011) Association between obesity and mortality after acute first-ever stroke: the obesity–stroke paradox. Stroke 42:30–36

Verma S, Li SH, Wang CH, Fedak PW, Li RK, Weisel RD, Mickle DA (2003) Resistin promotes endothelial cell activation: further evidence of adipokine-endothelial interaction. Circulation 108:736–740

Volpe M, Iaccarino G, Vecchione C, Rizzoni D, Russo R, Rubattu S, Condorelli G, Ganten U, Ganten D, Trimarco B, Lindpaintner K (1996) Association and cosegregation of stroke with impaired endothelium-dependent vasorelaxation in stroke prone, spontaneously hypertensive rats. J Clin Invest 98:256–261

Wallace JL, Reaves AB, Tolley EA, Oliphant CS, Hutchison L, Alabdan NA, Sands CW, Self TH (2013) Comparison of initial warfarin response in obese patients versus non-obese patients. J Thromb Thrombolysis 36(1):96–101

Wan Z, Mah D, Simtchouk S, Kluftinger A, Little JP (2015) Human adipose tissue conditioned media from lean subjects is protective against H_2O_2 induced neurotoxicity in human SH-SY5Y neuronal cells. Int J Mol Sci 16:1221–1231

Wang P, Miao CY (2015) NAMPT as a Therapeutic Target against Stroke. Trends Pharmacol Sci 36(12):891–905

Wang WH, Yu WH, Dong XQ, Zhang ZY, Zhu Q, Che ZH, Du Q, Wang H (2011) Plasma adiponectin as an independent predictor of early death after acute intracerebral hemorrhage. Clin Chim Acta 412(17–18):1626–1631

Wang C, Liu Y, Yang Q, Dai X, Wu S, Wang W et al (2013) Body mass index and risk of total and type-specific stroke in Chinese adults: results from a longitudinal study in China. Int J Stroke 8(4):245–250

Wannamethee SG, Shaper AG, Whincup PH, Lennon L, Sattar N (2013) Adiposity, adipokines, and risk of incident stroke in older men. Stroke 44(1):3–8

Wardlaw JM, Smith EE, Biessels GJ, Cordonnier C, Fazekas F, Frayne R, Lindley RI, O'Brien JT, Barkhof F, Benavente OR, Black SE, Brayne C, Breteler M, Chabriat H, Decarli C, de Leeuw FE, Doubal F, Duering M, Fox NC, Greenberg S, Hachinski V, Kilimann I, Mok V, Oostenbrugge R, Pantoni L, Speck O, Stephan BC, Teipel S, Viswanathan A, Werring D, Chen C, Smith C, van Buchem M, Norrving B, Gorelick PB, Dichgans M, STandards for ReportIng Vascular changes on nEuroimaging (STRIVE v1) (2013) Neuroimaging standards for research into small vessel disease and its contribution to ageing and neurodegeneration. Lancet Neurol 12:822–838

Weikert C, Westphal S, Berger K, Dierkes J, Möhlig M, Spranger J, Rimm EB, Willich SN, Boeing H, Pischon T (2008) Plasma resistin levels and risk of myocardial infarction and ischemic stroke. J Clin Endocrinol Metab 93(7):2647–2653

Whitlock G, Lewington S, Sherliker P, Clarke R, Emberson J, Halsey J, Qizilbash N, Collins R, Peto R, Emerging Risk Factors Collaboration (2009) Body-mass index and cause-specific mortality in 900 000 adults: collaborative analyses of 57 prospective studies. Lancet 373(9669):1083–1096

Wiesner G, Brown RE, Robertson GS, Imran SA, Ur E, Wilkinson M (2006) Increased expression of the adipokine genes resistin and fasting-induced adipose factor in hypoxic/ischaemic mouse brain. Neuroreport 17(11):1195–1198

Wijnhoud AD, Koudstaal PJ, Dippel DW (2011) The prognostic value of pulsatility index, flow velocity, and their ratio, measured with TCD ultrasound, in patients with a recent TIA or ischemic stroke. Acta Neurol Scand 124:238–244

Willie CK, Colino FL, Bailey DM, Tzeng YC, Binsted G, Jones LW, Haykowsky MJ, Bellapart J, Ogoh S, Smith KJ, Smirl JD, Day TA, Lucas SJ, Eller LK, Ainslie PN (2011) Utility of transcranial Doppler ultrasound for the integrative assessment of cerebrovascular function. J Neurosci Methods 196:221–237

Wing RR, Bolin P, Brancati FL, Bray GA, Clark JM, Coday M, Crow RS, Curtis JM, Egan CM, Espeland MA, Evans M, Foreyt JP, Ghazarian S, Gregg EW, Harrison B, Hazuda HP, Hill JO, Horton ES, Hubbard VS, Jakicic JM, Jeffery RW, Johnson KC, Kahn SE, Kitabchi AE, Knowler WC, Lewis CE, Maschak-Carey BJ, Montez MG, Murillo A, Nathan DM, Patricio J, Peters A, Pi-Sunyer X, Pownall H, Reboussin D, Regensteiner JG, Rickman AD, Ryan DH, Safford M, Wadden TA, Wagenknecht LE, West DS, Williamson DF, Yanovski SZ, Look AHEAD Research Group (2013) Cardiovascular effects of intensive lifestyle intervention in type 2 diabetes. N Engl J Med 369(2):145–154

Winter Y, Rohrmann S, Linseisen J, Lanczik O, Ringleb PA, Hebebrand J, Back T (2008) Contribution of obesity and abdominal fat mass to risk of stroke and transient ischemic attacks. Stroke 39(12):3145–3151

Woo D, Sekar P, Chakraborty R, Haverbusch MA, Flaherty ML, Kissela BM et al (2005) Genetic epidemiology of intracerebral hemorrhage. J Stroke Cerebrovasc Dis 14:239–243

Wormser D, Kaptoge S, Di Angelantonio E, Wood AM, Pennells L, Thompson A, Sarwar N, Kizer JR, Lawlor DA, Nordestgaard BG, Ridker P, Salomaa V, Stevens J, Woodward M, Sattar N, Collins R, Thompson SG, Whitlock G, Danesh J, Emerging Risk Factors Collaboration (2011) Separate and combined associations of body-mass index and abdominal adiposity with cardiovascular disease: collaborative analysis of 58 prospective studies. Lancet 377(9771):1085–1095

Woywodt A, Gerdes S, Ahl B, Erdbruegger U, Haubitz M, Weissenborn K (2012) Circulating endothelial cells and stroke: influence of stroke subtypes and changes during the course of disease. J Stroke Cerebrovasc Dis 21:452–458

Wu MH, Chio CC, Tsai KJ, Chang CP, Lin NK, Huang CC, Lin MT (2016) Obesity exacerbates rat cerebral ischemic injury through enhancing ischemic adiponectin-containing neuronal apoptosis. Mol Neurobiol 53:3702–3713

Wu Y, Wang X, Zhou X, Cheng B, Li G, Bai B (2017) Temporal expression of Apelin/Apelin receptor in ischemic stroke and its therapeutic potential. Front Mol Neurosci 10:1

Zhang L, Hou Y, Po SS (2015) Obstructive sleep apnoea and atrial fibrillation. Arrhythm Electrophysiol Rev 4(1):14–18

Zhao QJ, Sun M, Zhang XG, Wang LX (2012) Relationship between serum leptin levels and clinical outcomes of hypertensive intracerebral hemorrhage. Clin Exp Hypertens 34:161–164

Zhang J, Deng Z, Liao J, Song C, Liang C, Xue H, Wang L, Zhang K, Yan G (2013) Leptin Attenuates Cerebral Ischemia Injury through the Promotion of Energy Metabolism via the PI3K/Akt Pathway. J Cereb Blood Flow Metab 33 (4):567–574

Zimmermann C, Wimmer M, Haberl RL (2004) L-arginine mediated vasoreactivity in patients with a risk of stroke. Cerebrovasc Dis 17:128–133

Zvan B, Zaletel M, Pogacnik T, Kiauta T (2002) Testing of cerebral endothelium function with L-arginine after stroke. Int Angiol 21:256–259

Chapter 8
Multiple Sclerosis: Implications of Obesity in Neuroinflammation

Ana Margarida Novo and Sónia Batista

Abstract Since the discovery of the remarkable properties of adipose tissue as a metabolically active organ, several evidences on the possible link between obesity and the pathogenesis of multiple sclerosis (MS) have been gathered. Obesity in early life, mainly during adolescence, has been proposed as a relevant risk factor for late MS development. Moreover, once MS is initiated, obesity can contribute to increase disease severity by negatively influencing disease progress. Despite the fact that clinical data are not yet conclusive, many biochemical links have been recently disclosed. The "low-grade inflammation" that characterizes obesity can lead to neuroinflammation through different mechanisms, including choroid plexus and blood–brain barrier disruption. Furthermore, it is well known that resident immune cells of central nervous system and peripheral immune cells are involved in the pathogenesis of MS, and adipokines and neuropeptides such as neuropeptide Y may mediate the cross talk between them.

Keywords Multiple sclerosis • Obesity • Neuroinflammation • Adipokines • Neuropeptide Y • Mesenchymal stem cells

8.1 Introduction

Multiple sclerosis (MS) is a primary demyelinating and immune-mediated disease of the central nervous system (CNS), with both inflammatory and neurodegenerative characteristics (Trapp and Nave 2008).

A.M. Novo
Neurology Department, Centro Hospitalar e Universitário de Coimbra, Coimbra, Portugal

S. Batista (✉)
Neurology Department, Centro Hospitalar e Universitário de Coimbra, Coimbra, Portugal

Faculty of Medicine, University of Coimbra, Coimbra, Portugal

CNC-Center for Neuroscience and Cell Biology, University of Coimbra, Coimbra, Portugal
e-mail: soniarmbatista@huc.min-saude.pt

© Springer International Publishing AG 2017
L. Letra, R. Seiça (eds.), *Obesity and Brain Function*, Advances in Neurobiology 19, DOI 10.1007/978-3-319-63260-5_8

Despite the great advances in the understanding of the pathophysiology of this disease, the fact is that MS cause remains unknown. MS is a complex disease, resulting from the interaction of environmental risk factors and genetic susceptibility. The environmental risk factors that have shown the strongest and most consistent evidence of an association with MS are Epstein-Barr virus and cigarette smoking (Belbasis et al. 2015). However, in the last decades, also vascular risk factors and comorbidities, namely obesity, type 2 diabetes, and dyslipidemia, have been associated with MS progression (Marrie et al. 2010; Weinstock-Guttman et al. 2011a). Inflammation, endothelial dysfunction, and autonomic dysregulation have been proposed as possible pathophysiological links between these diseases (Minagar et al. 2006; Flachenecker et al. 1999).

Focusing attention specifically in obesity, the paradigm shift from the apparent passive role played by adipocytes to a metabolically active performance of adipocytes opened new horizons for the investigation of possible points of contact between obesity and immune-mediated diseases like MS.

There is also evidence that obesity interacts with genetic and environmental factors to increase MS susceptibility. In a study that included data from case-control studies in the United States and Sweden, having a body mass index (BMI) $\geq$ 27 kg/m^2 in young adulthood and carrying 1–2 risk alleles of HLA-DRB1*15 was associated with a sevenfold increased risk of MS compared to noncarriers with a BMI < 27 kg/m^2 (Hedström et al. 2014). Similarly, there was a significant interaction between absence of HLA-A*02 protective alleles and obesity, regardless of DRB1*15 status (Hedström et al. 2014).

In a separate study, regardless of HLA status, a substantial interaction was observed between adolescent obesity and past infectious mononucleosis with regard to MS risk (individuals with a BMI > 27 kg/m^2 and a history of infectious mononucleosis had over a sixfold increased risk of MS compared to individuals with a BMI < 27 kg/m^2 without history of the infection). Notably, this was much higher than the risk of MS due to BMI (OR = 1.3–1.7) or infectious mononucleosis (OR = 1.8–2.0) alone (Hedström et al. 2015).

8.2 Neuroinflammation in the Pathogenesis of MS

The concept of neuroinflammation has attracted generalized attention in the neurosciences field but there is still no clear consensual definition. Traditionally, the term neuroinflammation has been used to denote chronic, CNS-specific, inflammation-like responses produced by the activation of glial cells (microglia and astrocytes) that do not reproduce the classic characteristics of inflammation in the periphery, occurring in chronic neurodegenerative diseases such as Alzheimer's and Parkinson's disease (Streit et al. 2004).

A contemporary view extends the term of neuroinflammation to describe any inflammatory response in the CNS and proposes that the nature of inflammation, more precisely, whether the pathological process is driven by adaptive immune cells or by CNS-resident and/or potentially blood-derived innate immune cells, should be used to distinguish neuroinflammation (Heppner et al. 2015). Consequently, diseases of the CNS may be discriminated as being mainly driven by an innate immune element (glia and/or blood-derived innate immune cells) as occurs in neurodegenerative diseases such as Alzheimer disease, or by an adaptive immune component (B and T lymphocytes) as seen in MS (Heppner et al. 2015; Peterson and Fujinami 2007).

Neuroinflammation is present at all stages of MS, but it is more pronounced in acute phases (Dendrou et al. 2015). Early lesions show prominent infiltrates of peripheral immune cells and leakage of the blood–brain barrier (BBB). Macrophages dominate the infiltrate, followed by CD8$^+$ T cells, while lower numbers of CD4$^+$ T cells, B cells, and plasma cells are also found. The infiltration of immune cells from the periphery can occur by direct crossing of the BBB, from the subarachnoid space or from the choroid plexus across the blood–cerebrospinal fluid (CSF) barrier (Shechter et al. 2013).

In more advanced stages of the disease, the immune cell infiltration from the periphery wanes, perhaps due to adaptive immune cell exhaustion from chronic antigen exposure, but the chronic neuroinflammation continues. Diffuse inflammatory T cell and B cell infiltrates, microglia, and astrocyte activation are seen throughout the white matter (Dendrou et al. 2015; Frischer et al. 2009) and tertiary lymphoid-like structures are found in the meninges of the patients with secondary progressive MS, contributing to late-stage neuroinflammation in patients with this form of MS (Howell et al. 2011). Moreover, the CNS-resident innate immune cells, mainly microglia and astrocytes, remain in a chronic state of activation during all over the course of the disease, perpetuating the neuroinflammation (Heneka et al. 2014). Stimulated by the microglia, astrocytes can produce CC-chemokine ligand 2 (CCL2) and granulocyte–macrophage colony-stimulating factor (GM CSF), leading to further microglial activation (Mayo et al. 2014).

Whether neuroinflammation in MS is triggered in the periphery or in the CNS is still a matter of debate. In the CNS-extrinsic (outside-in) model, autoreactive T cells are activated in the peripheral lymphoid organs—potentially through molecular mimicry, bystander activation, or the co-expression of T cell receptors with different specificities—and infiltrate the CNS along with activated B cells and monocytes (Dendrou et al. 2015). This hypothesis is in accordance with the method used to induce the animal model of MS, the experimental autoimmune encephalomyelitis (EAE): pathogenic CD4$^+$ T helper 1 (TH1) cells and TH$_{17}$ cells are generated in the draining lymph nodes after the administration of emulsified CNS antigen. These pathogenic cells then enter the circulation and cross the BBB or the blood–CSF barrier to the CNS parenchyma and ultimately initiate the neuroinflammation cascade (Dendrou et al. 2015).

Alternatively, CNS-intrinsic events (inside-out model) may trigger MS onset, with the infiltration of autoreactive lymphocytes occurring as a secondary phenomenon (Dendrou et al. 2015). These CNS-intrinsic events might include a chronic latent viral infection of CNS, namely the recently identified *Multiple Sclerosis Associated Retro Virus* (MSRV) (Duperray et al. 2015), or a primary defect of oligodendrocytes causing its spontaneous death (Traka et al. 2016). These events would cause the subsequent activation of resident microglia and then a secondary recruitment of adaptive and innate immune cells from the periphery which characterizes MS pathogenesis (Hemmer et al. 2015).

8.3 From Obesity-Induced Inflammation to Neuroinflammation

Obesity results from an interplay between genetic predisposition and environmental factors resulting in an immune response and subsequent low-grade inflammation that affects numerous tissues, including the CNS (Lumeng and Saltiel 2011). When considering obesity and the subsequent neuroinflammation, the focus was long set on the hypothalamus, since this structure includes the arcuate nucleus where two distinct neuronal populations (NPY/AgRP and POMC/CART neurons) are involved in body weight regulation and energy balance (Thaler et al. 2013). Hypothalamic inflammation is involved in the onset and maintenance of the obese phenotype (Thaler et al. 2013). However, more recently, obesity-derived neuroinflammation has been shown to affect other brain structures such as the hippocampus, cortex, brainstem, cerebellum, and amygdala (Guillemot-Legris and Muccioli 2017). Several mechanisms by which obesity-induced peripheral inflammation can lead to neuroinflammation have been proposed (Fig. 8.1), and the majority of them involve the disruption of the barriers between the periphery and the CNS. Therefore, the CNS can be affected from the peripheral inflammation through different communication pathways, including the vagus nerve, the choroid plexus, and the BBB. This will contribute to obesity-induced neuroinflammation, although to a varying extent depending on numerous factors, namely obesity stage, the age of the subject, diet, and CNS structure examined (Guillemot-Legris and Muccioli 2017). A recent study using a db/db mice as a genetic model of obesity directly implicated BBB leakage as a contributing factor to obesity-derived neuroinflammation and cognitive deficits. Indeed, by reducing BBB leakage (using a PKCβ inhibitor), neuroinflammation and cognitive deficits in mice were rescued (Stranahan et al. 2016). The choroid plexus, the blood to CSF barrier, has been less studied than BBB with regard to obesity-induced alterations. The choroid plexus epithelial cells are responsible for the synthesis of CSF but it has recently been proposed that they also play a role in mediating the ingress of inflammatory cues (e.g., miRNA, proteins, lipids) from the periphery in the context of systemic inflammation (Balusu et al. 2016). This barrier was also found to display decreased mRNA expression for tight junction-associated proteins claudin-5 and -12 in rats with diet-induced obesity (Kanoski et al. 2010).

Another proposed mechanism for the neuroinflammation associated with obesity is the activation of several cell types in the CNS in response to changes in the levels of peripheral mediators such as leptin, insulin, free fatty acids, lipopolysaccharides, or cytokines (Guillemot-Legris and Muccioli 2017). In vitro experiments indicate that both microglia and astrocytes are activated by fatty acids and more specifically by

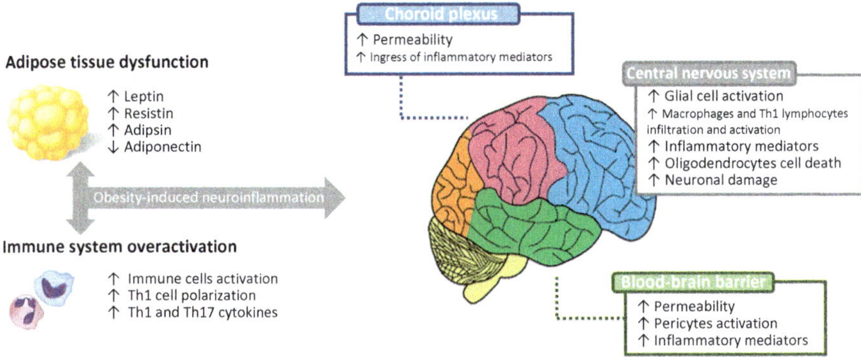

Fig. 8.1 Pathophysiological events likely to contribute to obesity-associated neuroinflammation

saturated fatty acids, leading to increased levels of pro-inflammatory cytokines and reactive oxygen species (Button et al. 2014). However, in vivo and in the hypothalamus, it seems that microglia is the key player, since microglia depletion rescues hypothalamus neuroinflammation and decreases food intake (Valdearcos et al. 2014). Pericytes are well known for their active role in BBB homeostasis. Upon activation by immune stimuli, they can produce pro-inflammatory mediators that will in turn disrupt the BBB permeability by destabilizing tight junction proteins (Pieper et al. 2014). These cells are also responsible for helping peripheral immune cell transmigration across the BBB (Pieper et al. 2013). Finally, the potential infiltration of peripheral immune cells into the CNS has been investigated. Buckman and colleagues, using mice transplanted with bone marrow from GFP⁺ mice, showed that 15 and 30 weeks of high-fat diet increased the recruitment of bone marrow-derived monocytes into the brain relative to chow-fed controls (Buckman et al. 2014). On the contrary, in another study using a similar model, mice fed a high-fat diet for 20 weeks did not display increased GFP⁺ myeloid cells in the brain (Baufeld et al. 2016). The reasons for these discrepancies are not yet clear, but could be related with differences in the cell sorting methods employed. In conclusion, regarding the infiltration of peripheral immune cells, the actual extent of recruitment of myeloid cells in the hypothalamus and cortex, their fate and their potential role in diet-induced obesity neuroinflammation are still unclear and need to be clarified (Guillemot-Legris and Muccioli 2017).

8.4 Obesity and Multiple Sclerosis: Epidemiological Links

Several studies have been conducted not only to investigate the prevalence of obesity in MS patients, but also to assess the possible role it may have in determining the risk of developing this immune-mediated disease and its associated neurological disability.

8.4.1 Prevalence of Obesity in MS

Data coming from epidemiological studies that aim to evaluate the prevalence of obesity in MS, compared to the general population, are indeed contradictory.

In a study involving 8.983 MS patients from the NARCOMS (North American Research Committee on Multiple Sclerosis) self-reported Registry, a quarter of participants were obese and 31.3% were overweight (Marrie et al. 2009). A posterior analysis showed that nearly 50% of MS patients from NARCOMS Registry were already overweight or obese at MS onset (Marrie et al. 2011). A survey study with a smaller population (123 MS patients) reported similar findings: 47.5% of patients were overweight and 25.8% were obese (Slawta et al. 2003).

However, in a cross-sectional study involving 4.703 veterans with MS, it was found a lower adjusted prevalence of obesity comparing with an historical cohort of veterans (20.1% versus 33.1%) (Khurana et al. 2009). Also, in another study, female patients with MS (but not male patients) showed a significant lower BMI compared to healthy controls (Markianos et al. 2013). On the other hand, in a study that analyzed the body composition in patients with MS, the mean body mass and BMI were lower in male patients than in male controls but the total amount of body fat was not found to differ between MS patients and controls, when males and females were grouped together (Sioka et al. 2011).

These inconsistent results may be attributed, at least in part, to the fact that in most of the studies the anthropometric data were self-reported by the participants.

8.4.2 Contribution of Obesity to the Risk of Developing MS

Regarding the possible role of obesity in the risk of developing MS, numerous publications over the last decade reported an association between early childhood and adolescent obesity and MS susceptibility. The first comprehensive study to examine this relationship was based on two cohorts of women from the Nurses' Health Study I and II ($n = 238.371$) (Munger et al. 2009). The researchers found that women with a BMI $\geq$ 30 kg/m^2 at age 18 had a 2.25-fold increased risk of developing MS compared to those with a BMI between 18.5 and 21 kg/m^2, after adjusting for age, latitude at age 15, ethnicity, and smoking. In line with these results, a population-based study involving 1.571 MS patients and 3.371 controls reported a twofold increased odds of MS in individuals with a BMI $\geq$ 30 kg/m^2; the association was significant in both females and males (Hedstrom et al. 2012). A study within Kaiser Permanente Northern California Region (Gianfrancesco et al. 2014) involving 1.235 MS cases and 697 controls showed that the twofold increased risk of MS in females with a BMI $\geq$ 30 kg/m^2 in one's twenties persisted after adjusting for history of infectious mononucleosis and genetic risk factors (HLA-DRB1*15 and established non-HLA risk alleles). However, no significant association was observed in males. A role for childhood obesity and risk of both pediatric and adult-onset MS had previously been found in two studies (Langer-Gould et al. 2013; Munger et al. 2013). Results demonstrated that obesity was associated with a significantly increased risk of Clinically Isolated Syndrome (CIS) and MS in girls but not in boys. Further, extremely obese girls had over three times the odds of developing MS compared to those at normal weight (Langer-Gould et al. 2013). In a Danish prospective study, school records from 302.043 individuals were used to assess if BMI at ages 7–13 years was associated with MS risk (Munger et al. 2013). The researchers found that, among girls, a one-unit increase in a BMI z-score was associated with an increased risk of MS (HR = 1.17–1.20), while the associations were attenuated in boys. However, a recent study reported contradictory findings, by demonstrating that BMI, during adolescence, rather than childhood, is critical in determining MS risk (Hedstrom et al. 2016). In this population-based case-control study (1.586 cases and 2.800 controls), individuals with adolescent obesity had a 90% increased risk of MS. Among participants who were not obese at age 20, no association was observed between body size at age 10 and subsequent MS risk. Interestingly, an interaction between HLA MS risk genes and adolescent, but not childhood, obesity was also observed (Hedstrom et al. 2016).

In summary, while there is a general consensus that obesity in adolescence/ young adulthood is associated with MS susceptibility, the association is less clear in male individuals and in childhood obesity.

Differences in findings may be attributed to several factors, namely: the use of silhouettes in some studies to assess body size and the fact that overweight individuals have been shown to have a more favorable perception of body silhouettes, potentially biasing results toward the null (Gianfrancesco and Barcellos 2016), recall bias with respect to weight and height when self-reported by participants, and lack of adjustment for socioeconomic status and for sun exposure and/or vitamin D serum levels. In fact, it is well known that obese individuals have lower vitamin D levels and less sun exposure, both due to behavioral factors and to the effect adipose tissue may have on the availability of vitamin D (Daly et al. 2012). Future studies aiming to clarify the strength of the association between obesity and the risk of MS development will necessarily have to rely on these covariates (Palavra et al. 2016).

8.4.3 Contribution of Obesity to the Neurological Disability Attributed to MS

In a longitudinal study, involving 269 subjects with relapsing-remitting MS (RRMS), over a 24-month time course, BMI was not predictive of disability, measured by Patient Determined Disease Steps (PDDS), a variant of the more widely used Kurtzke's Expanded Disability Status Scale (EDSS) (Pilutti et al. 2012). However, two other studies found that higher BMI objectively measured by the investigators, which is an advantage over the previous study, was independently associated with higher EDSS scores (Tettey et al. 2014a; Oliveira et al. 2014), but not with the hazard of relapse (Tettey et al. 2014b).

8.5 Obesity and Multiple Sclerosis: Potential Biological Links

Despite the complexity underlying the relationship between obesity and MS, recent lines of evidence have provided some clues about the possible biochemical mechanisms behind it. Clinical, experimental, and epidemiological data have suggested that the pathogenesis of MS might involve factors that link the immune system with the metabolic status and disorders of the lipid metabolism. Higher levels of aggressive and atherogenic molecules, particularly oxidized low-density lipoprotein (Ox-LDL) and small high-density lipoproteins (HDL), were found in MS patients' serum, compared to controls (Palavra et al. 2013). Interestingly, in this study, LDL content, especially Ox-LDL, showed a significant positive correlation with EDSS. The existence of an inter-dependence between lipids and vitamin D is also possible and it may result from a potential imbalance between pro-inflammatory effects of oxidized lipids on vascular endothelium and antioxidant and immuno-modulatory effects of vitamin D on the immune system (Palavra et al. 2016; Weinstock-Guttman et al. 2011b). In line with this, it was shown that plasma lipid profile could be improved by higher levels of vitamin D in serum (Mähler et al. 2012) and that a 12-week treatment with atorvastatin led to a clinically significant rise in 25-hydroxyvitamin D concentrations (Sathyapalan et al. 2010).

Another very exciting field of recent investigation comprises the role of adipokines in the pathogenesis of MS. Some studies have reported increased levels of leptin (Kraszula et al. 2012; Matarese et al. 2005; Messina et al. 2013) and resistin (Kraszula et al. 2012) and decreased levels of adiponectin in patients with RRMS (Kraszula et al. 2012; Musabak et al. 2011) in comparison with healthy controls, a profile also observed among subjects with obesity (Neuparth et al. 2013; Nieva-Vazquez et al. 2014).

8.5.1 Adipokines

8.5.1.1 Leptin

This adipokine has a well-established role in controlling innate and adaptive immune responses, modulating the immune system toward a pro-inflammatory profile. Leptin is found at the crossroad between inflammation and autoimmunity. Several lines of evidence indicate that this hormone is able to participate as a link between metabolism and MS and that it may act as an important player in the pathogenesis of this immune-mediated disease (Guerrero-García et al. 2016). Leptin receptor (LepR) is expressed in CD4+, CD8+, regulatory T cells (Treg), and Natural Killer (NK) cells and in monocytes/macrophages (Guerrero-García et al. 2016). Leptin induces differential effects in CD4+ T cells, increasing proliferation of naïve T cells and inhibiting memory T cells. Furthermore, leptin increases INF-γ and inhibits

IL-4 production in memory T cells (Lord et al. 2002). In patients with RRMS, the expression of LepR has been found significantly higher in CD8[+] T cells and monocytes from patients in relapse, comparing to patients in remission and healthy controls (Frisullo et al. 2007). Moreover, exogenous leptin treatment sustained STAT3 phosphorylation, but only in monocytes from patients in relapse, suggesting that LepR may be involved in the development of clinical relapses in MS (Frisullo et al. 2007). Leptin gene was one of the numerous genes of the neuroimmune endocrine axis whose transcription was increased in a gene microarray analysis of Th_1 lymphocytes from active MS lesions (Lock et al. 2002). Taken together, these data suggest that leptin and its receptor induce Th_1 cell and cytokine environment and favor the induction of inflammation in MS. Leptin-deficient ob/ob mice display impaired cell-mediated immunity and a reduced in vitro secretion, upon antigen stimulation, of the classical Th_1-type pro-inflammatory cytokines such as IL-2 and INF-γ and an increased production of IL-4, typical of the Th_2 regulatory phenotype (Matarese et al. 2001). It has been demonstrated that leptin-deficient *ob/ob* mice are resistant to induction of EAE, which was associated with a progressive decline in the survival of autoreactive CD4[+] T cells and reduced production of Th_1 and Th_{17} cytokines (Matarese et al. 2001). T cells demonstrated downregulation of Bcl-2, a survival protein, reduction in P-ERK1/2 and cell-cycle arrest associated with reduced degradation of cell-cycle inhibitor $p27^{kip1}$. These molecular events revealed a reduced activity of the nutrient/energy-sensing AKT/mammalian target of rapamycin pathway, which was restored in vivo by exogenous leptin replacement (Galgani et al. 2010). Also, the replacement of this hormone converted disease resistance to susceptibility in *ob/ob* mice, which was accompanied by a switch from a Th_2 to Th_1 pattern of cytokine release (Matarese et al. 2001). Another interesting data collected from the EAE model concerns the influence of leptin on the kinetics of EAE onset and clinical manifestations. It was found that a serum leptin increases before the clinical onset of EAE in disease-susceptible strains of mice, which correlated with disease susceptibility (Sanna et al. 2003). Indeed, acute starvation, which provoked a decrease in serum leptin, delayed disease onset and attenuated clinical symptoms in mice. In this study, immunohistochemical analysis revealed a parallel in situ production of leptin in inflammatory infiltrates and in neurons only during the acute/active phase of both chronic-progressive and relapsing-remitting EAE. Furthermore, leptin secretion by activated T cells sustained their proliferation in an autocrine loop, since anti-leptin receptor antibodies were able to inhibit the proliferative response of autoreactive T cells in vitro (Sanna et al. 2003). These data support the hypothesis that leptin is required for the induction and maintenance of an effective pro-inflammatory immune response in the CNS. In line with the previous results, blockade of leptin with anti-leptin antibodies or with a soluble mouse leptin receptor chimera, either before or after onset of EAE, improved clinical scores, slowed disease progression, and reduced disease relapses (De Rosa et al. 2006). Studies involving MS patients also displayed important results. While early studies reported comparable serum leptin levels between RRMS patients and healthy subjects (Batocchi et al. 2003; Chatzantoni et al. 2004), more recent works found significantly higher leptin levels in patients with RRMS ((Kraszula et al. 2012, Frisullo et al. 2007) and an inverse correlation between leptin concentration and the presence of the transcription factor Foxp3 in Treg cells was reported (Kraszula et al. 2012). Nevertheless, the finding of increased leptin secretion only in female patients with RRMS, but not in patients with CIS, led to the hypothesis that leptin may not have a pathogenic role from the early stages of the disease (Evangelopoulos et al. 2014). An interesting study that investigated the pregnancy-induced fluctuations of serum leptin levels in women with RRMS found that leptin increased during the third trimester (Neuteboom et al. 2009). A postdelivery drop in leptin levels was observed in both the MS and control groups and was associated with the occurrence of postpartum relapse. Additionally, leptin levels were found to be significantly higher in patients with secondary progressive MS (SPMS), but not in primary progressive MS (PPMS), compared to controls, suggesting that this adipokine is differentially involved in the PP and SP forms of the disease (Messina et al. 2013). In this study, there was no differences in leptin levels in patients before or during a relapse compared to the remission phase. However, this is not a universal finding, since a prior study reported that leptin levels increased before a clinical exacerbation in RRMS (Batocchi et al. 2003). A leptin increase in the CSF of MS patients was reported and it correlated with INF-γ secretion in this fluid (Matarese et al. 2005). T cells against human myelin basic protein (hMBP) from naïve-to-therapy RRMS patients produced immunoreactive leptin and upregulated the expression of the LepR after activation with hMBP (Gianfrancesco and Barcellos 2016). The sexual dimorphism in leptin levels (higher in females) is well established in normal subjects (Saad et al. 1997). Regarding the correlation of leptin to BMI in MS patients specifically, previous reports have been somewhat conflicting. While Matarese et al. (2005) reported a loss of this correlation in MS patients with low disability in relapse, Rotondi et al. (2013) reported that the correlation was maintained in MS patients with low and intermediate disability but lost in patients with high disability (EDSS score $\geq$ 3.5) in remission. The authors proposed that other factors, beyond body fat mass and sex, are involved in the modulation of leptin serum levels during the clinical progression of MS, namely its independent production by effector and regulatory CD4$^+$ T cells. Several studies have also investigated possible correlations between leptin levels and disease parameters, such as disease duration, EDSS, or number of clinical relapses, with negative results (Batocchi et al. 2003; Chatzantoni et al. 2004; Rotondi et al. 2013). However, a study with a larger sample found that leptin levels correlated positively with disease duration (but not to EDSS), in female patients with RRMS (Evangelopoulos et al. 2014). Regarding the effects of disease-modifying therapies for MS on leptin levels, one study reported a decrease two months after initiation of INFbeta-1a (Batocchi et al. 2003) and another study found a reduction in leptin and IL-6 levels after 6 and 12 months of treatment with INFbeta-1b in 12 patients with SPMS who did not show progression of disability (Angelucci et al. 2005). However, it was not observed a significant variation of this adipokine level after glatiramer acetate treatment (Carrieri et al. 2015). Corticosteroid administration has been shown to increase serum leptin levels, albeit only transiently (Rotondi et al. 2013).

In conclusion, previous reports have presented to some extent conflicting results, and the precise role of leptin in MS pathogenesis needs to be further clarified.

8.5.1.2 Adiponectin

A variety of anti-inflammatory properties have been reported for adiponectin, including a direct immunomodulatory effect on antigen-activated human T cells (Wilk et al. 2011). In 2008, Piccio and coworkers were able to demonstrate that calorie restriction ameliorated murine EAE, in association with higher adiponectin and decreased leptin levels. In a posterior study, the same group of investigators demonstrated a protective role of adiponectin in the EAE model for MS. Adiponectin deficient (ADPKO) mice developed more clinical and histologic severe EAE, with greater inflammation, demyelination, and axonal injury (Piccio et al. 2013). Also, lymphocytes from ADPKO mice proliferated more, produced higher amounts of INF-γ, IL-17, TNF-α, and IL-6, and transferred more severe EAE than wild-type lymphocytes. At EAE peak, ADPKO had reduced numbers of regulatory CD4$^+$ T cells and a defect of CD25$^+$ Foxp3$^+$ Treg suppressive function. During EAE recovery, IL-10 and TGF-β expression levels in the CNS were reduced in ADPKO mice. Treatment with globular adiponectin in vivo ameliorated EAE and was associated with an increase in Treg cells (Piccio et al. 2013). These data indicate that adiponectin is an important regulator of T-cell function, supporting a protective role of this adipokine in EAE. Altered adiponectin levels in MS patients have been reported in three studies, with two showing decreased adiponectin serum levels (Mayo et al. 2014; Moschen et al. 2007) and the other showing elevated CSF levels (Hietaharju et al. 2010). In fact, Hietaharju et al. (2010) found higher CSF concentrations of adiponectin and adipsin in twins with MS in remission compared to their asymptomatic twins, although these levels did not correlate with plasma levels. The authors hypothesized a possible intrathecal synthesis of adiponectin or increased transport across the BBB following enhanced systemic production. However, the sample size of this study was small. Consequently, there is a need for larger studies to clarify the significance of adiponectin levels in MS patients.

8.5.1.3 Resistin

Resistin can be produced by immunocompetent cells, including mononuclear cells in humans, and there is some evidence supporting its engagement in inflammatory conditions in vitro and in vivo (Guerrero-García et al. 2016). A few studies had attempted to discover the relationship between resistin and MS pathogenesis in patients with the disease. Emamgholipour et al. (2013) observed an elevation of resistin, leptin, and visfatin levels, as well as a decrease in the Foxp3 mRNA expression of T cells. Additionally, the authors found that these adipokines were positively correlated with levels of inflammatory mediators (TNF-α, IL-1β, and human sensitive C-reactive protein) and negatively correlated with Foxp3 expression in MS patients. Another study also reported higher levels of resistin in MS patients compared to the control group (Hossein-Nezhad et al. 2013), which adds evidence to support the role of resistin in MS, although more studies are needed.

8.5.1.4 Visfatin

This adipokine is not only produced by adipose tissue, hepatocytes, and skeletal muscle, but also by leucocytes and macrophages (Moschen et al. 2007). In the area of immune activities, visfatin induces the production of IL-6, TNF-α, and IL-1β in monocytes and increases the surface expression of costimulatory molecules CD54, CD40, and CD80 (Moschen et al. 2007). Almost nothing is known about the role of this adipokine in MS. Only one study reported higher visfatin levels in MS patients, which correlated with levels of inflammatory mediators (Emamgholipour et al. 2013).

8.5.1.5 Adipocyte-Fatty Acid-Binding Protein (A-FABP)

Serum A-FABP is produced by adipose tissue, monocytes, and macrophages, and its expression is enhanced by Toll-Like Receptor-2 (TLR-2) stimulation (Guerrero-García et al. 2016). Regarding MS, A-FABP levels are higher in SPMS, suggesting a possible role in the pathogenesis of this disease subtype (Messina et al. 2013). Also, A-FABP levels were found to be increased in patients with Pediatric-Onset MS and may play a role in the early stages of disease (Messina et al. 2013). Therefore, more studies are necessary to clarify the mechanisms by which A-FABP is involved in MS pathogenesis.

8.5.1.6 Adipsin

Adipsin was described as a molecular marker of obesity in rodents but its precise role on energy homeostasis and systemic metabolism remains unknown (Lo et al. 2014). Adipsin CSF levels are correlated with inflammation mediators but not with the presence of oligoclonal bands (Schmid et al. 2016). In a study of twins with MS, higher CSF concentrations of adipsin were found in twins with MS in remission compared to their asymptomatic twins and no correlation with adipsin plasma levels was observed (Hietaharju et al. 2010). Recently, an exploratory study conducted by Natarajan and coworkers found a significant correlation, in RRMS patients, between the serum levels of adipsin and EDSS scores, number of T1-weighted lesions (at baseline and over a two-year follow-up) and FLAIR lesions on MRI (Natarajan et al. 2015). The authors proposed that adipsin may exert predictive potential as a biomarker of neurodegeneration.

8.5.1.7   Chemerin

This adipokine is expressed in vascular endothelial cells in the meninges and in white-matter lesions of MS, whereas its receptor is expressed in infiltrating leukocytes, including plasmacytoid dendritic cells. These data suggest that chemerin is directly involved in the migration of peripheral cells into the CNS (Lande et al. 2008). Chemerin is a proteolytically regulated leukocyte chemoattractant when it binds to CheMoKine-Like Receptor-1 (CMKLR-1) (Graham et al. 2009). CMKLR-1 knockout mice exhibit reduced symptoms of EAE (Graham et al. 2009). In a recent study, it was found that chemerin levels are higher in MS patients with overweight or obesity compared with MS patients without obesity and controls (Tomalka-Kochanowska et al. 2014). These results suggest that obesity in patients with MS increases chemerin levels, leading to an increase in CNS-infiltrating cells, that may, in turn, contribute to disease severity.

8.5.1.8   Vaspin

Vaspin has been poorly studied in patients with MS. In this regard, Assadi et al. (2011) did not find significant difference in vaspin levels between MS patients and controls nor any correlation between vaspin levels and age, BMI, biochemical and bone mineral density measurements in patients with MS.

8.5.2   Neuropeptide Y (NPY)

Another important player in the relationship between obesity and MS may be NPY. This 36-amino acid C-terminally amidated neurotransmitter peptide is generally considered the most abundant peptide in the central and peripheral nervous system. NPY is involved in the complex networks controlling food intake and energy balance, integrating the functional projections from the arcuate nucleus toward other areas of the hypothalamus and also participating in the inputs received from the brainstem, cortical areas, and reward pathways (Simpson et al. 2009). NPY injected into the paraventricular nucleus is the most potent central appetite stimulant known, inducing feeding and obesity (Williams et al. 2000). Besides its central role in brain regulation of energy homeostasis, NPY is also distinguished by exhibiting pleiotropic functions in many other physiological systems, including the immune system (Palavra et al. 2016). Not only NPY can be produced by immune cells upon appropriate stimulation, but also specific Y receptor subtypes are expressed by immune cells. NPY is known to modulate immune cell trafficking, T helper cell differentiation, cytokine secretion, NK cell activity, phagocytosis, and the production of reactive oxygen species (Dimitrijevic and Stanojevic 2013). The concentration of NPY in CSF of MS patients was shown to be reduced (Maeda et al. 1994) and, in the EAE model, the intravenous administration of NPY in a dose-dependent fashion ameliorated the symptoms and severity of the disease (Bedoui et al. 2003). The beneficial effect of NPY on clinical EAE was linked with the decreased amounts of INF-γ secreted from autoreactive T lymphocytes when stimulated with the specific autoantigen, and the elevated IgG1-IgG2a ratio of autoantigen-specific antibodies, which is indicative of a response favoring a Th$_2$ phenotype. In the same sense, by treating rats with NPY during the induction phase of EAE, it was possible to delay the onset and reduce the severity of clinical manifestations, these effects being correlated with a reduction of the infiltration of the brain at disease peak by T cells and macrophages (Dimitrijevic and Stanojevic 2013). This suppressive effect of NPY over EAE seems to be mediated by Y1 receptors, because specific agonists have been demonstrated to inhibit the induction of the disease and, on the contrary, the usage of Y1 receptor antagonists led to an earlier onset of clinical signs (Bedoui et al. 2003). The importance of Y1 receptor signaling for disease control is also suggested in MS patients. Namely, an increased expression of dipeptidyl peptidase 4 (also known as CD26, the enzyme involved in degrading NPY and finishing its interaction with Y1 receptor) on T lymphocytes in MS patients may contribute to a decreased amount of intact NPY and therefore a reduced Y1 receptor signaling (Reinhold et al. 2002). Indeed, the pharmacological inhibition of CD26 activity successfully suppresses the clinical course of EAE, including an active TGF-β1 mediated anti-inflammatory effect on the CNS (Steinbrecher et al. 2001).

In conclusion, it will be relevant to explore the role of neuropeptides like NPY in relation to feeding behavior, whose mechanisms are regulated by orexigenic and anorexigenic hypothalamic neurons, because an adequate regulation of this neural circuitry might lead to an improvement in the inflammatory response in MS.

8.6 Transplantation of Adipose Tissue-Mesenchymal Stem Cells as Therapy for MS

Mesenchymal stem cells (MSCs) are a pleiotropic population of cells that are self-renewing and capable of differentiating into canonical cells of the mesenchyme, including adipocytes, chondrocytes, and osteocytes (Glenn and Whartenby 2014). The discovery of their immunosuppressive functions ushered in a new interest in MSCs as a promising tool to suppress inflammation and downregulate pathogenic immune responses in autoimmune diseases. Furthermore, the immunomodulatory and neuroprotective effects of adipose tissue-MSCs (AT-MSCs) make them possibly suitable candidates for stem cell-based MS therapy. In a study that compared the immune regulatory properties of AT-MSCs (Yousefi et al. 2013) in two independent routes of injection (intraperitoneal versus intravenous) in EAE mice, it was shown that the intraperitoneal route produced a more pronounced effect in maintaining the splenic CD4$^+$ CD25$^+$ Foxp3$^+$ T cell population and increase of IL-4 secretion. The intraperitoneal route also resulted in lower IFN-γ secretion and reduced cell infiltration in brain more effectively. However, the effects of AT-MSCs on downregulation of IL-17 secretion and alleviating the severity of clinical scores were similar in both routes of injection (Yousefi et al. 2013). To investigate whether MS affects the biologic properties of AT-MSCs and whether autologous AT-MSCs from MS-affected sources could serve as an effective source for stem cell therapy, a group of investigators isolated AT-MSCs from fat pads of mice with EAE and assessed the therapeutic efficacy of in vivo transplantation into other EAE mice. The AT-MSCs from EAE mice demonstrated increased expression of pro-inflammatory cytokines and chemokines, specifically an elevation in the expression of monocyte chemoattractant protein-1 and keratin chemoattractant. In vivo, infusion of wild-type AT-MSCs significantly ameliorated the disease course, autoimmune mediated demyelination, and cell infiltration through the regulation of the inflammatory responses. On the contrary, mice treated with autologous AT-MSCs showed no therapeutic improvement on the disease progression (Zhang et al. 2014). The impact of obesity on the therapeutic efficacy of AT-MSCs was also investigated (Strong et al. 2016). The therapeutic efficacy of AT-MSCs isolated from lean subjects (BMI < 25 kg/m^2) and obese subjects (BMI > 30 kg/m^2) were determined in murine EAE. Compared with the EAE disease-modifying effects of AT-MSCs from lean subjects, AT-MSCs from obese subjects consistently failed to alleviate clinical symptoms or inhibit inflammation in the CNS. When activated, AT-MSCs from obese subjects expressed higher mRNA levels of several pro-inflammatory cytokines compared with AT-MSCs from lean individuals. Additionally, conditioned media collected from the AT-MSCs of obese subjects markedly enhanced the proliferation and differentiation of T cells, whereas conditioned media from AT-MSCs of lean individuals did not. These results indicate that obesity reduces, or eliminates, the anti-inflammatory effects of human AT-MSCs, and therefore they may not be a suitable cell source for the treatment of autoimmune diseases (Strong et al. 2016). The majority of MSCs clinical trials are currently in phase 2 of development, during which safety and tolerability of treatment continue to be evaluated. Consequently, although AT-MSCs transplantation can be regarded as a potential source of treatment for MS, several studies now at clinical stages need to see whether these show a real benefit in practice, particularly in MS progressive stages (Guerrero-García et al. 2016). These works will contribute to the design of future trials conducted to establish whether MSCs transplantation comprises an effective therapy for patients with MS.

References

Angelucci F, Mirabella M, Caggiula M et al (2005) Evidence of involvement of leptin and IL-6 peptides in the action of interferon-beta in secondary progressive multiple sclerosis. Peptides 26(11):2289–2293

Assadi M, Salimipour H, Akbarzadeh S et al (2011) Correlation of circulating omentin-1 with bone mineral density in multiple sclerosis: the crosstalk between bone and adipose tissue. PLoS One 6(9):e24240

Balusu S, Van Wonterghem E, De Rycke R et al (2016) Identification of a novel mechanism of blood–brain communication during peripheral inflammation via choroid plexus-derived extracellular vesicles. EMBO Mol Med 8(10):1162–1183

Batocchi AP, Rotondi M, Caggiula M et al (2003) Leptin as a marker of multiple sclerosis activity in patients treated with interferon-beta. J Neuroimmunol 139(1–2):150–154

Baufeld C, Osterloh A, Prokop S et al (2016) High-fat diet-induced brain region- specific phenotypic spectrum of CNS resident microglia. Acta Neuropathol 132:361–375

Bedoui S, Miyake S, Lin Y et al (2003) Neuropeptide Y (NPY) suppresses experimental autoimmune encephalomyelitis: NPY1 receptor-specific inhibition of autoreactive Th1 responses in vivo. J Immunol 171:3451–3458

Belbasis L, Bellou V, Evangelou E et al (2015) Environmental risk factors and multiple sclerosis: an umbrella review of systematic reviews and meta-analyses. Lancet Neurol 4(3):263–273

Buckman LB, Hasty AH, Flaherty DK et al (2014) Obesity induced by a high-fat diet is associated with increased immune cell entry into the central nervous system. Brain Behav Immun 35:33–42

Button EB, Mitchell AS, Domingos MM et al (2014) Microglial cell activation increases saturated and decreases monounsaturated fatty acid content but both lipid species are proinflammatory. Lipids 49:305–316

Carrieri PB, Carbone F, Perna F et al (2015) Longitudinal assessment of immuno-metabolic parameters in multiple sclerosis patients during treatment with glatiramer acetate. Metabolism 64(9):1112–1121

Chatzantoni K, Papathanassopoulos P, Gourzoulidou E et al (2004) Leptin and its soluble receptor in plasma of patients suffering from remitting-relapsing multiple sclerosis (MS): in vitro effects of leptin on type-1 and type-2 cytokine secretion by peripheral blood mononuclear cells, T-cells and monocytes of MS patients. J Autoimmun 23(2):169–177

Daly RM, Gagnon C, Lu ZX et al (2012) Prevalence of vitamin D deficiency and its determinants in Australian adults aged 25 years and older: a national, population-based study. Clin Endocrinol 77(1):26–35

De Rosa V, Procaccini C, La Cava A et al (2006) Leptin neutralization interferes with pathogenic T cell autoreactivity in autoimmune encephalomyelitis. J Clin Invest 116(2):447–455

Dendrou CA, Fugger L, Friese MA (2015) Immunopathology of multiple sclerosis. Nat Rev Immunol 15(9):545–558

Dimitrijevic M, Stanojevic S (2013) The intriguing mission of neuropeptide Y in the immune system. Amino Acids 45:41–53

Duperray A, Barbe D, Raguenez G et al (2015) Inflammatory response of endothelial cells to a human endogenous retrovirus associated with multiple sclerosis is mediated by TLR4. Int Immunol 27(11):545–553

Emamgholipour S, Eshaghi SM, Hossein-Nezhad A et al (2013) Adipocytokine profile, cytokine levels and Foxp3 expression in multiple sclerosis: a possible link to susceptibility and clinical course of disease. PLoS One 8(10):e76555

Evangelopoulos ME, Koutsis G, Markianos M (2014) Serum leptin levels in treatment-naive patients with clinically isolated syndrome or relapsing-remitting multiple sclerosis. Autoimmune Dis 2014:486282

Flachenecker P, Wolf A, Krauser M et al (1999) Cardiovascular autonomic dysfunction in multiple sclerosis: correlation with orthostatic intolerance. J Neurol 246(7):578–586

Frischer JM, Bramow S, Dal-Bianco A et al (2009) The relation between inflammation and neurodegeneration in multiple sclerosis brains. Brain 132(5):1175–1189

Frisullo G, Mirabella M, Angelucci F et al (2007) The effect of disease activity on leptin, leptin receptor and suppressor of cytokine signalling-3 expression in relapsing-remitting multiple sclerosis. J Neuroimmunol 192(1–2):174–183

Galgani M, Procaccini C, De Rosa V et al (2010) Leptin modulates the survival of autoreactive CD4+ T cells through the nutrient/energy-sensing mammalian target of rapamycin signaling pathway. J Immunol 185(12):7474–7479

Gianfrancesco MA, Barcellos L (2016) Obesity and multiple sclerosis susceptibility: a review. J Neurol Neuromed 1(7):1–5

Gianfrancesco MA, Acuna B, Shen L et al (2014) Obesity during childhood and adolescence increases susceptibility to multiple sclerosis after accounting for established genetic and environmental risk factors. Obes Res Clin Pract 8(5):e435–e447

Glenn JD, Whartenby KA (2014) Mesenchymal stem cells: emerging mechanisms of immuno-modulation and therapy. World J Stem Cells 6(5):526–539

Graham KL, Zabel BA, Loghavi S et al (2009) Chemokinelike receptor-1 expression by central nervous system-infiltrating leukocytes and involvement in a model of autoimmune demyelinating disease. J Immunol 183(10):6717–6723

Guerrero-García JJ, Carrera-Quintanar L, López-Roa RI et al (2016) Multiple sclerosis and obesity: possible roles of adipokines. Mediat Inflamm 2016:4036232

Guillemot-Legris O, Muccioli GG (2017) Obesity induced neuroinflammation beyond the hypothalamus. Trends Neurosci 40(4):237–253

Hedstrom AK, Olsson T, Alfredsson L (2012) High body mass index before age 20 is associated with increased risk for multiple sclerosis in both men and women. Mult Scler 18(9):1334–1336

Hedström AK, Lima Bomfim I, Barcellos L et al (2014) Interaction between adolescent obesity and HLA risk genes in the etiology of multiple sclerosis. Neurology 82(10):865–872

Hedström AK, Lima Bomfim I, Hillert J et al (2015) Obesity interacts with infectious mononucleosis in risk of multiple sclerosis. Eur J Neurol 22(3):578–e38

Hedstrom AK, Olsson T, Alfredsson L (2016) Body mass index during adolescence, rather than childhood, is critical in determining MS risk. Mult Scler 22(7):878–883

Hemmer B, Kerschensteiner M, Korn T (2015) Role of the innate and adaptive immune responses in the course of multiple sclerosis. Lancet Neurol 14(4):406–419

Heneka MT, Kummer MP, Latz E (2014) Innate immune activation in neurodegenerative disease. Nat Rev Immunol 14(7):463–477

Heppner FL, Ransohoff RM, Becher B (2015) Immune attack: the role of inflammation in Alzheimer disease. Nat Rev Neurosci 16(6):358–372

Hietaharju A, Kuusisto H, Nieminen R et al (2010) Elevated cerebrospinal fluid adiponectin and adipsin levels in patients with multiple sclerosis: a Finnish co-twin study. Eur J Neurol 17(2):332–334

Hossein-Nezhad A, Varzaneh FN, Mirzaei K et al (2013) A polymorphism in the resistin gene promoter and the risk of multiple sclerosis. Minerva Med 104(4):431–438

Howell OW, Reeves CA, Nicholas R et al (2011) Meningeal inflammation is widespread and linked to cortical pathology in multiple sclerosis. Brain 134(9):2755–2771

Kanoski SE, Zhang Y, Zheng W et al (2010) The effects of a high-energy diet on hippocampal function and blood–brain barrier integrity in the rat. J Alzheimers Dis 21(1):207–219

Khurana SR, Bamer AM, Turner AP et al (2009) The prevalence of overweight and obesity in veterans with multiple sclerosis. Am J Phys Med Rehabil 88(2):83–91

Kraszula L, Jasińska A, Eusebio M et al (2012) Evaluation of the relationship between leptin, resistin, adiponectin and natural regulatory T cells in relapsing-remitting multiple sclerosis. Neurol Neurochir Pol 46(1):22–28

Lande R, Gafa V, Serafini B et al (2008) Plasmacytoid dendritic cells in multiple sclerosis: intracerebral recruitment and impaired maturation in response to interferon-β. J Neuropathol Exp Neurol 67(5):388–401

Langer-Gould A, Brara SM, Beaber BE et al (2013) Childhood obesity and risk of pediatric multiple sclerosis and clinically isolated syndrome. Neurology 80(6):548–552

Lo JC, Ljubicic S, Leibiger B et al (2014) Adipsin is an adipokine that improves β cell function in diabetes. Cell 158(1):41–53

Lock C, Hermans G, Pedotti R et al (2002) Gene-microarray analysis of multiple sclerosis lesions yields new targets validated in autoimmune encephalomyelitis. Nat Med 8(5):500–508

Lord GM, Matarese G, Howard JK et al (2002) Leptin inhibits the anti-CD3-driven proliferation of peripheral blood T cells but enhances the production of proinflammatory cytokines. J Leukoc Biol 72(2):330–338

Lumeng CN, Saltiel AR (2011) Inflammatory links between obesity and metabolic disease. J Clin Invest 121:2111–2117

Maeda K, Yasuda M, Kaneda H et al (1994) Cerebrospinal fluid (CSF) neuropeptide Y- and somatostatin-like immunoreactivities in man. Neuropeptides 27:323–332

Mähler A, Steiniger J, Bock M et al (2012) Is metabolic flexibility altered in multiple sclerosis patients? PLoS One 7:e43675

Markianos M, Evangelopoulos ME, Koutsis G et al (2013) Body Mass Index in Multiple Sclerosis: Associations with CSF Neurotransmitter Metabolite Levels. ISRN Neurol 2013:981070

Marrie RA, Horwitz R, Cutter G et al (2009) High frequency of adverse health behaviors in multiple sclerosis. Mult Scler 15(1):105–113

Marrie RA, Rudick R, Horwitz R et al (2010) Vascular comorbidity is associated with more rapid disability progression in multiple sclerosis. Neurology 74(13):1041–1047

Marrie RA, Horwitz R, Cutter G et al (2011) Association between comorbidity and clinical characteristics of MS. Acta Neurol Scand 124(2):135–141

Matarese G, Di Giacomo A, Sanna V et al (2001) Requirement for leptin in the induction and progression of autoimmune encephalomyelitis. J Immunol 166:5909–5916

Matarese G, Carrieri PB, La Cava A et al (2005) Leptin increase in multiple sclerosis associates with reduced number of CD4(+)CD25+ regulatory T cells. Proc Natl Acad Sci U S A 102(14):5150–5155

Mayo L, Trauger SA, Blain M et al (2014) Regulation of astrocyte activation by glycolipids drives chronic CNS inflammation. Nat Med 20(10):1147–1156

Messina S, Vargas-Lowy D, Musallam A et al (2013) Increased leptin and A-FABP levels in relapsing and progressive forms of MS. BMC Neurol 13:172

Minagar A, Jy W, Jimenez JJ et al (2006) Multiple sclerosis as a vascular disease. Neurol Res 28(3):230–235

Moschen AR, Kaser A, Enrich B et al (2007) Visfatin, an adipocytokine with proinflammatory and immunomodulating properties. J Immunol 178(3):1748–1758

Munger KL, Chitnis T, Ascherio A (2009) Body size and risk of MS in two cohorts of US women. Neurology 73(19):1543–1550

Munger KL, Bentzen J, Laursen B et al (2013) Childhood body mass index and multiple sclerosis risk: a long-term cohort study. Mult Scler 19(10):1323–1329

Musabak U, Demirkaya S, Genç G et al (2011) Serum adiponectin, TNF-α, IL-12p70, and IL-13 levels in multiple sclerosis and the effects of different therapy regimens. Neuroimmunomodulation 18(1):57–66

Natarajan R, Hagman S, Hämäläinen M et al (2015) Adipsin is associated with multiple sclerosis: a follow-up study of adipokines. Mult Scler Int 2015:371734

Neuparth MJ, Proença JB, Santos-Silva A et al (2013) Adipokines, oxidized low-density lipoprotein, and C reactive protein levels in lean, overweight, and obese Portuguese patients with type 2 diabetes. ISRN Obes 2013:142097

Neuteboom RF, Verbraak E, Voerman JSA et al (2009) Serum leptin levels during pregnancy in multiple sclerosis. Mult Scler 15(8):907–912

Nieva-Vazquez A, Perez-Fuentes R, Torres-Rasgado E et al (2014) Serum resistin levels are associated with adiposity and insulin sensitivity in obese Hispanic subjects. Metab Syndr Relat Disord 12(2):143–148

Oliveira SR, Simão AN, Kallaur AP et al (2014) Disability in patients with multiple sclerosis: influence of insulin resistance, adiposity, and oxidative stress. Nutrition 30(3):268–273

Palavra F, Marado D, Mascarenhas-Melo F et al (2013) New markers of early cardiovascular risk in multiple sclerosis patients: oxidized-LDL correlates with clinical staging. Dis Markers 34:341–348

Palavra F, Almeida L, Ambrósio AF et al (2016) Obesity and brain inflammation: a focus on multiple sclerosis. Obes Rev 17(3):211–224

Peterson LK, Fujinami RS (2007) Inflammation, demyelination, neurodegeneration and neuroprotection in the pathogenesis of multiple sclerosis. J Neuroimmunol 184(1–2):37–44

Piccio L, Stark JL, Cross AH (2008) Chronic calorie restriction attenuates experimental autoimmune encephalomyelitis. J Leukoc Biol 84(4):940–948

Piccio L, Cantoni C, Henderson JG et al (2013) Lack of adiponectin leads to increased lymphocyte activation and increased disease severity in a mouse model of multiple sclerosis. Eur J Immunol 43(8):2089–2100

Pieper C, Pieloch P, Galla HJ (2013) Pericytes support neutrophil transmigration via interleukin-8 across a porcine co-culture model of the blood–brain barrier. Brain Res 1524:1–11

Pieper C, Marek JJ, Unterberg M et al (2014) Brain capillary pericytes contribute to the immune defense in response to cytokines or LPS in vitro. Brain Res 1550:1–8

Pilutti LA, McAuley E, Motl RW (2012) Weight status and disability in multiple sclerosis: an examination of bi-directional associations over a 24-month period. Mult Scler Relat Disord 1:139–144

Reinhold D, Kahne T, Steinbrecher A et al (2002) The role of dipeptidyl peptidase IV (DP IV) enzymatic activity in T cell activation and autoimmunity. Biol Chem 383:1133–1138

Rotondi M, Batocchi AP, Coperchini F et al (2013) Severe disability in patients with relapsing-remitting multiple sclerosis is associated with profound changes in the regulation of leptin secretion. Neuroimmunomodulation 20(6):341–347

Saad MF, Damani S, Gingerich RL et al (1997) Sexual dimorphism in plasma leptin concentration. J Clin Endocrinol Metab 82(2):579–584

Sanna V, Di Giacomo A, La Cava A et al (2003) Leptin surge precedes onset of autoimmune encephalomyelitis and correlates with development of pathogenic T cell responses. J Clin Invest 111(2):241–250

Sathyapalan T, Shepherd J, Arnett C et al (2010) Atorvastatin increases 25-hydroxy vitamin D concentrations in patients with polycystic ovary syndrome. Clin Chem 56:1696–1700

Schmid A, Hochberg A, Berghoff M et al (2016) Quantification and regulation of adipsin in human cerebrospinal fluid (CSF). Clin Endocrinol 84(2):194–202

Shechter R, London A, Schwartz M (2013) Orchestrated leukocyte recruitment to immune-privileged sites: absolute barriers versus educational gates. Nat Rev Immunol 13(3):206–218

Simpson KA, Martin NM, Bloom SR (2009) Hypothalamic regulation of food intake and clinical therapeutic applications. Arq Bras Endocrinol Metabol 53:120–128

Sioka C, Fotopoulos A, Georgiou A et al (2011) Body composition in ambulatory patients with multiple sclerosis. J Clin Densitom 14(4):465–470

Slawta JN, Wilcox AR, McCubbin JA et al (2003) Health behaviors, body composition, and coronary heart disease risk in women with multiple sclerosis. Arch Phys Med Rehabil 84(12):1823–1830

Steinbrecher A, Reinhold D, Quigley L et al (2001) Targeting dipeptidyl peptidase IV (CD26) suppresses autoimmune encephalomyelitis and up-regulates TGF-beta 1 secretion in vivo. J Immunol 166:2041–2048

Stranahan AM, Hao S, Dey A et al (2016) Blood–brain barrier breakdown promotes macrophage infiltration and cognitive impairment in leptin receptor-deficient mice. J Cereb Blood Flow Metab 36:2108–2121

Streit WJ, Mrak RE, Griffin WS (2004) Microglia and neuroinflammation: a pathological perspective. J Neuroinflammation 1(1):14

Strong AL, Bowles AC, Wise RM et al (2016) Human adipose stromal/stem cells from obese donors show reduced efficacy in halting disease progression in the experimental autoimmune encephalomyelitis model of multiple sclerosis. Stem Cells 34(3):614–626

Tettey P, Simpson S Jr, Taylor B et al (2014a) An adverse lipid profile is associated with disability and progression in disability in people with MS. Mult Scler 20:1737–1744

Tettey P, Simpson S Jr, Taylor B et al (2014b) Adverse lipid profile is not associated with relapse risk in MS: results from an observational cohort study. J Neurol Sci 340:230–232

Thaler JP, Guyenet SJ, Dorfman MD et al (2013) Hypothalamic inflammation: marker or mechanism of obesity pathogenesis? Diabetes 62:2629–2634

Tomalka-Kochanowska J, Baranowska B, Wolinska-Witort E et al (2014) Plasma chemerin levels in patients with multiple sclerosis. Neuro Endocrinol Lett 35(3):218–223

Traka M, Podojil JR, McCarthy DP et al (2016) Oligodendrocyte death results in immune-mediated CNS demyelination. Nat Neurosci 19(1):65–74

Trapp BD, Nave KA (2008) Multiple sclerosis: an immune or neurodegenerative disorder? Annu Rev Neurosci 31:247–269

Valdearcos M, Robblee MM, Benjamin DI et al (2014) Microglia dictate the impact of saturated fat consumption on hypothalamic inflammation and neuronal function. Cell Rep 9:2124–2138

Weinstock-Guttman B, Zivadinov R, Mahfooz N et al (2011a) Serum lipid profiles are associated with disability and MRI outcomes in multiple sclerosis. J Neuroinflammation 4(8):127

Weinstock-Guttman B, Zivadinov R, Ramanathan M (2011b) Inter-dependence of vitamin D levels with serum lipid profiles in multiple sclerosis. J Neurol Sci 311:86–91

Wilk S, Scheibenbogen C, Bauer S et al (2011) Adiponectin is a negative regulator of antigen-activated T cells. Eur J Immunol 41(8):2323–2332

Williams G, Harrold JA, Cutler DJ (2000) The hypothalamus and the regulation of energy homeostasis: lifting the lid on a black box. Proc Nutr Soc 59(3):385–396

Yousefi F, Ebtekar M, Soleimani M et al (2013) Comparison of in vivo immunomodulatory effects of intravenous and intraperitoneal administration of adipose tissue mesenchymal stem cells in experimental autoimmune encephalomyelitis (EAE). Int Immunopharmacol 17(3):608–616

Zhang X, Bowles AC, Semon JA et al (2014) Transplantation of autologous adipose stem cells lacks therapeutic efficacy in the experimental autoimmune encephalomyelitis model. PLoS One 9(1):e85007

Part III
Brain Function After Bariatric Surgery

Chapter 9
Central Modulation of Energy Homeostasis and Cognitive Performance After Bariatric Surgery

Hans Eickhoff

Abstract In moderately or morbidly obese patients, bariatric surgery has been proven to be an effective therapeutic approach to control body weight and comorbidities. Surgery-mediated modulation of brain function via modified postoperative secretion of gut peptides and vagal nerve stimulation was identified as an underlying mechanism in weight loss and improvement of weight-related diseases. Increased basal and postprandial plasma levels of gastrointestinal hormones like glucagon-like peptide 1 and peptide YY that act on specific areas of the hypothalamus to reduce food intake, either directly or mediated by the vagus nerve, are observed after surgery while suppression of meal-induced ghrelin release is increased. Hormones released from the adipose tissue like leptin and adiponectin are also affected and leptin plasma levels are reduced in treated patients. Besides homeostatic control of body weight, surgery also changes hedonistic behavior in regard to food intake and cognitive performance involving the limbic system and prefrontal areas.

Keywords Bariatric surgery • Brain function • Cognitive performance • Energy homeostasis • Hedonic behavior

9.1 Introduction

Bariatric surgery is widely recognized as an effective therapeutic procedure to control severe and morbid obesity and weight-related diseases. Case-control studies, randomized controlled trials, and large meta-analyses have shown that surgery not only reduces body weight more efficiently than medical treatment but also

H. Eickhoff (✉)
Institute of Physiology, Institute for Biomedical Imaging and Life Sciences—IBILI,
Faculty of Medicine, University of Coimbra, Coimbra, Portugal

Obesity Center, Hospital da Luz de Setúbal, Setúbal, Portugal
e-mail: h.c.a.e@sapo.pt

© Springer International Publishing AG 2017
L. Letra, R. Seiça (eds.), *Obesity and Brain Function*, Advances in
Neurobiology 19, DOI 10.1007/978-3-319-63260-5_9

improves comorbidities like type 2 diabetes, hypertension, and sleep apnea, among others (Buchwald et al. 2004, 2009; Schauer et al. 2014; Sjöström 2008, 2013). However, until recently, surgical procedures have been regarded as either restrictive or malabsorptive, or a combination of both (Mingrone and Castagneto-Gissey, 2009; Sandoval 2011).

The surgical treatment of diet-resistant morbid obesity started in the 1950s with the jejunoileal bypass which excluded a large segment of the small intestine but created a blind loop prone to bacterial overgrowth. Other early and late complications included hypoproteinemia, vitamin deficiency, severe diarrhea, and liver disease (Drenick et al. 1976; Fikri and Cassella 1974; Halverson et al. 1978; Weismann and Johnson 1977) which led to the abandonment of the technique and its substitution by biliopancreatic diversion (Scopinaro 2006; Scopinaro et al. 1998) and gastric bypass (Mason et al. 1978; Mason and Ito 1969). Both techniques included the creation of a gastric pouch and a more or less extended bypass of the small intestine avoiding creating a blind segment by using a loop or a Roux-en-Y anastomosis to the gastric remnant. As early as 1981 a clinical study on gut hormones in patients submitted to either jejunoileal bypass or biliopancreatic diversion showed a postprandial elevation of an enteroglucagon from the distal gut in operated individuals in comparison to controls (Sarson et al. 1981). However, the complete scope of the impact of surgical procedures on the complex regulation of hunger and satiety and overall energy homeostasis only came into focus after the discovery and general acknowledgement of the importance of gut hormones and the enteroendocrine axis with regard to weight loss and the control of comorbidities (Bloom et al. 2005; Cummings et al. 2004; le Roux and Bloom 2005; Neary et al. 2004).

Gastrointestinal hormones like glucose-dependent insulinotropic polypeptide (GIP) and glucagon-like peptide-1 (GLP-1) have a powerful effect on insulin secretion after an oral glucose load and are produced in duodenal K-cells (GIP) and intestinal L-cells (GLP-1) that are present in increasing number in the distal jejunum, ileum, and colon (Dupre et al. 1973; Eissele et al. 1992; Gutzwiller et al. 1999; Kreymann et al. 1987; Polak et al. 1973; Turner et al. 1973). This so-called incretin effect explains the increased tolerance to oral glucose administration in comparison to intravenous administration that had been recognized since the late nineteenth century (Creutzfeldt 1979). Whereas GIP regulates primarily insulin and glucagon secretion, besides its effects on gastric emptying, GLP-1 has been shown to act as well in the central nervous system in specific areas of the hypothalamus (Taminato et al. 1977; Turton et al. 1996). Studies in rats and humans showed its inhibitory effects on food intake while enhancing satiety (Gutzwiller et al. 1999; Turton et al. 1996). Other gastrointestinal hormones that are modulated by bariatric surgical procedures include peptide tyrosine tyrosine (PYY) and ghrelin (Batterham et al. 2002; Cummings et al. 2002).

However, the transformation of the gastrointestinal tract through bariatric surgery does not act solely on gut peptides to alter the interplay between hunger and satiety. It also exerts its effects by stimulation of the vagal nerve and interferes in the striatal dopamine homeostasis. Moreover, as the regulation of food intake is also subject to control through higher nervous function, including learned social behavior and cognitive action, recent research has also demonstrated the effects of surgery in this domain.

9.2 Effects of Bariatric Surgery on Gut Peptide Hormones

9.2.1 *Glucagon-like Peptide 1, Oxyntomodulin, and Peptide YY*

The gut peptides glucagon-like peptide 1 (GLP-1), oxyntomodulin (OXM), and peptide YY (PYY) are released from intestinal endocrine L-cells in response to food intake (Ballantyne 2006; Murphy and Bloom 2004; Wynne and Bloom 2006). While GLP-1 and OXM are derived from cleavage of proglucagon by prohormone convertase 1 and 2, peptide YY belongs to the pancreatic peptide family and is structurally similar to neuropeptide Y (Lundberg et al. 1982; Tatemoto et al. 1982; Tatemoto and Mutt 1980); it binds to the Y2 receptor in the arcuate nucleus (ARC) of the hypothalamus and reduces food intake in rodents (Batterham et al. 2002; Chelikani et al. 2006) and humans (Batterham et al. 2003).

Modulation of postprandial secretion of GLP-1 after gastrointestinal surgery has been under intense scrutiny as it increases the secretion of insulin in pancreatic beta-cells (Mojsov et al. 1987; Orskov 1992), a possible mechanism for the improvement of type 2 diabetes after bariatric surgery (Kashyap et al. 2013; Vidal and Jiménez 2013). However, as shown in animal models, GLP-1 acts in the central nervous system as well, especially on GLP-1 receptors (GLP-1R) in the paraventricular nucleus (PVN) of the hypothalamus (Larsen et al. 1997; Turton et al. 1996), and inhibits food intake in humans (Gutzwiller et al. 1999; Verdich et al. 2001). GLP-1Rs are also present in the dorsal raphe nucleus of the brainstem and their activation might induce hypophagia via an increase in hypothalamic serotonin signaling (Anderberg et al. 2017). Oxyntomodulin, besides its effects on gastric acid secretion, gastroduodenal motility, and gastric emptying (Schjoldager et al. 1989), also acts centrally and its intracerebroventricular (ICV) infusion induces hypophagia in rats (Dakin et al. 2001). Peripheral infusion in humans reduces ad libitum food intake and preprandial levels of the orexigenic hormone ghrelin (Cohen et al. 2003). Apparently, OXM acts through dual activation of the GLP-1R and, additionally, by signaling via the glucagon receptor (Kosinski et al. 2012).

In obese patients submitted to gastric bypass, a surgical technique that creates a small gastric pouch with a gastrojejunal anastomosis, postprandial GLP-1 and PYY levels in peripheral blood increased significantly after surgery in comparison to preoperative values (Borg et al. 2006; le Roux et al. 2007; Morínigo et al. 2006). Sleeve gastrectomy, a surgical technique which includes a resection of the larger curvature of the stomach without bypassing any segment of the gastrointestinal tract, also induces a similar increase in postprandial GLP-1 and PYY (Nannipieri et al. 2013; Papamargaritis et al. 2013; Romero et al. 2012). Basic research including bariatric surgery performed in obese and non-obese rodents support the observation that both sleeve gastrectomy and gastric bypass have a comparable effect on postprandial gastrointestinal hormone levels (Cummings et al. 2012; Eickhoff et al. 2015). Gastric banding, a bariatric procedure that does not involve any resection or bypassing of the gastrointestinal tract and creates only a small pouch in the upper stomach with a restricted outlet, does not induce a similar rise in postprandial plasma levels of GLP-1 and PYY (le Roux et al. 2006).

Blood oxygen level-dependent (BOLD) functional magnetic resonance imaging (fMRI) is able to demonstrate brain activity in response to fasting and feeding in specific regions of interest (ROIs) including amygdala, caudate, insula, nucleus accumbens, orbitofrontal cortex, and putamen. In fasting subjects exposed to images of food, BOLD signals decreased after intravenous administration of GLP-1 and PYY, either isolated or together. The signal change after combined administration of both hormones was comparable to that seen in subjects after an ad libitum meal intake which strongly suggests a role for both hormones (De Silva et al. 2011). Comparing the effects of surgical techniques with (gastric bypass) and without (gastric banding) the transformation of the gastrointestinal tract, fMRI showed that the exposure to images of food, and particularly high calorie food, produced significantly lower BOLD values in ROIs after gastric bypass than after gastric banding. Simultaneously, the increase in GLP-1 and PYY plasma levels after food intake was significantly more pronounced after gastric bypass in the same study population (Scholtz et al. 2014).

Recently, however, discussion arose regarding the importance of postprandial GLP-1 rise for appetite control after bariatric surgery (Vidal et al. 2016), notwithstanding the favorable effect of exogenous GLP-1R agonists on food intake and weight loss (Davies et al. 2015; Pi-Sunyer et al. 2015). In animal studies, sleeve gastrectomy and gastric bypass in GLP-1 receptor-deficient mice showed similar outcomes regarding weight loss, reduced food intake, and glucose homeostasis, in comparison to wild-type mice (Wilson-Pérez et al. 2013; Ye et al. 2014).

Nonetheless, the administration of the somatostatin analogue octreotide, a nonspecific inhibitor of gut hormone release, induces an increase in food intake in patients submitted to gastric bypass while decreasing postprandial levels of GLP-1 and PYY (Goldstone et al. 2016; le Roux et al. 2007), indicating an effect for both hormones, at least while acting together, in patients submitted to surgery. In a rodent model, the ICV infusion of exendin (9-39), a GLP-1R antagonist, blocked the inhibitory action on food intake normally seen after ICV administration of GLP-1. In untreated animals, the sole infusion of exendin (9-39) increased the food intake in already satiated rats, as well as the response to the orexigenic neuropeptide Y (Turton et al. 1996). In rats submitted to gastric bypass, the blockade of GLP-1R by ICV infusion of exendin (9-39) induced a steady increase in food intake, feed efficiency, body weight, and adiposity. Interestingly, in the same study, the ICV infusion of an Y2 receptor antagonist without simultaneous blockade of the GLP-1R did not have any effect on body weight or food intake, although the influence of PYY on hunger and satiety is well established (Ye et al. 2014). Besides a technical issue, brought forward by the authors as an explanation, ICV infusion is possibly less effective regarding the blockade of PYY mediated decrease in food intake in comparison to direct injection into the arcuate nucleus and the locus of Y2 receptors (Abbott et al. 2005b). Additionally, as shown in another rodent model, gastric bypass surgery downregulates neuropeptide Y itself (Romanova et al. 2004) which possibly induces the simultaneous downregulation of the Y2 receptor. This finding might explain the reduced effect of the ICV infusion of the Y2 receptor antagonist on the effect of PYY mediated decrease in food intake.

9.2.2 Cholecystokinin

Cholecystokinin (CCK) is secreted from duodenal and jejunal I-cells in response to a meal (Liddle et al. 1985) and exerts effects on short-term satiety via receptors on the vagus nerve, brainstem, and hypothalamus (Blevins et al. 2000; Garlicki et al. 1990; Kopin et al. 1999). Besides its direct effects on the control of hunger and satiety, CCK mediates the stimulating effect of intraduodenal fat hydrolysis on GLP-1 and PYY secretion, as well as the inhibition of ghrelin secretion (Beglinger et al. 2010; Degen et al. 2007). After bariatric surgery, postprandial plasma levels of CCK are not significantly affected by banded gastroplasty, a surgical technique that simply reduces the gastric lumen (Foschi et al. 2004). However, sleeve gastrectomy, a technique that includes the surgical resection of the larger curvature and the gastric fundus, induces a more pronounced postprandial increase in plasma CCK, in comparison to gastric bypass, a procedure that includes duodenal exclusion, the main locus of CCK secretion (Lee et al. 2011; Peterli et al. 2012).

9.2.3 Ghrelin

In contrast to the previously described hormones, ghrelin has a markedly orexigenic effect and its ICV or peripheral administration increases food intake, body weight, and adiposity in rats, apparently through the activation of the growth hormone receptor on neuropeptide Y neurons in the ARC of the hypothalamus (Kohno et al. 2003; Tschöp et al. 2000; Wren et al. 2000). Ghrelin is secreted mainly from the gastric fundus but can also be found in the remaining gastrointestinal tract in rats and humans in decreasing levels towards the colon (Date et al. 2000; Kojima et al. 1999). Endogenous ghrelin also stimulates food intake in humans (Wren et al. 2001) but fasting plasma ghrelin levels are reduced in obese individuals in comparison to lean controls (Tschöp et al. 2001). Regarding diurnal variation, fasting ghrelin levels tend to be elevated and decrease markedly in response to a meal in healthy individuals (Cummings 2001) while this variation appears to be impaired in obese patients (English et al. 2002).

The effect of bariatric surgery on ghrelin secretion has been extensively studied. Fasting and meal reduced plasma ghrelin levels are generally reduced after gastric bypass in comparison to lean and obese controls, while diet-induced weight loss increases fasting and circadian ghrelin secretin (Cummings et al. 2002; Frühbeck et al. 2004; Korner et al. 2006; Tritos et al. 2003), possibly related to the difficulty that patients experience to maintain weight loss after dieting. In other studies somewhat contradictory results have been observed regarding fasting ghrelin values after gastric bypass, although the authors did not measure postprandial hormone levels which might have shown the characteristic reduction after food intake (Borg et al. 2006; Faraj et al. 2003; Holdstock et al. 2003). Conversely, surgical techniques that involve the resection of the gastric fundus induce a sustained reduction of both fasting and postprandial ghrelin secretion. This observation was confirmed in patients submitted to sleeve gastrectomy (Karamanakos et al. 2008; Langer et al. 2005; Yousseif et al. 2014), biliopancreatic diversion with duodenal switch (Kotidis et al. 2006), and gastric bypass with additional resection of the gastric fundus (Chronaiou et al. 2012).

Isolated postoperative ghrelin levels might have limited prognostic value, at least in patients submitted to gastric bypass, regarding the extent of weight loss after surgery and do not necessarily correspond to satiety experienced by operated patients (Christou et al. 2005). However, patients with poor weight loss after gastric bypass (excess body mass index loss (EBL) < 50%) showed less reduction in postprandial ghrelin levels expressed as area under the curve than patients with good results after surgery (EBL > 60%) (Dirksen et al. 2013). Moreover, in fMRI studies after the intraperitoneal injection of ghrelin in anesthetized mice provoked an increase in signal intensity in ROIs in the hypothalamus associated with the regulation of food intake, in comparison to animals treated with saline. This observation was accompanied by an increased food intake after the end of the anesthesia. An opposite effect was obtained after injection of PYY, both regarding signal intensity and eating behavior (Kuo et al. 2007). In humans, fMRI using BOLD effect sizes with peak voxels show activation of brain areas related to hedonic feeding behavior after exogenous ghrelin administration in comparison to controls, in particular the bilateral amygdala and left orbitofrontal cortex (OFC) and pulvinar gyrus, which was also associated with self-reported hunger (Malik et al. 2008). These findings support the association between ghrelin-mediated activation of specific areas of the brain and the increase in food intake in vivo but also show that ghrelin levels alone are not responsible for the modulation of satiety after surgery, underscoring the multifactorial origin of surgical induced weight reduction.

Not surprisingly, narrowing of the gastric inlet alone and creating a small gastric pouch either through gastric banding or vertical banded gastroplasty did not significantly reduce and might even increase fasting and postprandial ghrelin levels, similarly to diet-induced weight loss (Frühbeck et al. 2004; Korner et al. 2006; Nijhuis et al. 2004).

9.3 Modulation of the Brain-Adipose Tissue Axis After Surgery

Leptin is secreted by white adipose tissue and is encoded by the *ob* gene (Zhang et al. 1994). Its exogenous administration decreases food intake in rodents, particularly in animals with deficient leptin secretion or an altered leptin receptor (Campfield et al. 1995; Sahu 1998). The leptin receptor is densely localized in the ARC of the hypothalamus and signaling involves activation of proopiomelanocortin (POMC) neurons (Balthasar et al. 2004). An initial study in humans showed a small dose-dependent effect at 4 weeks in lean and obese subjects and at 24 weeks in obese subjects (Heymsfield et al. 1999) but these observations were not confirmed by later research in obese patients (Liu et al. 2013; Zelissen et al. 2005). Also its short-term administration in non-obese humans did not affect energy metabolism or food intake (Mackintosh and Hirsch 2001). However, exogenous leptin administration is extremely efficient in patients that suffer from a rare congenital leptin deficiency characterized by extreme hyperphagia and adiposity and very low plasma leptin levels (Montague et al. 1997). Nevertheless, in general leptin levels are increased in obese subjects in comparison to lean individuals and correlate to total body fat mass (Considine et al. 1996), except for acute fasting states where a decrease in circulating leptin occurs, out of proportion regarding small fat mass loss (Chan and Mantzoros 2003).

Adiponectin is another hormone secreted by adipose tissue that apparently interferes with the regulation of hunger and satiety. It is reported to stimulate food intake by activating specific receptors in the ARC nucleus in rodents (Kadowaki et al. 2008; Kubota et al. 2007) and has anti-inflammatory properties (Ouchi et al. 1999, 2000). Chronic ICV infusion of high doses of adiponectin reduced food intake slightly, without affecting whole body metabolic rate (Bassi et al. 2012). However, conflicting research showed a decrease in body weight mediated by increased energy expenditure in mice, with no effect on food intake (Qi et al. 2004). While in patients with insulin resistance and obesity plasma adiponectin tends to be low (Hotta et al. 2000; Weyer et al. 2001), diet-induced weight loss increases circulating hormone levels (Yang et al. 2001).

Studies in patients submitted to bariatric surgery revealed a postoperative increase in plasma adiponectin and a reduction in circulating leptin, both after gastric bypass and sleeve gastrectomy (Faraj et al. 2003; Gumbau et al. 2014; Holdstock et al. 2003). After sleeve gastrectomy in rodents, leptin levels decreased (Stefater et al. 2010), but not more than in pair-fed animals or in animals submitted to gastric banding (Kawasaki et al. 2015) which is consistent with the hypothesis that leptin correlates with body fat mass rather than being an independent player in the regulation of body weight. Furthermore, research investigating the hypothesis that low leptin levels after gastric bypass might contribute to insufficient weight loss did not reveal any difference in body weight change in patients that received exogenous leptin during 16 weeks, in comparison to controls (Korner et al. 2013).

Gastric bypass in leptin deficient *ob/ob* mice did not produce the same effect on body weight and insulin sensitizing if compared to the same operation in mice with diet-induced obesity (DIO), yet exogenous administration of leptin was able to restore the effect of gastric bypass in leptin deficient animals (Hao et al. 2015). As such, leptin seems to be necessary for the effects of bariatric surgery on the regulation of body weight, but not in a dose-dependent manner. Interestingly, in type 2 diabetic patients submitted to biliopancreatic diversion, remitters showed higher circulating leptin levels preoperatively and increased plasma adiponectin at 5 years, in comparison to non-remitters. Leptin levels decreased in both groups but remitters had values almost three times as high 5 years after surgery (Adami et al. 2016).

Leptin secretion after bariatric surgery might be suppressed under the influence of high GLP-1. Patients submitted to gastric bypass that suffered from persistent postoperative nausea and vomiting showed significantly increased fasting plasma GLP-1 and decreased circulating leptin levels. Adipose tissue samples incubated with recombinant human GLP-1 showed significantly reduced leptin concentrations in culture media in comparison to samples without GLP-1 (Al-Rasheid et al. 2014). Another mechanism for the reduction of postoperative leptin secretion is possibly linked to reduced circulating free fatty acids after surgery as suggested by a study in patients submitted to biliopancreatic diversion (Raffaelli et al. 2015).

9.4 The Role of the Vagal Nerve

Besides the direct effect of gut hormones on specific structures of the brain, and particularly the brainstem, stimulation of the vagus nerve, either mechanically or by gut hormones, plays an important role in the regulation of appetite and satiety, particularly through afferent signaling (de Lartigue and Diepenbroek 2016; Kentish and Page 2015; Kral et al. 2009). In a small study truncal vagotomy alone has been shown to be effective in reducing food intake and body weight in obese patients (Gortz et al. 1990). Recent research in Wistar rats made obese by a palatable high fat cafeteria diet showed that animals submitted to subdiaphragmatic truncal vagotomy experienced a decrease in body weight and adiposity in comparison to pair-fed rats, achieving and maintaining a body weight similar to animals in the control group fed standard rat chow. Furthermore, fasting and postprandial insulin secretion normalized in operated animals (Balbo et al. 2016). However, these findings contrast with evidence that CCK and leptin inhibit short-term feeding behavior inducing the ending of a meal via afferent vagal fibers, in opposition to the orexigenic effect of ghrelin (Owyang and Heldsinger 2011). Additionally, PPY and GLP-1 activate vagal afferents to decrease food intake and the effects of their peripheral or intraperitoneal administration are ablated after subdiaphragmatic truncal vagotomy or transection of the brainstem-hypothalamic pathway in rodents (Abbott et al. 2005a).

Nevertheless, there is evidence that "over-activation" of afferent fibers in obesity might lead to a loss of its effects through processes that involve synaptic plasticity

(Blasi 2016). As such, truncal vagotomy possibly interrupts this unfavorable feedback mechanism. Moreover, efferent vagal fibers are equally sectioned during this procedure inducing not only reduced gastrin secretion and gastric motility, but possibly also a decrease in ghrelin secretion, as can be shown after anticholinergic blockade of the vagal tone in human volunteers (Veedfald et al. 2016).

Interestingly, intermittent vagal nerve blockade (vBloc) therapy, a novel mini-invasive surgical procedure for treatment of obesity and associated diseases, has been shown to decrease body weight and ameliorate type 2 diabetes and cardiovascular risk factors in obese patients (Apovian et al. 2017; Shikora et al. 2016). Leads are placed on the anterior and posterior abdominal vagal trunks and connected to a rechargeable pacemaker-like neuroregulator that delivers current during at least 12 h per day. Possible mechanisms of action include vagal-mediated increase in brainstem receptors for leptin and CCK while plasma concentrations of gut peptides including gastrin remained unchanged as shown in a study in rodents, which is consistent with an exclusive activation of afferent fibers and blockade of efferent signals (Johannessen et al. 2017).

Gastric banding, a surgical procedure that creates a small gastric pouch by an outlet obstruction in the upper stomach without any anatomical transformation of the gastrointestinal tract, besides the anatomical restriction also reduces food intake and body weight in rats through a mechanism that involves stimulation of afferent vagal fibers which could be reversed by capsaicin-mediated lesioning of sensory fibers. Brainstem activity in the nucleus tractus solitarius (NTS) and the parabrachial nucleus assessed by Fos protein elevation after band insufflation decreased in capsaicin-treated animals (Aneta Stefanidis et al. 2016). However, capsaicin might not be selective for afferent vagal fibers alone but exert effects on motor fiber and the brainstem as well (Travagli and Anselmi 2016). In contrast, damage to efferent and afferent vagal structures and vagal-hindbrain communication can be shown in rats submitted to gastric bypass in comparison to sleeve gastrectomy or sham-operated animals using a retrograde tracer. A reorganization of vagal circuits in the hindbrain was observed after sleeve gastrectomy leading to an increased density of vagal afferents whereas a decreased density occurred after gastric bypass, together with intensified microglia activation (Ballsmider et al. 2015) that is also observed after truncal vagotomy (Gallaher et al. 2012). However, afferent vagal signaling after gastric bypass from mid and lower intestines via the celiac branch still contributes to hypophagia and weight loss after surgery as has been shown in rats submitted to gastric bypass with or without sectioning of the celiac branch of the vagus nerve (Hao et al. 2014).

In addition to the effects on sensory vagal signaling, diet-induced obesity also affects efferent vagal neurons. In rodents exposed to high-fat diet for 3 months, neurons from the dorsal motor nucleus of the vagus (DMV) were less excitable and less responsive to GLP-1 and CCK than controls. However, in animals submitted to gastric bypass, these effects were reversed whereas morphological changes persisted, possibly attributable rather to diet than to obesity itself (Browning et al. 2013).

9.5 Striatal Dopamine Homeostasis and Opioid Receptor Modulation After Surgery

The regulation of body weight, food intake, hunger, and satiety is not only subject to homeostatic signals but also modulated by hedonic and reward driven behavior. The mesolimbic dopamine pathway is involved in substance abuse and in excessive food intake beyond satiation by mechanisms that include conditioning or reward learning (Hall et al. 2014; Lutter and Nestler 2009; Nummenmaa et al. 2012) and also essential in the quest for nutritional value (McCutcheon 2015). Consequently, absence of DA in genetically modified mice lead to early death from starvation which could be prevented by its exogenous administration (Szczypka et al. 2001), which underscores the central role of DA signaling in the context of food intake. It can be shown that mesolimbic dopamine expression is subject to activation (ghrelin) and inhibition (e.g., leptin and insulin) by gut and adipose tissue peptides that are modulated by bariatric surgery (Palmiter 2007). Using fMRI, increased dopaminergic signaling into the nucleus accumbens can also be shown after the presentation of highly palatable food to obese patients in comparison to controls, associated with possibly dopamine-mediated activation of the dorsal striatum (Rothemund et al. 2007). However, the presence of dopamine D_2 receptors in any brain region assessed by positron emission tomography (PET) with selective radioligands was not distinctive in obese or lean subjects. On the other hand, µ-opioid receptors in regions linked to reward processing like the ventral striatum, the insula, and thalamus were decreased in obese patients, suggestive of downregulation of the reward system that might lead to compensatory overeating (Karlsson et al. 2015).

In DIO mice submitted to gastric bypass, DA levels in the dorsal striatum were increased 4 weeks after surgery in comparison to controls, and associated with reduced body weight, adiposity, and food intake (Reddy et al. 2014) but dopaminergic transmission markers (dopamine transporter and tyrosine hydroxylase) in the midbrain and preference for high-fat diet were reduced after gastric bypass surgery in Sprague-Dawley rats (Barkholt et al. 2016). In another experimental study, gastric bypass upregulated striatal D_1 receptors in operated animals, particularly under high-fat feeding conditions whereas diet-induced weight loss had no effect. Vagotomy attenuated the effect of surgery on striatal dopamine expression, a finding that underscores the role of afferent vagal fibers on eating behavior. Apparently, gastric bypass increases the production of oleoylethanolamide (OEA) in the lower intestine, which increases striatal dopamine levels via activation of afferent vagal fibers through the peroxisome proliferator-associated receptor-α (PPAR-α) (Hankir et al. 2017).

Regarding sweet taste and glucose intake, the effect of duodenal-jejunal bypass on sweet-seeking behavior was assessed in a rodent model. After surgery, animals satiated by an intra-gastric glucose preload showed significantly reduced craving for the non-caloric sweetener sucralose. This was associated with an abolished DA expression in the dorsal but not in the ventral striatum in response to glucose in operated rats in comparison to sham surgery. However, optogenetic activation of D_1 receptor neurons in the dorsal striatum was able to override this effect and induced increased sweet-seeking behavior in rats submitted to duodenal-jejunal bypass (Han et al. 2015).

Modulation of food choice has been assessed after gastric bypass in obese subjects. Reduced reward value for sweet and fat taste was observed after surgery using a setting that involved progressive ratio tasking (Miras et al. 2012). In another study, a questionnaire measuring the motivation to consume highly palatable foods was used to assess the hedonic drive to consume sweet and fatty food. In parallel, actual food intake was assessed by a second survey focusing on food frequency. After surgery, a reduced hedonic drive for the consumption of highly palatable food was observed while dietary habits showed an increased intake in protein-rich food and vegetables, in detriment of snacks and sugar containing beverages (Miras et al. 2012). Similar changes were observed after sleeve gastrectomy (Coluzzi et al. 2016). Taken together with results of previously mentioned animal studies, these observations are suggestive of underlying mechanisms involving DA-mediated changes in striatal activation after surgery. However, results might be biased by the influence of dietary counseling during the pre- and postoperative management.

In human obese patients submitted to bariatric surgery, the hedonic and reward driven food intake has been assessed indirectly through fMRI, comparing the response to pictures of high and low-calorie food after gastric bypass and gastric banding, a purely restrictive surgical procedure. Besides increased stimulated plasma PYY levels, reduced BOLD activation in areas related to hedonic food intake and reward processing in the ventral (nucleus accumbens) and dorsal (putamen and caudate nucleus) was observed after gastric bypass (Scholtz et al. 2014). Additionally, signal intensity was also reduced in the orbitofrontal cortex, a prefrontal region associated with reward driven decision-making (Kringelbach and Radcliffe 2005). Another research group was also able to demonstrate reduced mesolimbic activation after visual and auditory exposition to high energy density food cues using fMRI in operated patients, which was associated with a reduced desire to eat. No change in response to low energy density food cues was observed (Ochner et al. 2011, 2012).

Striatal dopamine D_2 receptor expression, evaluated by PET scan, decreased after weight loss following both gastric bypass and sleeve gastrectomy and was associated with less hunger and improved satiety (Dunn et al. 2010). However, conflicting results were reported in another study in gastric bypass patients that showed an increased D_2 receptor availability, 6 weeks after surgery, using volumes of interests (VOIs) with a MRI and PET co-registration technique (Steele et al. 2010). Recent research using single-photon emission computed tomography (SPECT) in patients submitted to gastric bypass demonstrated an increase in dopamine $D_{2/3}$ receptor availability after 2 years of follow-up, yet still inferior to lean controls, with no change in the short-term (De Weijer et al. 2014; van der Zwaal et al. 2016).

However, unfavorable results of surgery related to addiction transfer due to the reward deficiency syndrome might also be linked to the dopaminergic mesolimbic system as an increased frequency of the dopamine D_2 receptor gene A1 allele can be found in morbidly obese patients and individuals with alike addictions (Blum and Bailey 2011). As presence of the A1 allele is associated with a decreased density of D_2 receptors (Pohjalainen et al. 1998; Thompson et al. 1997), this polymorphism might explain the reduced food reward leading to overeating in some obese patients and would be concordant with the association of a D_2 receptor availability and the results of weight loss surgery in some fMRI studies (De Weijer et al. 2014; van der Zwaal et al. 2016).

9.6 Short- and Long-Term Variation of Cognitive Function in Patients Submitted to Bariatric Surgery

Multiple studies have demonstrated an inverse correlation between obesity and cognitive function (Boeka and Lokken 2008; Elias et al. 2003; Fagundo et al. 2012; Gunstad et al. 2007). The Baltimore Longitudinal Study of Aging showed that obesity indexes were associated with a poorer performance in global cognitive function and domains like global screening measures, memory, and verbal fluency tasks but not in executive function and tests of attention and visuospatial ability (Gunstad et al. 2010). Similarly, cognitive function was impaired in patients suffering from obesity alone or as part of the metabolic syndrome in a study involving a Canadian First Nations population (Fergenbaum et al. 2009). Although evidence is mounting regarding the association between obesity and impaired cognitive function, study criteria have been heterogeneous and methodological issues might need to be addressed with appropriate assessment tools (Prickett et al. 2015).

Functional MRI assessing the BOLD response to a challenging cognitive task in patients with central obesity controls revealed alterations in the right superior frontal gyrus and the left middle frontal gyrus that were related to a decline in task performance and associated with increased waist circumference (Gonzales et al. 2014). In another study using PET to evaluate regional brain glucose metabolism, a significant negative correlation was found between BMI and metabolic activity in the prefrontal cortex and the cingulate gyrus that were associated with tests for memory and executive function. However, no such correlation was found during cognitive stimulation using numerical calculations (Volkow et al. 2009).

Nevertheless, abovementioned studies were not designed to address the cause-effect relation between obesity and impaired cognitive function or the inverse. As such, the assessment of cognitive function after weight loss achieved through bariatric surgery would be indicative of the reversibility of impaired function and support the thesis that obesity itself interferes with cognitive performance.

In a prospective study with longitudinal assessment, bariatric surgery patients showed a preoperative impairment in cognitive function including attention, executive function, memory, and language. Twelve weeks after surgery, memory performance improved in comparison to preoperative assessment and obese controls

(Gunstad et al. 2011). At 1 year after surgery, patients maintained improvement in memory performance, particularly those suffering from sleep apnea preoperatively and patients with a lower BMI, independent of patient age (Alosco et al. 2014a; Miller et al. 2013). To test the hypothesis that reduced postoperative inflammation might be associated with cognitive improvement, inflammatory markers like high-sensitivity C-reactive protein were assessed, without any significant correlation to observed results (Hawkins et al. 2015). However, decreased leptin and increased ghrelin were associated with improved executive function and attention (Alosco et al. 2015). The same group published data regarding reevaluation at 24 and 36 months that confirmed their previous observations on memory performance. Patients who regained weight between 2 and 3 years after surgery showed a decline in performance of attention (Alosco et al. 2014b, c) while preserved preoperative cognitive function was associated with a better outcome after surgery, possibly due to better compliance and adherence to therapeutic counseling (Spitznagel et al. 2013).

In contrast, a cross-sectional study comparing 50 patients who experienced significant weight loss after gastric bypass to an age and gender matched preoperative control group did not find any differences regarding overall cognitive performance and non-food related impulsivity (Georgiadou et al. 2014). Also another comparative cross-sectional study that included patients submitted to gastric bypass, sleeve gastrectomy, and gastric banding with an average postoperative BMI of 32.6 kg/m^2 compared to a preoperative group with a BMI > 40 kg/m^2 was unable to find differences in executive function between groups. However, in both groups cognitive performance was inferior to expected results according to normative data (Sousa et al. 2012).

An experimental study using obese rodents under a high-fat diet and lean controls compared the effects of sleeve gastrectomy and gastric bypass to caloric restriction. Metabolic health assessed by glucose metabolism and body fat distribution improved in all groups. However, rats submitted to sleeve gastrectomy showed an impairment in spatial learning tasks that was associated with hippocampal inflammation whereas pair-fed or animals submitted to gastric bypass improved task performance and microglial infiltration (Grayson et al. 2014). In contrast, brain structural abnormalities evaluated in human patients submitted to sleeve gastrectomy using MRI scans showed significant improvement after surgery. Decreased preoperative fractional anisotropy and gray and white matter densities as well as increased mean diffusivity in regions associated with food intake control and cognitive and emotional modulation recovered after surgery (Zhang et al. 2016).

In another study, neurocognitive evaluation before and 24 weeks after bariatric surgery solely revealed improvement in executive function using the Trail Making Test while fMRI showed postoperative normalization of increased cerebral glycolytic metabolism in the right posterior cingulate gyrus and right posterior lobe of the cerebellum, similar to lean subjects (Marques et al. 2014). As obesity-induced endothelial dysfunction might lead to cerebral hypoperfusion, oxidative stress, and compensatory hypermetabolism, this pathophysiological mechanism can possibly be reversed by weight loss (Spitznagel et al. 2015; Toda et al. 2014).

Taken together, data from abovementioned studies supports the hypothesis that bariatric surgery has a favorable effect on cognitive function associated with neural plasticity. Nevertheless, caution is recommended as much of our knowledge derives from repetitive testing in a small group of patients included in the Longitudinal Assessment of Bariatric Surgery project.

References

Abbott CR, Monteiro M, Small CJ, Sajedi A, Smith KL, Parkinson JRC, Ghatei MA, Bloom SR (2005a) The inhibitory effects of peripheral administration of peptide YY 3-36 and glucagon-like peptide-1 on food intake are attenuated by ablation of the vagal-brainstem-hypothalamic pathway. Brain Res 1044:127–131

Abbott CR, Small CJ, Kennedy AR, Neary NM, Sajedi A, Ghatei MA, Bloom SR (2005b) Blockade of the neuropeptide Y Y2 receptor with the specific antagonist BIIE0246 attenuates the effect of endogenous and exogenous peptide YY (3-36) on food intake. Brain Res 1043:139–144

Adami GF, Gradaschi R, Andraghetti G, Scopinaro N, Cordera R (2016) Serum leptin and adiponectin concentration in type 2 diabetes patients in the short and long term following biliopancreatic diversion. Obes Surg 26:2442–2448

Al-Rasheid N, Gray R, Sufi P, Marina-Gonzalez N, Al-Sayrafi M, Atherton E, Mohamed-Ali V (2014) Chronic elevation of systemic glucagon-like peptide-1 following surgical weight loss: association with nausea and vomiting and effects on adipokines. Obes Surg 25:386–391

Alosco ML, Cohen R, Spitznagel MB, Strain G, Devlin M, Crosby RD, Mitchell JE, Gunstad J (2014a) Older age does not limit postbariatric surgery cognitive benefits: a preliminary investigation. Surg Obes Relat Dis 10:1196–1201

Alosco ML, Galioto R, Spitznagel MB, Strain G, Devlin M, Cohen R, Crosby RD, Mitchell JE, Gunstad J (2014b) Cognitive function after bariatric surgery: evidence for improvement 3 years after surgery. Am J Surg 207:870–876

Alosco ML, Spitznagel MB, Strain G, Devlin M, Cohen R, Crosby RD, Mitchell JE, Gunstad J (2015) Improved serum leptin and ghrelin following bariatric surgery predict better postoperative cognitive function. J Clin Neurol 11:48–56

Alosco ML, Spitznagel MB, Strain G, Devlin M, Cohen R, Paul R, Crosby RD, Mitchell JE, Gunstad J (2014c) Improved memory function two years after bariatric surgery. Obesity (Silver Spring) 22:32–38

Anderberg RH, Richard JE, Eerola K, Ferreras LL, Nordbeck EB, Hansson C, Nissbrandt H, Berqquist F, Gribble FM, Reimann F, Wernstedt-Asterholm I, Lamy C, Skibicka KP (2017) Glucagon-like peptide-1 and its analogues act in the dorsal raphe and modulate central serotonin to reduce appetite and body weight. Diabetes 66(4):1062–1073. doi:10.2337/db16-0755

Aneta Stefanidis PD, Forrest N, Brown WA, Dixon JB, O'Brien PB, Kampe J, Oldfield BJ (2016) An investigation of the neural mechanisms underlying the efficacy of the adjustable gastric band. Surg Obes Relat Dis 12:828–838

Apovian CM, Shah SN, Wolfe BM, Ikramuddin S, Miller CJ, Tweden KS, Billington CJ, Shikora SA (2017) Two-year outcomes of vagal nerve blocking (vBloc) for the treatment of obesity in the ReCharge trial. Obes Surg 27:169–176

Balbo SL, Ribeiro RA, Mendes MC, Lubaczeuski C, Maller ACPA, Carneiro EM, Bonfleur ML (2016) Vagotomy diminishes obesity in cafeteria rats by decreasing cholinergic potentiation of insulin release. J Physiol Biochem 72:625–633

Ballantyne GH (2006) Peptide YY(1-36) and peptide YY(3-36). Part I: Distribution, release and actions. Obes Surg 16:651–658

Ballsmider LA, Vaughn AC, David M, Hajnal A, Di Lorenzo PM, Czaja K (2015) Sleeve gastrectomy and Roux-en-Y gastric bypass alter the gut-brain communication. Neural Plast 2015:601985

Balthasar N, Coppari R, McMinn J, Liu SM, Lee CE, Tang V, Kenny CD, McGovern RA, Chua SC, Elmquist JK, Lowell BB (2004) Leptin receptor signaling in POMC neurons is required for normal body weight homeostasis. Neuron 42:983–991

Barkholt P, Pedersen PJ, Hay-Schmidt A, Jelsing J, Hansen HH, Vrang N (2016) Alterations in hypothalamic gene expression following Roux-en-Y gastric bypass. Mol Metab 5:296–304

Bassi M, Do Carmo JM, Hall JE, Da Silva AA (2012) Chronic effects of centrally administered adiponectin on appetite, metabolism and blood pressure regulation in normotensive and hypertensive rats. Peptides 37:1–5

Batterham RL, Cohen MA, Ellis SM, Le Roux CW, Withers DJ, Frost GS, Ghatei MA, Bloom SR (2003) Inhibition of food intake in obese subjects by peptide YY 3-36. N Engl J Med 349:941–948

Batterham RL, Cowley MA, Small CJ, Herzog H, Cohen MA, Dakin CL, Wren AM, Brynes AE, Low MJ, Ghatei MA, Cone RD, Bloom SR (2002) Gut hormone PYY(3-36) physiologically inhibits food intake. Nature 418:650–654

Beglinger S, Drewe J, Schirra J, Göke B, D'Amato M, Beglinger C (2010) Role of fat hydrolysis in regulating glucagon-like peptide-1 secretion. J Clin Endocrinol Metab 95:879–886

Blasi C (2016) The role of the vagal nucleus Tractus Solitarius in the therapeutic effects of obesity surgery and other interventional therapies on type 2 diabetes. Obes Surg 26:3045–3057

Blevins JE, Stanley BG, Reidelberger RD (2000) Brain regions where cholecystokinin suppresses feeding in rats. Brain Res 860:1–10

Bloom S, Wynne K, Chaudhri O (2005) Gut feeling--the secret of satiety? Clin Med 5:147–152

Blum K, Bailey J (2011) Neuro-genetics of reward deficiency syndrome (RDS) as the root cause of "addiction transfer": a new phenomenon common after bariatric surgery. J Genet Syndr Gene Ther 2012:1–21

Boeka AG, Lokken KL (2008) Neuropsychological performance of a clinical sample of extremely obese individuals. Arch Clin Neuropsychol 23:467–474

Borg CM, le Roux CW, Ghatei MA, Bloom SR, Patel AG, SJB A (2006) Progressive rise in gut hormone levels after Roux-en-Y gastric bypass suggests gut adaptation and explains altered satiety. Br J Surg 93:210–215

Browning KN, Fortna SR, Hajnal A (2013) Roux-en-Y gastric bypass reverses the effects of diet-induced obesity to inhibit the responsiveness of central vagal motoneurones. J Physiol 591:2357–2372

Buchwald H, Avidor Y, Braunwald E, Jensen MD, Pories W, Fahrbach K, Schoelles K (2004) Bariatric surgery: a systematic review and meta-analysis. JAMA 292:1724–1737

Buchwald H, Estok R, Fahrbach K, Banel D, Jensen MD, Pories WJ, Bantle JP, Sledge I (2009) Weight and type 2 diabetes after bariatric surgery: systematic review and meta-analysis. Am J Med 122. 248–256.e5

Campfield LA, Smith FJ, Guisez Y, Devos R, Burn P (1995) Recombinant mouse OB protein: evidence for a peripheral signal linking adiposity and central neural networks. Science 269:546–549

Chan J, Mantzoros C (2003) The role of falling Leptin levels in the neuroendocrime and metabolic adaption to short-term starvation in healthy men. J Clin Invest 111:1409–1421

Chelikani PK, Haver AC, Reeve JR, Keire DA, Reidelberger RD (2006) Daily, intermittent intravenous infusion of peptide YY(3-36) reduces daily food intake and adiposity in rats. Am J Physiol Regul Integr Comp Physiol 290:R298–R305

Christou NV, Look D, McLean AP (2005) Pre- and post-prandial plasma ghrelin levels do not correlate with satiety or failure to achieve a successful outcome after Roux-en-Y gastric bypass. Obes Surg 15:1017–1023

Chronaiou A, Tsoli M, Kehagias I, Leotsinidis M, Kalfarentzos F, Alexandrides TK (2012) Lower ghrelin levels and exaggerated postprandial peptide-YY, glucagon-like peptide-1, and insulin responses, after gastric fundus resection, in patients undergoing Roux-en-Y gastric bypass: a randomized clinical trial. Obes Surg 22:1761–1770

Cohen MA, Ellis SM, Le Roux CW, Batterham RL, Park A, Patterson M, Frost GS, Ghatei MA, Bloom SR (2003) Oxyntomodulin suppresses appetite and reduces food intake in humans. J Clin Endocrinol Metab 88:4696–4701

Coluzzi I, Raparelli L, Guarnacci L, Paone E, Del Genio G, le Roux CW, Silecchia G (2016) Food intake and changes in eating behavior after laparoscopic sleeve gastrectomy. Obes Surg 26:2059–2067

Considine RV, Sinha MK, Heiman ML, Kriauciunas A, Stephens TW, Nyce MR, Ohannesian JP, Marco CC, McKee LJ, Bauer TL, Caro JF (1996) Serum immunoreactive-leptin concentrations in normal-weight and obese humans. N Engl J Med 334:292–295

Creutzfeldt W (1979) The incretin concept today. Diabetologia 16:75–85

Cummings BP, Bettaieb A, Graham JL, Stanhope KL, Kowala M, Haj FG, Chouinard ML, Havel PJ (2012) Vertical sleeve gastrectomy improves glucose and lipid metabolism and delays diabetes onset in UCD-T2DM rats. Endocrinology 153:3620–3632

Cummings DE (2001) A preprandial rise in plasma ghrelin levels suggests a role in meal initiation in humans. Diabetes 50:1714–1719

Cummings DE, Overduin J, Foster-Schubert KE (2004) Gastric bypass for obesity: mechanisms of weight loss and diabetes resolution. J Clin Endocrinol Metab 89:2608–2615

Cummings DE, Weigle DS, Frayo RS (2002) Plasma ghrelin levels after diet-induced weight loss or gastric bypass surgery. N Engl J Med 346:1623–1630

Dakin CL, Gunn I, Small CJ, Edwards CM, Hay DL, Smith DM, Ghatei MA, Bloom SR (2001) Oxyntomodulin inhibits food intake in the rat. Endocrinology 142:4244–4250

Date Y, Kojima M, Hosoda H, Sawaguchi A, Mondal MS, Suganuma T, Matsukura S, Kangawa K, Nakazato M (2000) Ghrelin, a novel growth hormone-releasing acylated peptide, is synthesized in a distinct endocrine cell type in the gastrointestinal tracts of rats and humans. Endocrinology 141:4255–4261

Davies MJ, Bergenstal R, Bode B, Kushner RF, Lewin A, Skjøth TV, Andreasen AH, Jensen CB, DeFronzo RA (2015) Efficacy of Liraglutide for weight loss among patients with type 2 diabetes. JAMA 314:687

de Lartigue G, Diepenbroek C (2016) Novel developments in vagal afferent nutrient sensing and its role in energy homeostasis. Curr Opin Pharmacol 31:38–43

De Silva A, Salem V, Long CJ, Makwana A, Newbould RD, Rabiner EA, Ghatei MA, Bloom SR, Matthews PM, Beaver JD, Dhillo WS (2011) The gut hormones PYY 3-36 and GLP-1 7-36 amide reduce food intake and modulate brain activity in appetite centers in humans. Cell Metab 14:700–706

De Weijer BA, Van De Giessen E, Janssen I, Berends FJ, Van De Laar A, Ackermans MT, Fliers E, La Fleur SE, Booij J, Serlie MJ (2014) Striatal dopamine receptor binding in morbidly obese women before and after gastric bypass surgery and its relationship with insulin sensitivity. Diabetologia 57:1078–1080

Degen L, Drewe J, Piccoli F, Gräni K, Oesch S, Bunea R, D'Amato M, Beglinger C (2007) Effect of CCK-1 receptor blockade on ghrelin and PYY secretion in men. Am J Physiol Regul Integr Comp Physiol 292:R1391–R1399

Dirksen C, Jørgensen NB, Bojsen-Møller KN, Kielgast U, Jacobsen SH, Clausen TR, Worm D, Hartmann B, Rehfeld JF, Damgaard M, Madsen JL, Madsbad S, Holst JJ, Hansen DL (2013) Gut hormones, early dumping and resting energy expenditure in patients with good and poor weight loss response after Roux-en-Y gastric bypass. Int J Obes 37:1452–1459

Drenick EJ, Ament ME, Finegold SM, Corrodi P, Passaro E (1976) Bypass enteropathy. Intestinal and systemic manifestations following small-bowel bypass. JAMA 236:269–272

Dunn JP, Cowan RL, Volkow ND, Feurer ID, Li R, Williams DB, Kessler RM, Abumrad NN (2010) Decreased dopamine type 2 receptor availability after bariatric surgery: preliminary findings. Brain Res 1350:123–130

Dupre J, Ross SA, Watson D, Brown JC (1973) Stimulation of insulin secretion by gastric inhibitory polypeptide in man. J Clin Endocrinol Metab 37:826–828

Eickhoff H, Louro TM, Matafome PN, Vasconcelos F, Seiça RM, Castro E, Sousa F (2015) Amelioration of glycemic control by sleeve Gastrectomy and gastric bypass in a lean animal model of type 2 diabetes: restoration of gut hormone profile. Obes Surg 25:7–18

Eissele R, Göke R, Willemer S, Harthus HP, Vermeer H, Arnold R, Göke B (1992) Glucagon-like peptide-1 cells in the gastrointestinal tract and pancreas of rat, pig and man. Eur J Clin Investig 22:283–291

Elias MF, Elias PK, Sullivan LM, Wolf PA, D'Agostino RB (2003) Lower cognitive function in the presence of obesity and hypertension: the Framingham heart study. Int J Obes Relat Metab Disord 27:260–268

English PJ, Ghatei MA, Malik IA, Bloom SR, Wilding JPH (2002) Food fails to suppress ghrelin levels in obese humans. J Clin Endocrinol Metab 87:2984

Fagundo AB, de la Torre R, Jiménez-Murcia S, Agüera Z, Granero R, Tárrega S, Botella C, Baños R, Fernández-Real JM, Rodríguez R, Forcano L, Frühbeck G, Gómez-Ambrosi J, Tinahones FJ, Fernández-García JC, Casanueva FF, Fernández-Aranda F (2012) Executive functions profile in extreme eating/weight conditions: from anorexia nervosa to obesity. PLoS One 7(8):e43382

Faraj M, Havel PJ, Phélis S, Blank D, Sniderman AD, Cianflone K (2003) Plasma acylation-stimulating protein, adiponectin, leptin, and ghrelin before and after weight loss induced by gastric bypass surgery in morbidly obese subjects. J Clin Endocrinol Metab 88:1594–1602

Fergenbaum JH, Bruce S, Lou W, Hanley AJG, Greenwood C, Young TK (2009) Obesity and lowered cognitive performance in a Canadian first nations population. Obesity (Silver Spring) 17:1957–1963

Fikri E, Cassella R (1974) Jejunoileal bypass for massive obesity: results and complications in fifty-two patients. Ann Surg 179:460–464

Foschi D, Corsi F, Pisoni L, Vago T, Bevilacqua M, Asti E, Righi I, Trabucchi E (2004) Plasma cholecystokinin levels after vertical banded gastroplasty: effects of an acidified meal. Obes Surg 14:644–647

Frühbeck G, Diez-Caballero A, Gil MJ, Montero I, Gómez-Ambrosi J, Salvador J, Cienfuegos JA (2004) The decrease in plasma ghrelin concentrations following bariatric surgery depends on the functional integrity of the fundus. Obes Surg 14:606–612

Gallaher ZR, Ryu V, Herzog T, Ritter RC, Czaja K (2012) Changes in microglial activation within the hindbrain, nodose ganglia, and the spinal cord following subdiaphragmatic vagotomy. Neurosci Lett 513:31–36

Garlicki J, Konturek PK, Majka J, Kwiecien N, Konturek SJ (1990) Cholecystokinin receptors and vagal nerves in control of food intake in rats. Am J Phys 258:E40–E45

Georgiadou E, Gruner-Labitzke K, Köhler H, de Zwaan M, Müller A (2014) Cognitive function and nonfood-related impulsivity in post-bariatric surgery patients. Front Psychol 5:1–7

Goldstone AP, Miras AD, Scholtz S, Jackson S, Neff KJ, Pénicaud L, Geoghegan J, Chhina N, Durighel G, Bell JD, Meillon S, le Roux CW (2016) Link between increased satiety gut hormones and reduced food reward after gastric bypass surgery for obesity. J Clin Endocrinol Metab 101:599–609

Gonzales MM, Kaur S, Eagan DE, Goudarzi K, Pasha E, Doan DC, Tanaka H, Haley AP (2014) Central adiposity and the functional magnetic resonance imaging response to cognitive challenge. Int J Obes 38:1193–1199

Gortz L, Bjorkman A C, Andersson H, Kral JG (1990) Truncal vagotomy reduces food and liquid intake in man. Physiol Behav 48:779–781

Grayson BE, Fitzgerald MF, Hakala-Finch AP, Ferris VM, Begg DP, Tong J, Woods SC, Seeley RJ, Davidson TL, Benoit SC (2014) Improvements in hippocampal-dependent memory and microglial infiltration with calorie restriction and gastric bypass surgery, but not with vertical sleeve gastrectomy. Int J Obes 38:349–356

Gumbau V, Bruna M, Canelles E, Guaita M, Mulas C, Basés C, Celma I, Puche J, Marcaida G, Oviedo M, Vázquez A (2014) A prospective study on inflammatory parameters in obese patients after sleeve gastrectomy. Obes Surg 24:903–908

Gunstad J, Lhotsky A, Wendell CR, Ferrucci L, Zonderman AB (2010) Longitudinal examination of obesity and cognitive function: results from the Baltimore longitudinal study of aging. Neuroepidemiology 34:222–229

Gunstad J, Paul RH, Cohen RA, Tate DF, Spitznagel MB, Gordon E (2007) Elevated body mass index is associated with executive dysfunction in otherwise healthy adults. Compr Psychiatry 48:57–61

Gunstad J, Strain G, Devlin MJ, Wing R, Cohen RA, Paul RH, Crosby RD, Mitchell JE (2011) Improved memory function 12 weeks after bariatric surgery. Surg Obes Relat Dis 7:465–472

Gutzwiller JP, Göke B, Drewe J, Hildebrand P, Ketterer S, Handschin D, Winterhalder R, Conen D, Beglinger C (1999) Glucagon-like peptide-1: a potent regulator of food intake in humans. Gut 44:81–86

Hall KD, Hammond RA, Rahmandad H (2014) Dynamic interplay among homeostatic, hedonic, and cognitive feedback circuits regulating body weight. Am J Public Health 104:1169–1175

Halverson JD, Wise L, Wazna MF, Ballinger WF (1978) Jejunoileal bypass for morbid obesity. A critical appraisal. Am J Med 64:461–475

Han W, Tellez LA, Niu J, Medina S, Ferreira TL, Zhang X, Su J, Tong J, Schwartz GJ, van den Pol A, de Araujo IE (2015) Striatal dopamine links gastrointestinal rerouting to altered sweet appetite. Cell Metab 23:103–112

Hankir MK, Seyfried F, Hintschich CA, Diep T-A, Kleberg K, Kranz M, Deuther-Conrad W, Tellez LA, Rullmann M, Patt M, Teichert J, Hesse S, Sabri O, Brust P, Hansen HS, de Araujo IE, Krügel U, Fenske WK (2017) Gastric bypass surgery recruits a gut PPAR-α-striatal D1R pathway to reduce fat appetite in obese rats. Cell Metab 25:1–10

Hao Z, Münzberg H, Rezai-Zadeh K, Keenan M, Coulon D, Lu H, Berthoud H-R, Ye J (2015) Leptin deficient ob/ob mice and diet-induced obese mice responded differently to Roux-en-Y bypass surgery. Int J Obes 39:798–805

Hao Z, Townsend RL, Mumphrey MB, Patterson LM, Ye J, Berthoud H-R (2014) Vagal innervation of intestine contributes to weight loss after Roux-en-Y gastric bypass surgery in rats. Obes Surg 24:2145–2151

Hawkins MAW, Alosco ML, Spitznagel MB, Strain G, Devlin M, Cohen R, Crosby RD, Mitchell JE, Gunstad J (2015) The association between reduced inflammation and cognitive gains after bariatric surgery. Psychosom Med 77:688–696

Heymsfield SB, Greenberg AS, Fujioka K, Dixon RM, Kushner R, Hunt T, Lubina JA, Patane J, Self B, Hunt P (1999) Recombinant leptin for weight loss in obese and lean adults. JAMA 282:1568–1575

Holdstock C, Engström BE, Ohrvall M, Lind L, Sundbom M, Karlsson FA (2003) Ghrelin and adipose tissue regulatory peptides: effect of gastric bypass surgery in obese humans. J Clin Endocrinol Metab 88:3177–3183

Hotta K, Funahashi T, Arita Y, Takahashi M, Matsuda M, Okamoto Y, Iwahashi H, Kuriyama H, Ouchi N, Maeda K, Nishida M, Kihara S, Sakai N, Nakajima T, Hasegawa K, Muraguchi M, Ohmoto Y, Nakamura T, Yamashita S, Hanafusa T, Matsuzawa Y (2000) Plasma concentrations of a novel, adipose-specific protein, adiponectin, in type 2 diabetic patients. Arterioscler Thromb Vasc Biol 20:1595–1599

Johannessen H, Revesz D, Kodama Y, Cassie N, Skibicka K, Barrett P, Dickson S, Holst J, Rehfeld J, van der Plasse G, Adan R, Kulseng B, Ben-Menachem E, Zhao CM, Chen D (2017) Vagal blocking for obesity control: a possible mechanism-of-action. Obes Surg 27:177–185

Kadowaki T, Yamauchi T, Kubota N (2008) The physiological and pathophysiological role of adiponectin and adiponectin receptors in the peripheral tissues and CNS. FEBS Lett 582:74–80

Karamanakos SN, Vagenas K, Kalfarentzos F, Alexandrides TK (2008) Weight loss, appetite suppression, and changes in fasting and postprandial ghrelin and peptide-YY levels after Roux-en-Y gastric bypass and sleeve gastrectomy: a prospective, double blind study. Ann Surg 247:401–407

Karlsson HK, Tuominen L, Tuulari JJ, Hirvonen J, Parkkola R, Helin S, Salminen P, Nuutila P, Nummenmaa L (2015) Obesity is associated with decreased μ-opioid but unaltered dopamine D2 receptor availability in the brain. J Neurosci 35:3959–3965

Kashyap SR, Bhatt DL, Wolski K, Watanabe RM, Abdul-Ghani M, Abood B, Pothier CE, Brethauer S, Nissen S, Gupta M, Kirwan JP, Schauer PR (2013) Metabolic effects of bariatric surgery in patients with moderate obesity and type 2 diabetes: analysis of a randomized control trial comparing surgery with intensive medical treatment. Diabetes Care 36:2175–2182

Kawasaki T, Ohta M, Kawano Y, Masuda T, Gotoh K, Inomata M, Kitano S (2015) Effects of sleeve gastrectomy and gastric banding on the hypothalamic feeding center in an obese rat model. Surg Today 45:1560–1566

Kentish SJ, Page AJ (2015) The role of gastrointestinal vagal afferent fibres in obesity. J Physiol 593:775–786

Kohno D, Gao H-Z, Muroya S, Kikuyama S, Yada T (2003) Ghrelin directly interacts with neuropeptide-Y-containing neurons in the rat arcuate nucleus: Ca2+ signaling via protein kinase a and N-type channel-dependent mechanisms and cross-talk with leptin and orexin. Diabetes 52:948–956

Kojima M, Hosoda H, Date Y, Nakazato M, Matsuo H, Kangawa K (1999) Ghrelin is a growth-hormone-releasing acylated peptide from stomach. Nature 402:656–660

Kopin AS, Mathes WF, McBride EW, Nguyen M, Al-Haider W, Schmitz F, Bonner-Weir S, Kanarek R, Beinborn M (1999) The cholecystokinin-a receptor mediates inhibition of food intake yet is not essential for the maintenance of body weight. J Clin Invest 103:383–391

Korner J, Conroy R, Febres G, McMahon DJ, Conwell I, Karmally W, Aronne LJ (2013) Randomized double-blind placebo-controlled study of leptin administration after gastric bypass. Obesity 21:951–956

Korner J, Inabnet W, Conwell IM, Taveras C, Daud A, Olivero-Rivera L, Restuccia NL, Bessler M (2006) Differential effects of gastric bypass and banding on circulating gut hormone and leptin levels. Obesity (Silver Spring) 14:1553–1561

Kosinski JR, Hubert J, Carrington PE, Chicchi GG, Mu J, Miller C, Cao J, Bianchi E, Pessi A, Sinharoy R, Marsh DJ, Pocai A (2012) The glucagon receptor is involved in mediating the body weight-lowering effects of oxyntomodulin. Obesity (Silver Spring) 20:1566–1571

Kotidis EV, Koliakos GG, Baltzopoulos VG, Ioannidis KN, Yovos JG, Papavramidis ST (2006) Serum ghrelin, leptin and adiponectin levels before and after weight loss: comparison of three methods of treatment—a prospective study. Obes Surg 16:1425–1432

Kral JG, Paez W, Wolfe BM (2009) Vagal nerve function in obesity: therapeutic implications. World J Surg 33:1995–2006

Kreymann B, Williams G, Ghatei MA, Bloom SR (1987) Glucagon-like peptide-1 7-36: a physiological incretin in man. Lancet 2:1300–1304

Kringelbach ML, Radcliffe J (2005) The human orbitofrontal cortex: linking reward to hedonic experience. Nat Rev Neurosci 6:691–702

Kubota N, Yano W, Kubota T, Yamauchi T, Itoh S, Kumagai H, Kozono H, Takamoto I, Okamoto S, Shiuchi T, Suzuki R, Satoh H, Tsuchida A, Moroi M, Sugi K, Noda T, Ebinuma H, Ueta Y, Kondo T, Araki E, Ezaki O, Nagai R, Tobe K, Terauchi Y, Ueki K, Minokoshi Y, Kadowaki T (2007) Adiponectin stimulates AMP-activated protein kinase in the hypothalamus and increases food intake. Cell Metab 6:55–68

Kuo Y-T, Parkinson JRC, Chaudhri OB, Herlihy AH, So P-W, Dhillo WS, Small CJ, Bloom SR, Bell JD (2007) The temporal sequence of gut peptide CNS interactions tracked in vivo by magnetic resonance imaging. J Neurosci 27:12341–12348

Langer FB, Reza Hoda MA, Bohdjalian A, Felberbauer FX, Zacherl J, Wenzl E, Schindler K, Luger A, Ludvik B, Prager G (2005) Sleeve gastrectomy and gastric banding: effects on plasma ghrelin levels. Obes Surg 15:1024–1029

Larsen PJ, Tang-Christensen M, Holst JJ, Orskov C (1997) Distribution of glucagon-like peptide-1 and other preproglucagon-derived peptides in the rat hypothalamus and brainstem. Neuroscience 77:257–270

le Roux CW, Aylwin SJB, Batterham RL, Borg CM, Coyle F, Prasad V, Shurey S, Ghatei M a, Patel AG, Bloom SR (2006) Gut hormone profiles following bariatric surgery favor an anorectic state, facilitate weight loss, and improve metabolic parameters. Ann Surg 243:108–114

le Roux CW, Bloom SR (2005) Why do patients lose weight after Roux-en-Y gastric bypass? J Clin Endocrinol Metab 90:591–592

le Roux CW, Welbourn R, Werling M, Osborne A, Kokkinos A, Laurenius A, Lönroth H, Fändriks L, Ghatei MA, Bloom SR, Olbers T (2007) Gut hormones as mediators of appetite and weight loss after Roux-en-Y gastric bypass. Ann Surg 246:780–785

Lee W-J, Chen C-Y, Chong K, Lee Y-C, Chen S-C, Lee S-D (2011) Changes in postprandial gut hormones after metabolic surgery: a comparison of gastric bypass and sleeve gastrectomy. Surg Obes Relat Dis 7:683–690

Liddle RA, Goldfine ID, Rosen MS, Taplitz RA, Williams JA (1985) Cholecystokinin bioactivity in human plasma. J Clin Invest 75:1144–1152

Liu AG, Smith SR, Fujioka K, Greenway FL (2013) The effect of leptin, caffeine/ephedrine, and their combination upon visceral fat mass and weight loss. Obesity 21:1991–1996

Lundberg JM, Tatemoto K, Terenius L, Hellström PM, Mutt V, Hökfelt T, Hamberger B (1982) Localization of peptide YY (PYY) in gastrointestinal endocrine cells and effects on intestinal blood flow and motility. Proc Natl Acad Sci U S A 79:4471–4475

Lutter M, Nestler EJ (2009) Homeostatic and hedonic signals interact in the regulation of food intake. J Nutr 139:629–632

Mackintosh RM, Hirsch J (2001) The effects of leptin administration in non-obese human subjects. Obes Res 9:462–469

Malik S, McGlone F, Bedrossian D, Dagher A (2008) Ghrelin modulates brain activity in areas that control appetitive behavior. Cell Metab 7:400–409

Marques EL, Halpern A, Mancini MC, De Melo ME, Horie NC, Buchpiguel CA, Coutinho AMN, Ono CR, Prando S, Santo MA, Cunha-Neto E, Fuentes D, Cercato C (2014) Changes in neuropsychological tests and brain metabolism after bariatric surgery. J Clin Endocrinol Metab 99:E2347–E2352

Mason EE, Ito C (1969) Gastric bypass. Ann Surg 170:329–339

Mason EE, Printen KJ, Blommers TJ, Scott DH (1978) Gastric bypass for obesity after ten years experience. Int J Obes 2:197–206

McCutcheon JE (2015) The role of dopamine in the pursuit of nutritional value. Physiol Behav 152:408–415

Miller LA, Crosby RD, Galioto R, Strain G, Devlin MJ, Wing R, Cohen RA, Paul RH, Mitchell JE, Gunstad J (2013) Bariatric surgery patients exhibit improved memory function 12 months postoperatively. Obes Surg 23:1527–1535

Mingrone G, Castagneto-Gissey L (2009) Mechanisms of early improvement/resolution of type 2 diabetes after bariatric surgery. Diabetes Metab 35:518–523

Miras AD, Jackson RN, Jackson SN, Goldstone AP, Olbers T, Hackenberg T, Spector AC, Le Roux CW (2012) Gastric bypass surgery for obesity decreases the reward value of a sweet-fat stimulus as assessed in a progressive ratio task. Am J Clin Nutr 96:467–473

Mojsov S, Weir GC, Habener JF (1987) Insulinotropin: glucagon-like peptide I (7-37) co-encoded in the glucagon gene is a potent stimulator of insulin release in the perfused rat pancreas. J Clin Invest 79:616–619

Montague CT, Farooqi IS, Whitehead JP, Soos MA, Rau H, Wareham NJ, Sewter CP, Digby JE, Mohammed SN, Hurst JA, Cheetham CH, Earley AR, Barnett AH, Prins JB, O'Rahilly S (1997) Congenital leptin deficiency is associated with severe early-onset obesity in humans. Nature 387:903–908

Morínigo R, Moizé V, Musri M, Lacy AM, Navarro S, Marín JL, Delgado S, Casamitjana R, Vidal J (2006) Glucagon-like peptide-1, peptide YY, hunger, and satiety after gastric bypass surgery in morbidly obese subjects. J Clin Endocrinol Metab 91:1735–1740

Murphy KG, Bloom SR (2004) Gut hormones in the control of appetite. Exp Physiol 89:507–516

Nannipieri M, Baldi S, Mari A, Colligiani D, Guarino D, Camastra S, Barsotti E, Berta R, Moriconi D, Bellini R, Anselmino M, Ferrannini E (2013) Roux-en-Y gastric bypass and sleeve gastrectomy: mechanisms of diabetes remission and role of gut hormones. J Clin Endocrinol Metab 98(11):4391–4399

Neary NM, Goldstone AP, Bloom SR (2004) Appetite regulation: from the gut to the hypothalamus. Clin Endocrinol 60:153–160

Nijhuis J, van Dielen FMH, Buurman WA, Greve JWM (2004) Ghrelin, leptin and insulin levels after restrictive surgery: a 2-year follow-up study. Obes Surg 14:783–787

Nummenmaa L, Hirvonen J, Hannukainen JC, Immonen H, Lindroos MM, Salminen P, Nuutila P (2012) Dorsal striatum and its limbic connectivity mediate abnormal anticipatory reward processing in obesity. PLoS One 7(2):e31089

Ochner CN, Kwok Y, Conceição E, Pantazatos SP, Puma LM, Carnell S, Teixeira J, Hirsch J, Geliebter A (2011) Selective reduction in neural responses to high calorie foods following gastric bypass surgery. Ann Surg 253:502–507

Ochner CN, Stice E, Hutchins E, Afifi L, Geliebter A, Hirsch J, Teixeira J (2012) Relation between changes in neural Responsivity and reductions in desire to eat high-calorie foods following gastric bypass surgery. Neuroscience 209:128–135

Orskov C (1992) Glucagon-like peptide-1, a new hormone of the entero-insular axis. Diabetologia 35:701–711

Ouchi N, Kihara S, Arita Y, Maeda K, Kuriyama H, Okamoto Y, Hotta K, Nishida M, Takahashi M, Nakamura T, Yamashita S, Funahashi T, Matsuzawa Y (1999) Novel modulator for endothelial adhesion molecules: adipocyte-derived plasma protein adiponectin. Circulation 100:2473–2476

Ouchi N, Kihara S, Arita Y, Okamoto Y, Maeda K, Kuriyama H, Hotta K, Nishida M, Takahashi M, Muraguchi M, Ohmoto Y, Nakamura T, Yamashita S, Funahashi T, Matsuzawa Y (2000) Adiponectin, an adipocyte-derived plasma protein, inhibits endothelial NF-kappaB signaling through a cAMP-dependent pathway. Circulation 102:1296–1301

Owyang C, Heldsinger A (2011) Vagal control of satiety and hormonal regulation of appetite. J Neurogastroenterol Motil 17:338–348

Palmiter RD (2007) Is dopamine a physiologically relevant mediator of feeding behavior? Trends Neurosci 30:375–381

Papamargaritis D, le Roux CW, Sioka E, Koukoulis G, Tzovaras G, Zacharoulis D (2013) Changes in gut hormone profile and glucose homeostasis after laparoscopic sleeve gastrectomy. Surg Obes Relat Dis 9:192–201

Peterli R, Steinert RE, Woelnerhanssen B, Peters T, Christoffel-Courtin C, Gass M, Kern B, von Fluee M, Beglinger C (2012) Metabolic and hormonal changes after laparoscopic Roux-en-Y gastric bypass and sleeve gastrectomy: a randomized, prospective trial. Obes Surg 22:740–748

Pi-Sunyer X, Astrup A, Fujioka K, Greenway F, Halpern A, Krempf M, Lau DCW, le Roux CW, Violante Ortiz R, Jensen CB, Wilding JPH (2015) A randomized, controlled trial of 3.0 mg of Liraglutide in weight management. N Engl J Med 373:11–22

Pohjalainen T, Rinne JO, Någren K, Lehikoinen P, Anttila K, Syvälahti EK, Hietala J (1998) The A1 allele of the human D2 dopamine receptor gene predicts low D2 receptor availability in healthy volunteers. Mol Psychiatry 3:256–260

Polak JM, Bloom SR, Kuzio M, Brown JC, Pearse AG (1973) Cellular localization of gastric inhibitory polypeptide in the duodenum and jejunum. Gut 14:284–288

Prickett C, Brennan L, Stolwyk R (2015) Examining the relationship between obesity and cognitive function: a systematic literature review. Obes Res Clin Pract 9:93–113

Qi Y, Takahashi N, Hileman SM, Patel HR, Berg AH, Pajvani UB, Scherer PE, Ahima RS (2004) Adiponectin acts in the brain to decrease body weight. Nat Med 10:524–529

Raffaelli M, Iaconelli A, Nanni G, Guidone C, Callari C, Fernandez Real JM, Bellantone R, Mingrone G (2015) Effects of biliopancreatic diversion on diurnal leptin, insulin and free fatty acid levels. Br J Surg 102:682–690

Reddy IA, Wasserman DH, Ayala JE, Hasty AH, Abumrad NN, Galli A (2014) Striatal dopamine homeostasis is altered in mice following Roux-en-Y gastric bypass surgery. ACS Chem Neurosci 5:943–951

Romanova IV, Ramos EJB, Xu Y, Quinn R, Chen C, George ZM, Inui A, Das U, Meguid MM (2004) Neurobiologic changes in the hypothalamus associated with weight loss after gastric bypass. J Am Coll Surg 199:887–895

Romero F, Nicolau J, Flores L, Casamitjana R, Ibarzabal A, Lacy A, Vidal J (2012) Comparable early changes in gastrointestinal hormones after sleeve gastrectomy and Roux-en-Y gastric bypass surgery for morbidly obese type 2 diabetic subjects. Surg Endosc 26:2231–2239

Rothemund Y, Preuschhof C, Bohner G, Bauknecht HC, Klingebiel R, Flor H, Klapp BF (2007) Differential activation of the dorsal striatum by high-calorie visual food stimuli in obese individuals. NeuroImage 37:410–421

Sahu A (1998) Leptin decreases food intake induced by melanin-concentrating hormone (MCH), galanin (GAL) and neuropeptide Y (NPY) in the rat. Endocrinology 139:4739–4742

Sandoval D (2011) Bariatric surgeries: beyond restriction and malabsorption. Int J Obes (Lond) 35(Suppl 3):S45–S49

Sarson DL, Scopinaro N, Bloom SR (1981) Gut hormone changes after jejunoileal (JIB) or biliopancreatic (BPB) bypass surgery for morbid obesity. Int J Obes 5:471–480

Schauer PR, Bhatt DL, Kirwan JP, Wolski K, Brethauer SA, Navaneethan SD, Aminian A, Pothier CE, Kim ESH, Nissen SE, Kashyap SR (2014) Bariatric surgery versus intensive medical therapy for diabetes—3-year outcomes. N Engl J Med 370:2002–2013

Schjoldager B, Mortensen PE, Myhre J, Christiansen J, Holst JJ (1989) Oxyntomodulin from distal gut. Role in regulation of gastric and pancreatic functions. Dig Dis Sci 34:1411–1419

Scholtz S, Miras AD, Chhina N, Prechtl CG, Sleeth ML, Daud NM, Ismail NA, Durighel G, Ahmed AR, Olbers T, Vincent RP, Alaghband-Zadeh J, Ghatei MA, Waldman AD, Frost GS, Bell JD, le Roux CW, Goldstone AP (2014) Obese patients after gastric bypass surgery have lower brain-hedonic responses to food than after gastric banding. Gut 63:891–902

Scopinaro N (2006) Biliopancreatic diversion: mechanisms of action and long-term results. Obes Surg 16:683–689

Scopinaro N, Adami GF, Marinari GM, Gianetta E, Traverso E, Friedman D, Camerini G, Baschieri G, Simonelli A (1998) Biliopancreatic diversion. World J Surg 22:936–946

Shikora SA, Toouli J, Herrera MF, Kulseng B, Brancatisano R, Kow L, Pantoja JP, Johnsen G, Brancatisano A, Tweden KS, Knudson MB, Billingto CJ, Billingto CJ (2016) Intermittent vagal nerve block for improvements in obesity, cardiovascular risk factors, and glycemic control in patients with type 2 diabetes mellitus: 2-year results of the VBLOC DM2 study. Obes Surg 26:1021–1028

Sjöström L (2008) Bariatric surgery and reduction in morbidity and mortality: experiences from the SOS study. Int J Obes (Lond) 32(Suppl 7):S93–S97

Sjöström L (2013) Review of the key results from the Swedish obese subjects (SOS) trial—a prospective controlled intervention study of bariatric surgery. J Intern Med 273:219–234

Sousa S, Ribeiro O, Horácio JG, Faísca L (2012) Funções executivas em sujeitos candidatos e submetidos a cirurgia bariátrica. Psicol Saúde Doenças 13:389–398

Spitznagel MB, Garcia S, Miller LA, Strain G, Devlin M, Wing R, Cohen R, Paul R, Crosby R, Mitchell JE, Gunstad J (2013) Cognitive function predicts weight loss after bariatric surgery. Surg Obes Relat Dis 9:453–459

Spitznagel MB, Hawkins M, Alosco M, Galioto R, Garcia S, Miller L, Gunstad J (2015) Neurocognitive effects of obesity and bariatric surgery. Eur Eat Disord Rev 23:488–495

Steele KE, Prokopowicz GP, Schweitzer MA, Magunsuon TH, Lidor AO, Kuwabawa H, Kumar A, Brasic J, Wong DF (2010) Alterations of central dopamine receptors before and after gastric bypass surgery. Obes Surg 20:369–374

Stefater MA, Pérez-Tilve D, Chambers AP, Wilson-Pérez HE, Sandoval DA, Berger J, Toure M, Tschöp M, Woods SC, Seeley RJ (2010) Sleeve gastrectomy induces loss of weight and fat mass in obese rats, but does not affect leptin sensitivity. Gastroenterology 138:2426–2436

Szczypka MS, Kwok K, Brot MD, Marck BT, Matsumoto AM, Donahue BA, Palmiter RD (2001) Dopamine production in the caudate putamen restores feeding in dopamine-deficient mice. Neuron 30:819–828

Taminato T, Seino Y, Goto Y, Inoue Y, Kadowaki S (1977) Synthetic gastric inhibitory polypeptide. Stimulatory effect on insulin and glucagon secretion in the rat. Diabetes 26:480–484

Tatemoto K, Carlquist M, Mutt V (1982) Neuropeptide Y—a novel brain peptide with structural similarities to peptide YY and pancreatic polypeptide. Nature 296:659–660

Tatemoto K, Mutt V (1980) Isolation of two novel candidate hormones using a chemical method for finding naturally occurring polypeptides. Nature 285:417–418

Thompson J, Thomas N, Singleton A, Piggott M, Lloyd S, Perry EK, Morris CM, Perry RH, Ferrier IN, Court JA (1997) D2 dopamine receptor gene (DRD2) Taq1 a polymorphism: reduced dopamine D2 receptor binding in the human striatum associated with the A1 allele. Pharmacogenetics 7:479–484

Toda N, Ayajiki K, Okamura T (2014) Obesity-induced cerebral hypoperfusion derived from endothelial dysfunction: one of the risk factors for Alzheimer's disease. Curr Alzheimer Res 11:733–744

Travagli RA, Anselmi L (2016) Vagal neurocircuitry and its influence on gastric motility. Nat Rev Gastroenterol Hepatol 13:389–401

Tritos NA, Mun E, Bertkau A, Grayson R, Maratos-Flier E, Goldfine A (2003) Serum ghrelin levels in response to glucose load in obese subjects post-gastric bypass surgery. Obes Res 11:919–924

Tschöp M, Smiley DL, Heiman ML (2000) Ghrelin induces adiposity in rodents. Nature 407:908–913

Tschöp M, Weyer C, Tataranni PA, Devanarayan V, Ravussin E, Heiman ML (2001) Circulating ghrelin levels are decreased in human obesity. Diabetes 50:707–709

Turner DS, Shabaan A, Etheridge L, Marks V (1973) The effect of an intestinal polypeptide fraction on insulin release in the rat in vitro and in vivo. Endocrinology 93:1323–1328

Turton MD, O'Shea D, Gunn I, Beak SA, Edwards CM, Meeran K, Choi SJ, Taylor GM, Heath MM, Lambert PD, Wilding JP, Smith DM, Ghatei MA, Herbert J, Bloom SR (1996) A role for glucagon-like peptide-1 in the central regulation of feeding. Nature 379:69–72

van der Zwaal EM, de Weijer BA, van de Giessen EM, Janssen I, Berends FJ, van de Laar A, Ackermans MT, Fliers E, la Fleur SE, Booij J, Serlie MJ (2016) Striatal dopamine D2/3 receptor availability increases after long-term bariatric surgery-induced weight loss. Eur Neuropsychopharmacol 26:1190–1200

Veedfald S, Plamboeck A, Deacon CF, Hartmann B, Knop FK, Vilsboll T, Holst JJ (2016) Cephalic phase secretion of insulin and other entero-pancreatic hormones in humans. Am J Physiol Gastrointest Liver Physiol 310:G43–G51

Verdich C, Flint A, Gutzwiller J-P, Naslund E, Beglinger C, Hellstrøm PM, Long SJ, Morgan LM, Holst JJ, Astrup A (2001) A meta-analysis of the effect of glucagon-like peptide-1 (7–36) amide on ad libitum energy intake in humans. J Clin Endocrinol Metab 86:4382–4389

Vidal J, De Hollanda A, Jiménez A (2016) GLP-1 is not the key mediator of the health benefits of metabolic surgery. Surg Obes Relat Dis 12:1225–1229

Vidal J, Jiménez A (2013) Diabetes remission following metabolic surgery: is GLP-1 the culprit? Curr Atheroscler Rep 15:357

Volkow ND, Wang G-J, Telang F, Fowler JS, Goldstein RZ, Alia-Klein N, Logan J, Wong C, Thanos PK, Ma Y, Pradhan K (2009) Inverse association between BMI and prefrontal metabolic activity in healthy adults. Obesity (Silver Spring) 17:60–65

Weismann RE, Johnson RE (1977) Fatal hepatic failure after jejunoileal bypass: clinical and laboratory evidence of prognostic significance. Am J Surg 134:253–258

Weyer C, Funahashi T, Tanaka S, Hotta K, Matsuzawa Y, Pratley RE, Tataranni PA (2001) Hypoadiponectinemia in obesity and type 2 diabetes: close association with insulin resistance and hyperinsulinemia. J Clin Endocrinol Metab 86:1930–1935

Wilson-Pérez HE, Chambers AP, Ryan KK, Li B, Sandoval DA, Stoffers D, Drucker DJ, Pérez-Tilve D, Seeley RJ (2013) Vertical sleeve gastrectomy is effective in two genetic mouse models of glucagon-like peptide 1 receptor deficiency. Diabetes 62:2380–2385

Wren AM, Seal LJ, Cohen MA, Brynes AE, Frost GS, Murphy KG, Dhillo WS, Ghatei MA, Bloom
    SR (2001) Ghrelin enhances appetite and increases food intake in humans. J Clin Endocrinol
    Metab 86:5992–5995
Wren AM, Small CJ, Ward HL, Murphy KG, Dakin CL, Taheri S, Kennedy AR, Roberts GH,
    Morgan DG, Ghatei MA, Bloom SR (2000) The novel hypothalamic peptide ghrelin stimulates
    food intake and growth hormone secretion. Endocrinology 141:4325–4328
Wynne K, Bloom SR (2006) The role of oxyntomodulin and peptide tyrosine-tyrosine (PYY) in
    appetite control. Nat Clin Pract Endocrinol Metab 2:612–620
Yang WS, Lee WJ, Funahashi T, Tanaka S, Matsuzawa Y, Chao CL, Chen CL, Tai TY, Chuang
    LM (2001) Weight reduction increases plasma levels of an adipose-derived anti-inflammatory
    protein, adiponectin. J Clin Endocrinol Metab 86:3815–3819
Ye J, Hao Z, Mumphrey MB, Townsend RL, Patterson LM, Stylopoulos N, Münzberg H, Morrison
    CD, Drucker DJ, Berthoud H-R (2014) GLP-1 receptor signaling is not required for reduced
    body weight after RYGB in rodents. Am J Physiol Regul Integr Comp Physiol 306:R352–R362
Yousseif A, Emmanuel J, Karra E, Millet Q, Elkalaawy M, Jenkinson AD, Hashemi M, Adamo
    M, Finer N, Fiennes AG, Withers DJ, Batterham RL (2014) Differential effects of laparoscopic
    sleeve gastrectomy and laparoscopic gastric bypass on appetite, circulating acyl-ghrelin, pep-
    tide YY3-36 and active GLP-1 levels in non-diabetic humans. Obes Surg 24:241–252
Zelissen PMJ, Stenlof K, Lean MEJ, Fogteloo J, Keulen ETP, Wilding J, Finer N, Rössner S,
    Lawrence E, Fletcher C, McCamish M, Author Group (2005) Effect of three treatment sched-
    ules of recombinant methionyl human leptin on body weight in obese adults: a randomized,
    placebo-controlled trial. Diabetes Obes Metab 7:755–761
Zhang Y, Ji G, Xu M, Cai W, Zhu Q, Qian L, Zhang YE, Yuan K, Liu J, Li Q, Cui G, Wang H,
    Zhao Q, Wu K, Fan D, Gold MS, Tian J, Tomasi D, Liu Y, Nie Y, Wang G-J (2016) Recovery
    of brain structural abnormalities in morbidly obese patients after bariatric surgery. Int J Obes
    40:1558–1565
Zhang Y, Proenca R, Maffei M, Barone M, Leopold L, Friedman JM (1994) Positional cloning of
    the mouse obese gene and its human homologue. Nature 372:425–432

Part IV
Neuroimaging in Obesity

Chapter 10
Functional Neuroimaging in Obesity Research

Liliana Letra, Daniela Pereira, and Miguel Castelo-Branco

Abstract Functional neuroimaging is beginning to yield valuable insights into the neurobiological underpinnings of the effects of obesity on neural circuits. Functional magnetic resonance imaging (fMRI), positron emission tomography (PET), and single-photon emission computed tomography (SPECT) studies have been used to identify aberrant activation patterns in regions implicated in reward (e.g., striatum, orbitofrontal cortex, insula), emotion and memory (e.g., amygdala, hippocampus), sensory and motor processing (e.g., insula, precentral gyrus), and cognitive control and attention (e.g., prefrontal cortex, cingulate) in obese individuals. Although a great amount of research using these techniques has already unveiled the influence of different neural response patterns on obesogenic behaviors, in this chapter we will, otherwise, try to highlight the effects of obesity on specific neuronal circuits and discuss recent developments in fMRI-based neurofeedback approaches as an alternative in obesity treatment.

Keywords Functional neuroimaging • fMRI • PET • SPECT • Obesity • Neurofeedback

L. Letra
Institute of Physiology, Institute for Biomedical Imaging and Life Sciences—IBILI, Faculty of Medicine, University of Coimbra, Coimbra, Portugal

Neurology Department, Centro Hospitalar do Baixo Vouga, Aveiro, Portugal

D. Pereira (✉)
IBILI-Institute for Biomedical Imaging and Life Sciences, Faculty of Medicine, University of Coimbra, Coimbra, Portugal

Neuroradiology Unit - Medical Imaging Department, Centro Hospitalar e Universitário de Coimbra, Coimbra, Portugal
e-mail: danielajardimpereira@gmail.com

M. Castelo-Branco
IBILI-Institute for Biomedical Imaging and Life Sciences, Faculty of Medicine, University of Coimbra, Coimbra, Portugal

© Springer International Publishing AG 2017
L. Letra, R. Seiça (eds.), *Obesity and Brain Function*, Advances in Neurobiology 19, DOI 10.1007/978-3-319-63260-5_10

10.1 Introduction

It is well established that obesity is associated with disrupted cognitive, affective, and other neurobehavioral circuits. The bidirectional interplay between adipose tissue and the central nervous system has been targeted in several neuroimaging studies that have provided, in the last decade, exciting new developments. In this context, functional magnetic resonance imaging (fMRI), positron emission tomography (PET), single-photon emission computed tomography (SPECT), and functional near-infrared spectroscopy (fNIRS) have been valuable tools in unveiling the neurobiology beneath this association.

fMRI most commonly uses blood-oxygen level dependent (BOLD) effect as an endogenous contrast mechanism based on paramagnetic properties of deoxyhemoglobin. Since its inception, fMRI has been widely used in cognitive and behavioral neuroscience to assess brain activity during task performance or resting state. Stimuli can be presented in a block design, a simpler and statistically more efficient method to detect effect even at the individual level such as in clinical practice; or as an event-related design, randomized and with higher specificity and flexibility, despite lower statistical power. Data analysis can also be performed differently according to the investigator's particular hypothesis and previous knowledge, considering whole-brain BOLD contrast or focusing in selected regions-of-interest (ROI's). More recently, fMRI also offers the possibility of addressing interactions between brain areas, building up networks with functional or effective (causal) connectivity studies. All this diversity in fMRI methodology is, simultaneously, the origin of its flexible and wide application and the justification for the contradictory results we will report along this chapter.

In the last decade, fMRI has largely replaced PET-FDG (fluorodeoxyglucose) in the cognitive neuroscience scene, mainly due to its higher spatial resolution. However, nuclear medicine imaging techniques, including PET and SPECT, still have an important role in functional neuroimaging such as providing information about biodistribution of particular molecules in brain tissue through administration of specific radioactive tracers and subsequent gamma rays detection. Dopamine (DA) receptor levels, for example, can be inferred with these techniques. Since the dopaminergic circuit is linked with reward neural mechanisms, PET/SPECT imaging is particularly attractive to study behavioral changes related to eating disorders.

A drawback of both fMRI and nuclear medicine techniques is the low temporal resolution. fNIRS, a noninvasive vascular-based technology, offers not only a similar temporal resolution as fMRI but also portability, ease of application, and low cost. It can measure the cortical concentration changes of both oxygenated and deoxygenated hemoglobin based on their intrinsic optical absorption. However, the limited spatial resolution, the impossibility of exploring deeper brain structures, and the lower signal-to-noise ratio (SNR) do not allow fNIRS to take out the gold-standard status of fMRI in functional neuroimaging. In the obesity context, literature reporting the use of fNIRS is scarce and thus premature to weave considerations on the results.

Although a great amount of research using the techniques described has already shed light on the influence of aberrant neural responsivity and functioning on obesogenic behaviors, in this chapter we will, otherwise, focus on the effects of the so-called *adiposopathy* on specific neuronal circuits and discuss recent developments in neurofeedback approaches in obesity treatment.

10.2 Neuroimaging of Obesity

From childhood to adulthood elevated body mass index (BMI) has been linked to cerebral macrostructural changes which include reduced total brain volume as well as more localized gray matter volume decrease in several areas: hippocampal formation, prefrontal and orbitofrontal cortices, striatum, cerebellum, cuneus/precuneus, thalamus, and brainstem. Moreover, obesity seems to be positively correlated with white matter atrophy, though only few studies have addressed this association, reporting rather inconsistent findings (Figley et al. 2016). As a result of the advances in neuroimaging techniques in the last decades, particularly in functional imaging, more research has focused on the impact of obesity on the brain's reward, attentional, cognitive, and behavioral control circuits. The main findings from neuroimaging research in obesity-related brain activity patterns and neural network dynamic modulation will be described in the following paragraphs.

10.2.1 fMRI

Cross-sectional studies evaluating BOLD response to palatable food are by far the most common study design found in the literature. Overall, data indicate that, when compared to lean matched controls, overweight/obese individuals show a different pattern of neural responses and network connectivity when presented with food stimuli. This includes greater activation in specific brain regions: (1) visual and anterior cingulate cortices, (2) precuneus, and (3) anterior insula, frontal operculum, postcentral gyrus, and rolandic operculum, which are involved in visual processing, encoding of stimulus salience, and somatosensory processing, respectively. In addition, these individuals also show an elevated striatal response to anticipatory food-related cues and decreased striatal response during its consumption (Burger and Berner 2014). However, this type of study shows limitations in clarifying whether these aberrant neural responses are cause or consequence of obesogenic behaviors and obesity. In fact, most researchers share the opinion that these functional networks are highly dynamic and thus continuously modulated not only by the neurobiological environment but also by the influence of repetitive patterns of behavior.

Adipose tissue dysfunction is accompanied by an altered pattern of neuroendocrine secretion. Part of the molecules involved may impact findings from fMRI studies evaluating the response to palatable foods. Examples are *leptin*, able to reduce insula and striatopallidal and increase prefrontal cortex activation. This could in theory reduce appetitive drive and increase inhibitory control although one should take into account the leptin resistant state typically found in obese (Baicy et al. 2007; Farooqi et al. 2007); *ghrelin*, an orexigenic hormone which levels fail to decrease in the postprandial period of obese individuals, activates the striatum, amygdala, orbitofrontal cortex, and anterior insula in response to food stimuli (Malik et al. 2008), while anorexigenic peptides such as peptide YY (*PYY*) and glucagon-like peptide 1 (*GLP-1*) reduce BOLD signal in those same regions in fasted normal-weight individuals. However, obese individuals may present with lower circulating PYY and elevated GLP-1 levels, suggesting GLP-1 insensitivity, and thus may not fully benefit from their inhibitory action on areas associated with rewarding aspects of food (Burger and Berner 2014). Interestingly, insulin has also been associated with the modulation of food-related neural networks translated into reduced activation in the fusiform gyrus (bilaterally) and the right hippocampus, temporal superior cortex and middle frontal cortex when a visual food cue is presented (Guthoff et al. 2010). Moreover, resting state fMRI studies have demonstrated a positive correlation between functional connectivity strength in the left orbitofrontal cortex and right putamen and fasting insulin levels as well as a negative correlation between activity in these regions and insulin sensitivity (Kullmann et al. 2012). In obesity, with the development of insulin resistance, the pattern and intensity of connections within neural networks, such as the default mode network and temporal lobe network, are affected and likely influence brain function and ultimately, behavioral patterns. To our knowledge no functional neuroimaging data exist regarding *adiponectin*, the most abundant adipokine in plasma and with an important role in energy expenditure and insulin sensitivity. It would be of great interest to deepen our knowledge on the effects of obesity on brain networks by modulating the circulating levels of hormones and other molecules secreted by the adipose tissue and to evaluate their potential implication not only in food behavior but also in the long-term cognitive detrimental effects associated with this disease.

10.2.2   *PET and SPECT*

PET and SPECT have been very useful tools in metabolic imaging research once they can specifically trace the uptake of a substrate, binding of a neurotransmitter or detect blood flow and oxygen uptake using highly selective radioactive probes. Cerebral glucose metabolism measured by PET-FDG has been extensively used to characterize metabolic brain responses in physiological and several pathological situations. This tracer is transported into cells and phosphorylated but, unlike glucose, does not undergo glycolysis and therefore accumulates in the brain in proportion with cerebral glucose consumption. In obese, global brain glucose metabolism seems to be greater than in lean subjects, though some authors point significant differences arising in more specific areas such as the parietal somatosensory cortex, upper cerebellum, and precuneus (Wang et al. 2002), while others report that this increase is exclusively dependent on the insulin stimulus (Tuulari et al. 2013), or the existence of an inverse correlation between BMI and baseline brain glucose metabolism in prefrontal regions and in the anterior cingulate gyrus instead (Volkow et al. 2009). It is obvious that our comprehension on how obesity influences brain glucose metabolism, namely brain insulin resistance, is far from being fully understood. Longitudinal studies would help to address the mechanisms by which weight changes can modulate cerebral glucose and insulin pathways, although some reports already exist on the beneficial effects of bariatric surgery on brain FGD-PET profile of these patients (Tuulari et al. 2013).

Nuclear imaging techniques have also been used to address the involvement of serotoninergic, dopaminergic, and opioid pathways not only in binge eating, bulimia, and anorexia but also in obesity not associated with eating disorders, which constitutes the majority of cases. Serotonin (5-HT) is a key neuromodulator involved in fundamental cerebral functions such as appetite and mood. In obesity, the levels of brain 5-HT transporters have been inconsistently reported as up- or downregulated, and this is in part due to the inclusion of obese patients with eating disorders in these cohorts (Haahr et al. 2012; Kuikka et al. 2001; Koskela et al. 2007; Wu et al. 2017). In fact, although some selective serotonin reuptake inhibitors (SSRIs) can reduce weight (Halford et al. 2007) and lorcaserin, a selective 5-HT_{2C} agonist, has been approved for the treatment of obesity, the inconsistent response to these drugs indicates the existence of interindividual variations in serotonergic activity or sensitivity (Bello and Liang 2011; Nigro et al. 2013) and underlines the importance of other pathways in the pathophysiology of obesity.

DA is one of the most important neurotransmitters involved in the modulation of prefrontal activity, thus regulating executive functions and eating behavior, and consequently constitutes a molecule of great interest in this field of research. It is a central player in the mesolimbic reward circuit although the association between striatal dopamine D2 receptor (DRD$_2$) levels or its transporter (DAT) availability and obesity has been inconsistently demonstrated (de Weijer et al. 2011; Chen et al. 2008; van de Giessen et al. 2013). A meta-analysis including five studies on D$_2$R availability according to BMI using [^{11}C]raclopride has failed to establish a strong association between these two variables, though it is recognized that DA is closely related with food intake (Guo et al. 2014). On the other hand, there is considerable evidence on the interactions between opioid and dopamine systems, since the μ-opioid receptor (MOR) is able to modulate the mesolimbic dopamine system in ventral tegmental area and striatum, two key areas implicated in reward processing. Besides having significantly lower MOR availability when compared with lean controls, obese subjects are also prone to disruptions in the DA-MOR interaction (Tuominen et al. 2015), suggesting that research on treatment of obesity should follow more holistic strategies, taking into consideration the complex etiopathogeny of this metabolic disorder.

Finally, some studies have otherwise investigated regional cerebral blood flow (rCBF) changes in those with high BMI using SPECT imaging. Willeumier et al. (2011) have demonstrated for the first time that overweight individuals presented decreased rCBF in the prefrontal cortex and anterior cingulate gyrus when compared to lean matched controls. These changes may negatively impact behavioral responses, but whether this deficit precedes weight gain or is an acquired feature of obesity is still elusive. In fact, this remains the main unanswered question in obesity functional neuroimaging research, and thus an important gap that should be explored in more comprehensive conceptual frameworks underlying the design of future studies.

10.3 Neurofeedback Based on rt-fMRI: A Promising Therapeutic Approach in Obesity

Neurofeedback, which may be viewed as operant conditioning of brain activity, was traditionally performed based on electroencephalography (EEG) with beneficial effects both on behavioral and imaging measures (Levesque et al. 2006; Kouijzer et al. 2009). However, this approach is limited by the low spatial resolution and inaccessibility to deep brain structures. Neurofeedback based on real-time fMRI (rt-fMRI), with its improved anatomical precision, has recently emerged as a possible solution for this drawback. Simultaneously, it offers the opportunity of studying the neurophysiological basis of behavior or disease and the neuroplastic changes induced by therapeutic intervention.

Introduced by Cox two decades ago (Cox et al. 1995), rt-fMRI provides noninvasive online assessment to brain function. In turn, this allows targeting brain regions or networks, while the individual is inside the scanner, and the implementation of experimental paradigms including neurofeedback. Instant and continuously updated BOLD contrast signals are fed back to subjects, who try to voluntarily modulate their own brain activity on the defined target, adapting and learning strategies to achieve task success. When proved the efficiency of those self-provided strategies, they may subsequently be transferred to more practical brain-computer interfaces (BCI's), such as EEG or fNIRS, and also applied in real-life situations as they may possibly outlast neurofeedback sessions.

Rt-fMRI neurofeedback still represents a technical challenge (namely related to the delay of feedback, image resolution and SNR) and methodological improvements are ongoing (see Weiskopf 2012 for a revision on this topic). The feedback modality, training structure, and type of physiological target are also being debated, as well as the therapeutic value of this technique and transfer strategies. However, preliminary data in healthy individuals and in disorders such as depression, Parkinson's disease, stroke, chronic pain, addition, and schizophrenia (Ruiz et al. 2014) are encouraging. Individuals were able to autoregulate their brain activity through this method and this learned regulation of localized brain regions seems to affect brain dynamics.

Taking into account the substantial knowledge about pathophysiology of obesogenic behavior produced by neuroimaging in the last decades, as previously reviewed on this chapter, rt-fMRI neurofeedback seems to be an attractive therapeutic tool in obesity, promoting positive reinforcement and self-efficacy (Bartholdy et al. 2013). To our knowledge, only three studies addressed rt-fMRI neurofeedback using food stimuli on the context of eating disorders.

On the first exploratory study, 11 lean and 10 obese healthy male individuals (Frank et al. 2012) attempted to regulate BOLD signal in anterior insular cortex (AIC), known to be involved in gustatory perception and reward elicited by food cues. Since AIC is also involved in emotional processing, participants were instructed to think about something subjectively positive or negative, with any strategy, to try to regulate the feedback signal. After six training sessions distributed on 2 days, obese participants showed enhanced regulation ability, while the four lean individuals were not able to regulate at all. From those who were successful in self-regulation, lean participants had stronger functional connectivity during upregulation of the AIC and medial temporal and medial cingulate cortex. That is, obese people seem to have greater ability to upregulate AIC, which in lean people is probably achieved at the expense of increased network connectivity. This preliminary study inaugurated the research on fMRI-BCI training in obese, which may be based on AIC, in control regions (such as dorsolateral prefrontal cortex—dlPFC) or in a whole network related to reward and gustatory perception.

The same investigation group explored precisely the interplay between executive functions and reward processing, traduced by the functional connectivity between dlPFC and ventromedial prefrontal cortex (vmPFC), respectively, in a recent neurofeedback study (Spetter et al. 2017). Eight participants with overweight or obesity were trained for 4 weeks to increase dlPFC-vmPFC functional connectivity during visualization of appetitive high and unhealthy high-calorie food pictures. All participants were able to upregulate functional connectivity between dlPFC and vmPFC with an incremental increase across three consecutive runs, but not across four individual training sessions, suggesting that the training effect did not endure between days, at least with the number of sessions applied. Furthermore, in a food choice behavioral task performed on each session, there was a tendency to choose less unhealthy food post-training, that contrast with a trend to greater food intake, measured with three types of snacks offered to the subjects. Interestingly, dlPFC activity (but not vmPFC) was increased during upregulation contrasted to passive viewing and also when comparing first to the fourth session, possibly indicating that self-regulation was dependent on executive control and its influence in the reward system. In line with the previous paper (Frank et al. 2012), insula emerged once more as a key area in appetitive and emotional processing that participants were able to self-regulate, showing enhanced activation of the inferior frontal gyrus(IFG)/insula during upregulation vs. passive viewing.

The third neurofeedback study with food stimuli was performed in a healthy population (10 females) and had an innovative approach involving downregulation training instead of the more common (and easier) upregulation (Ihssen et al. 2017). Another novel feature of this experiment was the use of the cue size as feedback (decreased size corresponding to successful downregulation), rather than adding a symbolic indicator of brain activity (typically a thermometer). The authors called this new methodology "motivational neurofeedback." Target brain areas were individually defined trough a functional localizer run and included mainly limbic and subcortical structures, which were successfully downregulated by participants during neurofeedback runs. Moreover, in contrast with findings reported by Frank et al. (2012), prefrontal cortex "control regions" had not increased activation during regulation. That is, deactivation of motivational areas, particularly the amygdala and left insula, achieved with neurofeedback did not seem to be dependent on "top-down" control. More important, behavioral data showed a significant hunger reduction after training, which was positively related with right amygdala downregulation, suggesting some efficacy of the neurofeedback protocol.

Conceptually and according to these promising preliminary data, rt-fMRI neurofeedback is a potential therapeutic tool in eating disorders and simultaneously an interface to understand pathological neural mechanisms implicated in obesogenic behavior. Neurofeedback protocols individually tailored (using functional localizer runs), with innovative stimuli and mainly focussed on reward circuits, executive structures, and motivational areas, must be tested at a larger scale (both in terms of number of sessions and participants number) and contemplating transfer strategies to more practical EEG-based BCI tools.

10.4 Conclusion and Final Remarks

Neuroimaging research has undoubtedly contributed to a more thorough understanding of regional neural activity abnormalities associated with obesity, although the understanding of its dynamics and causal interactions are still poorly understood. Functional neuroimaging techniques have witnessed, in the last years, tremendous advances, and if initially they were almost exclusively used as research tools in cognitive neuroscience and neuropsychology, at present they are increasingly deployed in other areas of research, such is the case of obesity. Unfortunately, the methods currently used to determine body adipose tissue content did not keep up with neuroimaging developments. Most of the studies still use BMI as the main or the only adiposity measure, although it does not distinguish lean from adipose tissue neither reflects the degree of adipose tissue dysfunction. More accurate techniques such as bioelectrical impedance analysis should be used in the future to overcome methodological incongruities in this field of research (Verney et al. 2016).

References

Baicy K, London ED, Monterosso J, Wong ML, Delibasi T, Sharma A, Licinio J (2007) Leptin replacement alters brain response to food cues in genetically leptin-deficient adults. Proc Natl Acad Sci U S A 104(46):18276–18279
Bartholdy S, Musiat P, Campbell IC, Schmidt U (2013) The potential of neurofeedback in the treatment of eating disorders: a review of the literature. Eur Eat Disord Rev 21(6):456–463
Bello NT, Liang NC (2011) The use of serotonergic drugs to treat obesity—is there any hope? Drug Des Devel Ther 5:95–109

Burger KS, Berner LA (2014) A functional neuroimaging review of obesity, appetitive hormones and ingestive behavior. Physiol Behav 136:121–127

Wu C-H, Chang C-S, Yang YK, Shen L-H, Yao W-J (2017) Comparison of brain serotonin transporter using [I-123]-ADAM between obese and non-obese young adults without an eating disorder. PLoS One 12(2):e0170886

Chase R. Figley, Judith S. A. Asem, Erica L. Levenbaum, Susan M. Courtney, (2016) Effects of Body Mass Index and Body Fat Percent on Default Mode, Executive Control, and Salience Network Structure and Function. Frontiers in Neuroscience 10

Chen PS, Yang YK, Yeh TL, Lee IH, Yao WJ, Chiu NT, Lu RB (2008) Correlation between body mass index and striatal dopamine transporter availability in healthy volunteers—a SPECT study. NeuroImage 40:275–279

Cox RW, Jesmanowicz A, Hyde JS (1995) Real-time functional magnetic resonance imaging. Magnet Reson Med 33:230–236

Farooqi IS, Bullmore E, Keogh J, Gillard J, O'Rahilly S, Fletcher PC (2007) Leptin regulates striatal regions and human eating behavior. Science 317:1355

Figley CR, Asem JSA, Levenbaum EL, Courtney SM (2016) Effects of body mass index and body fat percent on default mode, executive control, and salience network structure and function. Front Neurosci 10:234. doi:10.3389/fnins.2016.00234

Frank S, Lee S, Preissl H, Schultes B, Birbaumer N, Veit R (2012) The obese brain athlete: self-regulation of the anterior insula in adiposity. PLoS One 7(8):3–8

van de Giessen E, Hesse S, Caan MW, Zientek F, Dickson JC, Tossici-Bolt L, Sera T, Asenbaum S, Guignard R, Akdemir UO, Knudsen GM, Nobili F, Pagani M, Vander Borght T, Van Laere K, Varrone A, Tatsch K, Booij J, Sabri O (2013) No association between striatal dopamine transporter binding and body mass index: a multi-center European study in healthy volunteers. NeuroImage 64:61–67

Guo J, Simmons WK, Herscovitch P, Martin A, Hall KD (2014) Striatal dopamine D2-like receptor correlation patterns with human obesity and opportunistic eating behavior. Mol Psychiatry 19:1078–1084

Guthoff M, Grichisch Y, Canova C, Tschritter O, Veit R, Hallschmid M et al (2010) Insulin modulates food-related activity in the central nervous system. J Clin Endocrinol Metab 95:748–755

Haahr ME, Rasmussen PM, Madsen K et al (2012) Obesity is associated with high serotonin 4 receptor availability in the brain reward circuitry. Neuroimage 61:884–888

Halford JC, Harrold JA, Boyland EJ, Lawton CL, Blundell JE (2007) Serotonergic drugs: effects on appetite expression and use for the treatment of obesity. Drugs 67:27–55

Ihssen N, Sokunbi MO, Lawrence AD, Lawrence NS, Linden DEJ (2017) Neurofeedback of visual food cue reactivity: a potential avenue to alter incentive sensitization and craving. Brain Imaging Behav 11(3):915–924

Koskela AK, Keski-Rahkonen A, Sihvola E, Kauppinen T, Kaprio J, Ahonen A et al (2007) Serotonin transporter binding of [123I] ADAM in bulimic women, their healthy twin sisters, and healthy women: a SPECT study. BMC Psychiatry 7:19

Kouijzer MEJ, de Moor JMH, Gerrits BJL, Congedo M, van Schie HT (2009) Neurofeedback improves executive functioning in children with autism spectrum disorders. Res Autism Spectrum Disord 3:145–162

Kuikka JT, Tammela L, Karhunen L, Rissanen A, Bergström KA, Naukkarinen H et al (2001) Reduced serotonin transporter binding in binge eating women. Psychopharmacology 155:310–314

Kullmann S, Heni M, Veit R, Ketterer C, Schick F, Häring HU et al (2012) The obese brain: association of body mass index and insulin sensitivity with resting state network functional connectivity. Hum Brain Mapp 33:1052–1061

Kyle S. Burger, Laura A. Berner, (2014) A functional neuroimaging review of obesity, appetitive hormones and ingestive behavior. Physiology & Behavior 136:121–127

Levesque J, Beauregard M, Mensour B (2006) Effect of neurofeedback training on the neural substrates of selective attention in children with attention-deficit/hyperactivity disorder: a functional magnetic resonance imaging study. Neurosci Lett 394:216–221

Maartje S. Spetter, Rahim Malekshahi, Niels Birbaumer, Michael Lührs, Albert H. van der Veer, Klaus Scheffler, Sophia Spuckti, Hubert Preissl, Ralf Veit, Manfred Hallschmid, (2017) Volitional regulation of brain responses to food stimuli in overweight and obese subjects: A real-time fMRI feedback study. Appetite 112:188–195

Malik S, McGlone F, Bedrossian D, Dagher A (2008) Ghrelin modulates brain activity in areas that control appetitive behavior. Cell Metab 7:400–409

Nigro SC, Luon D, Baker WL (2013) Lorcaserin: a novel serotonin 2C agonist for the treatment of obesity. Curr Med Res Opin 29:839–848

Ruiz S, Buyukturkoglu K, Rana M, Birbaumer N, Sitaram R (2014) Real-time fMRI brain computer interfaces: self-regulation of single brain regions to networks. Biol Psychol 95:4–20

Spetter MS, Malekshahi R, Birbaumer N, Lührs M, van der Veer AH, Scheffler K, Spuckti S, Preissl H, Veit R, Hallschmid M (2017) Volitional regulation of brain responses to food stimuli in overweight and obese subjects: a real-time fMRI feedback study. Appetite 112:188–195

Tuominen L, Tuulari J, Karlsson H, Hirvonen J, Helin S, Salminen P, Parkkola R, Hietala J, Nuutila P, Nummenmaa L (2015) Aberrant mesolimbic dopamine-opiate interaction in obesity. NeuroImage 15(122):80–86

Tuulari JJ, Karlsson HK, Hirvonen J, Hannukainen JC, Bucci M, Helmiö M, Ovaska J, Soinio M, Salminen P, Savisto N, Nummenmaa L, Nuutila P (2013) Weight loss after bariatric surgery reverses insulin-induced increases in brain glucose metabolism of the morbidly obese. Diabetes 62(8):2747–2751

Verney J, Metz L, Chaplais E, Cardenoux C, Pereira B, Thivel D (2016) Bioelectrical impedance is an accurate method to assess body composition in obese but not severely obese adolescents. Nutr Res 36(7):663–670

Volkow ND, Wang GJ, Telang F, Fowler JS, Goldstein RZ, Alia-Klein N, Logan J, Wong C, Thanos PK, Ma Y, Pradhan K (2009) Inverse association between BMI and prefrontal metabolic activity in healthy adults. Obesity (Silver Spring) 17:60–65

Wang GJ, Volkow ND, Felder C, Fowler JS, Levy AV, Pappas NR, Wong CT, Zhu W, Netusil N (2002) Enhanced resting activity of the oral somatosensory cortex in obese subjects. Neuroreport 13:1151–1155

de Weijer BA, van de Giessen E, van Amelsvoort TA, Boot E, Braak B, Janssen IM, van de Laar A, Fliers E, Serlie MJ, Booij J (2011) Lower striatal dopamine D2/3 receptor availability in obese compared with non-obese subjects. EJNMMI Res 1:37

Weiskopf N (2012) Real-time fMRI and its application to neurofeedback. NeuroImage 62:682–692

Willeumier KC, Taylor DV, Amen DG (2011) Elevated BMI is associated with decreased blood flow in the prefrontal cortex using SPECT imaging in healthy adults. Obesity 19(5):1095–1097

CPSIA information can be obtained
at www.ICGtesting.com
Printed in the USA
LVOW05*2109210218
567424LV00001B/62/P